Unless Recalled Earlier

DATE DUE

WATER

A COMPREHENSIVE TREATISE

Volume 5

Water in

Disperse Systems

WATER
A COMPREHENSIVE TREATISE

Edited by Felix Franks

WATER
A COMPREHENSIVE TREATISE

Edited by Felix Franks
Unilever Research Laboratory
Sharnbrook, Bedford, England

Volume 5
Water in
Disperse Systems

PLENUM PRESS • NEW YORK AND LONDON

Library of Congress Cataloging in Publication Data

Franks, Felix
  Water in disperse systems.

  (His Water: a comprehensive treatise, v. 5)
  Bibliography: p.
  1. Colloids. 2. Suspensions (Chemistry) I. Title.
QD169.W3F7 vol. 5 [QD549] 553'.7'08s [541'.34514]
ISBN 0-306-37185-5   74-17190

© 1975 Plenum Press, New York
A Division of Plenum Publishing Corporation
227 West 17th Street, New York, N.Y. 10011

United Kingdom edition published by Plenum Press, London
A Division of Plenum Publishing Company, Ltd.
4a Lower John Street, London W1R 3PD, England

Printed in the United States of America

Preface

This volume is the last in the series comprising "Water—A Comprehensive Treatise." It was originally planned to combine aqueous solutions of macromolecules and disperse systems in one volume, but largely because of the extensive coverage required by recent developments in aqueous solutions of proteins and synthetic polymers I decided to separate topics dealing with water in disperse systems.

The systems treated in the present volume are of a complex nature so that the theoretical frameworks established earlier in Volume 1 and utilized in Volumes 2 and 3 cannot at the present time be applied. On the other hand the systems discussed in Volumes 4 and 5 in particular, border on the many biological and technological areas where important attributes are related to the common factor—water. Included among such diverse problem areas are food processing and preservation, cryopreservation, paper and textile finishing, membrane processes, hemodynamics, etc. It is to be hoped that in days to come some of the results and principles discussed in these five volumes can be applied to improve our understanding of the complex interactions in medically and industrially important spheres of scientific activity.

An age seems to have passed since the concept of creating this treatise was first discussed, and since work began on Volume 1, much has happened in the science of Water; some of the recent developments were highlighted at this year's Gordon Research Conference in Plymouth, N.H. They extend from a better understanding of the interactions between pairs of water molecules, through the fascinating studies of supercooled water and the detailed nature of ion hydration to new thinking on the role of the solvent in maintaining native protein and polysaccharide conformations.

Once again I acknowledge my debt of gratitude to Joyce Johnson for playing a major part in the various stages leading to publication, and also to the contributing authors for their close cooperation. Finally, I should like to dedicate the five volumes to my wife Hedy and my daughters Carolyn and Suzanne, in gratitude for the patience and tolerance which they have (at most times) shown during the past five years when manuscripts, type-scripts, proofs, and indexes have invaded our home.

December, 1974 Felix Franks

Contents

Chapter 1

The Influence of Hydration on the Stability of Hydrophobic Colloidal Systems

D. Eagland

Chapter 2

Properties of Water in Capillaries and Thin Films

J. Clifford

Chapter 3

Hydration and the Stability of Foams and Emulsions

M. C. Phillips

Chapter 4

Clay–Water Systems

Erik Forslind and Arvid Jacobsson

Chapter 5

Adsorption of Water on Well-Characterized Solid Surfaces

A. C. Zettlemoyer, F. J. Micale, and K. Klier

Chapter 6

Specific Interactions of Water with Biopolymers

Herman J. C. Berendsen

Contents of Volume 1: The Physics and Physical Chemistry of Water

Contents of Volume 2: Water in Crystalline Hydrates; Aqueous Solutions of Simple Nonelectrolytes

Contents of Volume 3: Aqueous Solutions of Simple Electrolytes

Contents of Volume 4: Aqueous Solutions of Amphiphiles and Macromolecules

Volume 5:

Water in Disperse Systems

The Influence of Hydration on the Stability of Hydrophobic Colloidal Systems

D. Eagland

School of Studies in Chemistry, University of Bradford
Bradford, Yorkshire, England

1. INTRODUCTION

The surface of a solid becomes electrically charged when placed in contact with an aqueous medium; the potential thus acquired at the interface is responsible for the preferential attraction of ions of opposite charge toward the surface and the establishment of an "electrical double layer" at the solid/liquid interface. Close approach of two such charged surfaces, together with their associated double layers, will result in the generation of a repulsive force between the surfaces. Colloidal systems are characterized primarily by the extremely large interface which exists between the particulate or disperse phase and the bulk liquid or dispersion medium; the repulsive force between the surfaces is therefore of primary importance for the resistance to flocculation, which would otherwise occur spontaneously under the influence of longer range van der Waals attractive forces.

These major factors are quite clearly primarily responsible for the stability or otherwise of a colloidal dispersion; however, both are modified in varying degrees by a variety of other contributing agents; these will be briefly reviewed here, with particular emphasis upon the role of water and hydration structures in such modification.

2. REPULSIVE FORCES BETWEEN COLLOIDAL PARTICLES— THE ELECTRICAL DOUBLE LAYER

2.1. The Gouy–Chapman Model of the Double Layer

Early theories of the electrical double layer[334] envisaged the layer as being caused by simple separation of opposite charges along two parallel, adjacent planes; hence the double layer could be regarded as a parallel plate condenser of molecular dimensions. Subsequent work by Gouy[295] and Chapman[105] revealed the shortcomings of the model and laid the basis of our present understanding of the double layer, anticipating the similar work of Debye[147,148] on the charge distribution surrounding an ion in aqueous solution by almost a decade.

The primary conditions for the Gouy–Chapman model of the double layer are that the concentration of ions within the dispersion medium must be low in order that interaction between ions may be neglected; the charge on the solid surface is assumed to be uniformly distributed; ions in the dispersion medium are treated as point charges of negligible volume; and the dispersion medium is regarded as a continuum, influencing the double layer only through its permittivity ε.

The electric potential ψ and the density of the space charge ϱ in the dispersion medium are related by the Poisson equation

$$\nabla^2 \psi = -4\pi\varrho/\varepsilon \tag{1}$$

The space charge density is the sum total of the point charge positive and negative ions

$$\varrho = (n_+'z_+ - n_-'z_-)e \tag{2}$$

where $n_{\pm}'$ are the local concentrations (number cm^{-3}) of the ions, $z_{\pm}$ are their valences, and e is the electronic charge. The Coulombic interaction of the ions with the solid surface is subject to interference from their thermal motions; hence the resulting statistical distributions are described by the Boltzmann law:

$$n_{\pm}' = n \exp(\mp z_{\pm}e\psi/kT) \tag{3}$$

where n is the bulk concentration of the ions, far from the interface, k is the Boltzmann constant, and T is the absolute temperature.

Combining eqns. (1)–(3) results in the Poisson–Boltzmann equation

$$\frac{d^2\psi}{dx^2} = -\frac{4\pi e}{\varepsilon}\left(nz_+ \exp\frac{-z_+e\psi}{kT} - nz_- \exp\frac{z_-e\psi}{kT}\right) \tag{4}$$

and in its spherical form

$$\frac{d^2\psi}{dx^2} = \frac{1}{r^2}\frac{d}{dr}\left(r^2\frac{d\psi}{dr}\right) \tag{5}$$

where r is defined as the radial distance.

The boundary conditions for solution of this equation are given by $\lim_{r\to\infty}\psi = 0$ and $\psi = \psi_a$ when $r = a$, where a may be defined as the radius of the spherical colloidal particle.

The Gouy–Chapman model of the double layer is based on the assumption that $\psi_a = \psi_0$, i.e., the Poisson–Boltzmann distribution of the ions surrounding the particle commences from the surface of the particle ($\psi = \psi_0$); this assumption, however, at relatively high surface potentials (a surface potential of 200 mV for a solid surface in contact with 0.01 mol dm^{-3} NaCl) would apparently yield a concentration of sodium ions at the particle surface of approximately 300 mol dm^{-3}. The error arises from the approximation made in the Gouy–Chapman analysis of the double layer that all the ions can be treated as point changes.

The more recent theories of Stern[648] and Grahame[298] allow for the finite size of the ions which occupy the region of closest approach to the particle surface and thus are much less susceptible to the thermal disturbance of the dispersion medium. Such ions are much more firmly held by the colloidal surface, which leads to the division of the double layer into a "fixed" or Stern region, close to the colloid surface and of a finite thickness δ, and a "diffuse" region to which the Gouy–Chapman model might still be considered to apply.

2.1.1. The Diffuse Double Layer of Electric Charge

It seems from the previous discussion that a more realistic assessment of the significance of ψ_a should be the correlation that $\psi_a = \psi_\delta$, rather than ψ_0, where ψ_δ is the potential at a distance δ from the particle surface. It should be mentioned that electrokinetic phenomena (Section 2.1.4) such as electrophoresis give rise to a further interpretation of ψ_a; here the liquid layer immediately adjacent to the particle surface is considered stationary with respect to the surface; thus the concept of a shear surface must be invoked to discriminate between moving and stationary liquid layers. The potential associated with this shear surface is known as the zeta (ζ) potential.

For small potentials and uni-univalent electrolytes ($z_+ = z_- = z$) $z\psi \ll 25$ mV (a situation which is usually likely to occur relatively far into the diffuse double layer) it is sufficient to consider only the first two terms of the exponential series; thus

$$\frac{1}{\phi} \frac{d}{d\phi} \left(\phi^2 \frac{d\eta}{d\phi} \right) = \eta \tag{6}$$

where $\phi = ze\psi_a/kT$ and $\eta = \varkappa a$, $\varkappa$ being given by the equation

$$\varkappa^2 = 4\pi e^2 (2nz^2)/\varepsilon kT \tag{7}$$

The reciprocal of $\varkappa$ has the dimensions of length and is known as the thickness of the double layer.

Solution of eqn. (5) for the case where $z\psi \gg 25$ mV requires the use of computer facilities and the reader is referred to tables by Loeb *et al.*[444] for further details.

Much of the work reported in the literature upon the structure of the diffuse double layer is based upon a hypothetical model of a colloidal plate of infinite extent, i.e., a colloidal sphere of infinite radius, rather than a spherical particle of a finite radius. Such a device simplifies the theoretical calculations and therefore will be used in the ensuing discussion, but the reader's attention is drawn to the distinction.

Application of the Poisson–Boltzmann distribution to a plane surface, and considering ion distribution in a perpendicular direction (x) from the surface only, gives

$$d^2\psi/dx^2 = -(8\pi zne/\varepsilon) \sinh(ze\psi/kT) \tag{8}$$

The solution of this equation within the boundary conditions $\psi = \psi_0$ when $x = 0$, and $\psi = 0$, $d\psi/dx = 0$ when $x = \infty$, gives

$$\psi = \frac{2kT}{ze} \ln \frac{1 + f \exp(-\varkappa x)}{1 - f \exp(-\varkappa x)} \tag{9}$$

where

$$f = \frac{\exp(ze\psi_0/2kT) - 1}{\exp(ze\psi_0/2kT) + 1} \tag{10}$$

The double layer thickness is directly proportional to the square root of the ionic concentration in the dispersion medium and to the ionic valence; thus the potential decrease with distance is accelerated at higher electrolyte concentrations; the expression for $\varkappa$ also contains a term involving the square of the ionic valence; thus polyvalent counterions (the ion of charge sign opposite to that of the surface) produce a very rapid decrease in potential with distance (Figs. 1 and 2).

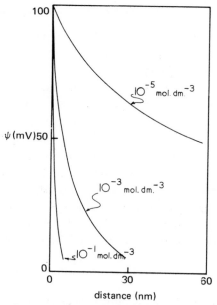

Fig. 1. The variation of potential with distance as a function of the concentration of a uni-univalent electrolyte.

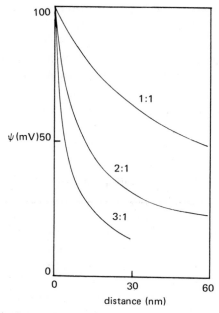

Fig. 2. The variation of potential with distance at a constant concentration of 10^{-5} mol dm^{-3} for a uni-univalent (1:1), a di-univalent (2:1), and a tri-univalent (3:1) electrolyte.

2.1.2. *Corrections to the Gouy–Chapman Model of the Diffuse Double Layer*

Two assumptions of the Gouy–Chapman model—that all the ions distributed in the diffuse region of the double layer can be treated as unpolarizable point changes, and that the permittivity of the continuum is unchanged upon closer approach to the solid–liquid interface—are acceptable only if the electrolyte concentration in the dispersion medium is very low and the potential at the surface is weak.

Measurements of the capacitance of the double layer have shown that the permittivity of the region of the diffuse layer closest to the solid surface is decreased due to dielectric saturation caused by the high field strength and structuring effects in the solvent continuum. Hasted and co-workers[325] have shown that for aqueous solutions of a single strong electrolyte of concentrations ranging from 10^{-5} to 2×10^{-3} mol cm^{-3} the permittivity (at constant temperature and pressure) is given by the expression

$$\varepsilon = \varepsilon_0 + \delta_+ c_+ + \delta_- c_- = \varepsilon_0 + 2\delta c \tag{11}$$

where ε_0 is the permittivity of the pure solvent in the absence of an applied electric field. For any given electrolyte, experiment only yields δ, but Hasted *et al.*, on the basis of assumptions concerning the hydrated structure of the ions, have split 2δ into δ_+ and δ_-; comparison of δ for electrolytes having a common ion enabled the authors to tabulate δ_+ and δ_- for a series of ions at zero electric field. Table I lists the values of the ionic hydration numbers and illustrates the fact that, since hydration of the cation is in general more pronounced than that of the anion, $|\delta_+| > |\delta_-|$.

Dielectric saturation at high field strengths must also be taken into account; according to Mandel,[471] ε decreases with the square of the field strength,

$$\varepsilon = \varepsilon_0 - hE^2 \tag{12}$$

TABLE I. Hydration Number of Ions[a]

Cation	Hydration number	Anion	Hydration number
Li$^+$	5	F$^-$	4
Na$^+$	5	Cl$^-$	1
K$^+$	4	Br$^-$	1
Rb$^+$	3	I$^-$	1

[a] From data of Bockris *et al.*[68]

For fields weaker than $\sim 1 \times 10^6$ V cm^{-1}, $h = 3 \times 10^{-7}$ V^{-2} cm^{-2}.[72,470] Combination of eqns. (11) and (12) yields the expression

$$\varepsilon = \varepsilon_0 + \delta_+ c_+ + \delta_- c_- - hE^2 \tag{13}$$

When the permittivity is not constant the Poisson equation takes the form

$$4\pi\varrho = -\operatorname{div}(\varepsilon \operatorname{grad} \psi) \tag{14}$$

$$= -\varepsilon \, V\psi - \operatorname{grad} \psi \operatorname{grad} \varepsilon \tag{15}$$

Thus for a spherical particle eqn. (15) becomes

$$\varepsilon \frac{d^2\psi}{dr^2} + \frac{2\varepsilon}{r}\frac{d\psi}{dr} + \frac{d\varepsilon}{dr}\frac{d\psi}{dr} = -4\pi z F(c_+ - c_-) \tag{16}$$

With the substitutions

$$\phi = zF\psi/kT \qquad \text{and} \qquad \eta = \varkappa r$$

eqn. (16) becomes

$$\varepsilon \frac{d^2\phi}{d\eta^2} + \frac{d\varepsilon}{d\eta}\frac{d\phi}{d\eta} + \frac{2\varepsilon}{\eta}\frac{d\phi}{d\eta} = \frac{\varepsilon_0(c_+ + c_-)}{2c} \tag{17}$$

Since the diffuse double layer is a charged and polarized medium each volume element is subject to a force of both mechanical and electrical origin and Devillez et al.[188] have shown by a local thermodynamic approach that the difference between the Kelvin pressure P and the Helmholtz pressure P_0, i.e., the pressure that would be present in the same volume in the absence of an electric field for equal concentration and temperature, is given for the bulk solution by the equation

$$P_0 - P = \frac{E^2}{8\pi} + \frac{1}{8\pi}\int_{0,c}^{E^2}\left[\sum c\left(\frac{\partial\varepsilon}{\partial c}\right)_{T,E} - \varepsilon\right]dE^2 \tag{18}$$

The same authors have shown that the distribution of the constituent ions in a charged and polarized medium is given by the expression

$$\ln\frac{\gamma_0 c}{\gamma_0' c'} = -V_0^*(P_0 - P) + \frac{1}{8\pi RT}\int_{0,c}^{E^2}\left(\frac{\partial\varepsilon}{\partial c}\right)_{T,E}dE^2 - \frac{zF\psi}{RT} \tag{19}$$

where γ_0 is the activity coefficient relative to molar fractions in the dissymmetric system of reference at field equal to zero and γ_0' is the activity coefficient relative to mole fractions in the bulk of the solution, c and c'

are the corresponding concentrations, and V_0^* is the molar standard volume at temperature T of the electrolyte component extrapolated to zero pressure.

Equations (13) and (19) enable a solution of the Poisson–Boltzmann equation (16) to be obtained under the boundary conditions of $r = a$, $\psi = \psi_\delta$, $\eta = \varkappa a$, $\phi = \phi_\delta$; and $\psi = 0$, $r = \infty$, and $\phi \rightarrow 0$ when $\eta \rightarrow \infty$. The solution indicates that for a constant potential the charge of the surface layer increases less rapidly with increasing electrolyte concentration than predicted by the (concentration)$^{1/2}$ term of the Gouy–Chapman theory; the effect of dielectric saturation increases with the potential at the surface and hence a decrease of charge for a given surface potential occurs. The influence of the anion (with the condition of negative potential at the surface) is shown by the authors to be negligible, but the effect of δ_+ (the cation hydration constant) is to decrease the surface layer charge for a constant potential; thus the more polarizable ions tend to concentrate in the layer in the order

$$q_{H^+} < q_{Li^+} < q_{Na^+}$$

where q is the charge in the diffuse layer per unit surface area.

The effect of molar volume is apparently negligible for small ions but the authors suggest that in the case of a large cation, taking the tetraethylammonium ion as an example, the effect is to lower the ion concentration in the diffuse layer. Unfortunately, no systematic study appears to have been reported using the series of tetraalkylammonium ions from the methyl up to the butyl homolog; it would be of considerable interest to examine the molar volume contribution of cations throughout this series since the differences might be expected to follow a Stokes law molar contribution unless the hydration structure known to surround the larger cations[211] is involved in the contribution.

Levine and Bell[429] have also obtained a modified Poisson–Boltzmann equation for the potential distribution in the diffuse double layer associated with a colloidal plate in an aqueous medium; these authors discuss the influence of ion size, the variation of permittivity, the image–self-atmosphere potentials, the compressibility of the aqueous medium, and the so-called cavity potentials ξ_{cav} and ξ'_{cav}, where ξ_{cav} is due to removal of diffuse layer charge distribution from the spherical exclusion volume of the hydrated ion and ξ'_{cav} is due to a reduction in the effective charge of the ion.

The linear form of the Poisson–Boltzmann equation thus obtained is of the form

$$\left[d\left(\frac{z\psi_0}{kT} \right) \middle/ dx \right]^2 = 4\varkappa^2 \left(\sinh \frac{z\psi_0}{kT} \right)^2 [1 + C(\infty)] \tag{20}$$

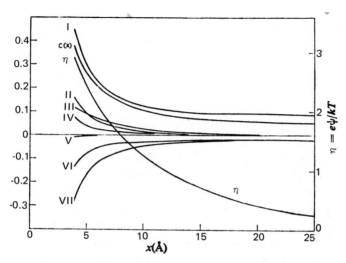

Fig. 3. Combination of the various factors in the overall correction term $C(x)$. (I) Cavity effect (ϕ_{cav}); (II) polarization energy of the ions and the dependence of ε on concentration; (III) self-atmosphere–image; (IV) dielectric saturation; (V) electrostriction; (VI) ionic volume effect; (VII) cavity effect (ϕ'_{cav}). (See Ref. 429.)

At large distances of $x(\infty)$ from the interface the collective correction term $C(\infty)$ includes the contributions from the correction terms outlined previously. The authors have calculated the variation of each contribution with distance from the interface for $ze\psi_0/kT = 4$, electrolyte concentration 10^{-1} mol dm^{-3}, and $\delta = 3.9$ Å as the distance of closest approach of the hydrated ions to the surface. Figure 3 shows that for potentials of 50–75 mV the corrections to the Poisson–Boltzmann equation are not large, due to the mutual cancellation of the various effects, but that the largest effects are caused by the cavity potentials and the ion cavity effects.

2.1.3. The Fixed Region of the Double Layer—The Stern Region

The major failing of the Gouy–Chapman model of the electrical double layer is the impossibly high concentration of ions which would result at the solid/liquid interface from the assumption that ions can be treated as unpolarizable point charges. Stern[648] and subsequently Grahame[298] overcame this fundamental difficulty by postulating that the finite size of ions placed limitations upon the numbers which may be adsorbed at the solid/liquid interface. From this distinction arises the concept of two discrete regions in the double layer, a region of thickness δ, adjacent to the interface, within which the ion size must be taken into account and the ions

are not subject to the normal thermal motions which affect the ions present
in the outermost diffuse region. The inner region takes its name from the
original proponent of the model and is known as the Stern layer, but is
itself subdivided in the model of Grahame into two further regions. (i) The
innermost region, immediately adjacent to the interface, of thickness β,
contains a layer of specifically adsorbed ions; such ions no longer have
their hydration shells intact and since anions, with the exception of the
fluoride ion, are more easily dehydrated, the innermost adsorbed layer
consists of anions irrespective of the sign of the charge on the interface
itself. This does not mean that the innermost region is devoid of water,
but water molecules present, because of the strong dipolar forces acting
between them and the interface, are not regarded as being associated with
the anions. The outermost surface of this region is termed the inner Helm-
holtz plane. (ii) The outer Helmholtz region of thickness ι, where $\beta + \iota = \delta$,
contains the hydrated counterions (ions whose sign of charge is opposite
to that of the surface); the outermost surface of this region, the outer
Helmholtz plane, is coincident with the thickness δ of the Stern region of
the double layer. Figure 4 outlines schematically the various regions of the
double layer and Fig. 5 illustrates the corresponding variation of potential
with distance from the surface.

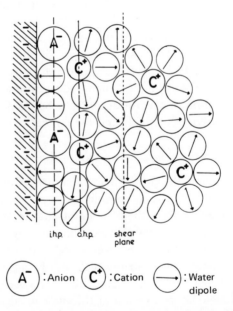

Fig. 4. A schematic representation of the electrical double layer.

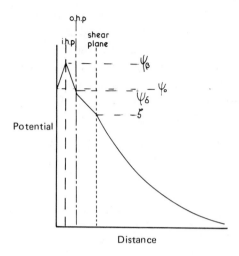

Fig. 5. Variation of potential corresponding to Fig. 4.

For the limiting case when no specific adsorption is assumed the adsorption of counterions has been described by Verwey and Overbeek[680] in the form of a Langmuir-type adsorption isotherm:

$$\sigma_s = \frac{N_1 ze}{1 + (N/M) \exp[(-ze\psi_\delta + A)kT]} \tag{21}$$

where σ_s is the surface charge density of the counterions, N is the number of adsorption centers per cm², M is the molecular weight of the solvent, N is the Avogadro number, and A is the specific chemical adsorption potential of the counterions.

In the case of the Grahame model, the number of anions per cm² in the inner Helmholtz plane is

$$n_{iHp} = 2rn \exp(-W/kT) \tag{22}$$

where r is the radius of the unhydrated anion and W is the work done in bringing an ion to the inner Helmholtz plane.

Levine and Matijevic[432] have more recently considered the relation between surface potential ψ_0 and the surface charge density Γ of stabilizing ions (the iodide ion) for the model of a negatively charged silver iodide–aqueous 1:1 electrolyte interface with and without adsorption of counterions. The model assumes that excess I^- ions are situated at the AgI solid/solution interface and that univalent cations are adsorbed up to the inner

Helmholtz plane; the Stern region is assumed to behave as a homogeneous dielectric medium (permittivity ε_1) of integral capacity K. The electrostatic potentials associated with the AgI surface, the inner Helmholtz plane, and the outer Helmholtz plane are ψ_0, ψ_β, and ψ_δ, respectively; σ_β is the surface charge density of adsorbed counterions, and the total charge of the diffuse layer per unit cross section is σ_δ. Applying the Gouy–Chapman theory to the diffuse layer, the standard relation

$$\sigma_\delta = -(\varepsilon \varkappa k T/2\pi e) \sinh(e\psi_d/2kT) \tag{23}$$

is obtained.

For the Stern region

$$\psi_0 = \psi_\beta + \frac{\beta}{\delta} \frac{\Gamma}{K} \tag{24}$$

and

$$\psi_\beta = \psi_\delta + \frac{i}{\delta} \frac{\Gamma + \sigma_\beta}{K} \tag{25}$$

Since for the condition of electrical neutrality

$$\Gamma + \sigma_\beta + \sigma_\delta = 0 \tag{26}$$

we have

$$\psi_0 = (1/K)(\Gamma + \sigma_\beta) - (2kT/e)\ln\theta \tag{27}$$

where

$$\theta = -\frac{2\pi e}{\varepsilon \varkappa kT}(\Gamma + \sigma_\beta) + \left[1 + \left(\frac{2\pi e}{\varepsilon \varkappa kT}\right)^2 (\Gamma + \sigma_\beta)^2\right]^{1/2} \tag{28}$$

If the iodide ion is assumed to be in thermodynamic equilibrium and the isoelectric point of AgI particles is taken as $pI = 10.6$, then according to the Nernst equation,

$$\psi_0 = 2.303(kT/e)(pI - 10.6) \tag{29}$$

In order to determine the relationship between ψ_0 and Γ, σ_β must be eliminated by introducing the adsorption isotherm of the counterions. The authors employ the form of the Stern adsorption isotherm[430] that allows for the discreteness-of-charge effect and the size of the hydrated counterions in the adsorbed layer; thus the chemical potential of an adsorbed counterion becomes

$$\bar{\mu}_M^{\;1} = kT[\ln v - p\ln(1 - pv/N_s] + \text{const} \tag{30}$$

where $v = \sigma_\beta/e$ is the density of adsorbed counterions, p is the number of adsorption sites occupied by partially hydrated counterions, and N_s is the number of adsorbed water molecules.

Making use of the dimensionless ratio r, where $r = \sigma_\beta/\sigma_\beta{}^\circ$ and $\sigma_\beta{}^\circ$ is a standard charge density in the discreteness-of-charge theory [equal to $(kT/e)(K\delta/\beta\iota)$ in this case], the authors deduce the relationship

$$r - \ln r + p \ln(1 - pbr) = 2 \ln \theta + \frac{\iota}{\delta} \; \frac{e}{kTK} \; (\Gamma + r\sigma_\beta{}^\circ)$$

$$- 2 \ln \varkappa + \frac{e^2\varkappa}{2\varepsilon kT} \left(\frac{1}{1 + \varkappa a} \right) + C \quad (31)$$

where b is the dimensionless quantity $\sigma_\beta{}^\circ/N_s e$, C is a constant which includes the chemical part of the adsorption energy of the counterions, and a is the diameter of an ion in the electrolyte. The first and third terms of this equation represent the effects of discreteness of charge and counterion size.

This equation is solved by assigning values to the parameters N_s, $\iota\delta$, and a; the adsorption energy parameter C is determined by choosing typical values for the potential ψ_δ at the outer Helmholtz plane; for example, at $r = 1$ and an electrolyte concentration of 0.1 mol dm^{-3}, $N_s = \frac{1}{2} \times 10^{15}$ cm^{-2}, $\iota/\delta = \frac{1}{3}$, and $a = 0.5$ nm. The value of N_s corresponds to an area of 0.20 nm^2 per adsorption site which, according to Parsons[563,564] and Damashin,[138] implies that water adsorbed into the inner Helmholtz region is arranged, on average, in clusters of two or three molecules; p values of one and two imply, respectively, areas of 0.20 nm^2 and 0.40 nm^2 per adsorbed (partially hydrated) counterion.

Experimental data[455,458] show that the integral capacity K of the inner region decreases with increase in the magnitude of Γ and Levine and Matijević[432] assume the relationship

$$K = 20 + J(\Gamma - \Gamma_0) \quad (32)$$

where Γ_0 is the value of Γ at $r = 1$ and J is given values of 0(5)25. Figure 6 illustrates the relationship between Γ and ψ_0 for three electrolyte concentrations, 0.1, 0.01, and 0.001 mol dm^{-3}; the extent of counterion adsorption is indicated by the short lines on each curve; thus the approximately linear relationship between Γ and ψ_0 observed experimentally is reproduced by a variation of K with Γ.

Bell and Levine[44] have recently considered the problem of the effect of the ions adsorbed into the Stern region of the double layer and of the

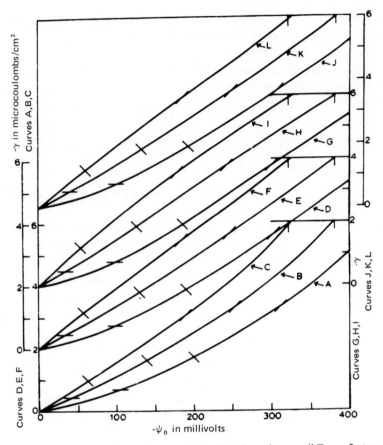

Fig. 6. Plots of (Γ, ψ_0) at 25°C in the case of specific adsorption. $e\psi_d/kT = -2$ at $r = 1$. Each group of three curves corresponds to $c = 0.001$, 0.01, and $0.1 \, \text{mol dm}^{-3}$. For groups (A, B, C), (D, E, F), and (G, H, I), $p = 2$, $C = 0.987$, and $J = 0$, 15, and 25 respectively, in eqn. (31). For group (J, K, L), $p = 1$, $C = 1.05$, and $J = 15$. (See Ref. 432.)

difference in dielectric properties of the inner region compared with that of the bulk electrolyte upon the self-atmosphere potential of an ion placed close to the Stern region but within the diffuse double layer. These authors divide the inner Stern region into regions of differing dielectric behavior, that between the inner Helmholtz plane and the solid surface being of permittivity ε_1 and that between the inner and outer Helmholtz planes being of permittivity ε_2. The authors suggest that a layer of water molecules within the inner Helmholtz layer whose polarizability differs from that of the bulk medium, due to boundary effects, makes no marked difference to the image effect at an uncharged dielectric interface, particularly at low electrolyte concentrations and longer distances from the interfaces.

Levine[427] has recently considered the modifications necessary to the Grahame model of the Stern region for a plane charged mercury/electrolyte interface when adsorption of ions into the inner Helmholtz layer is known to occur. The adsorption isotherm for univalent anions situated in the inner region is of the form

$$\ln \Theta + g(\Theta) = F(P, T) + (e\psi_A/kT) + \ln a_- \tag{33}$$

where Θ is the fraction of total coverage by specifically adsorbed anions at the mercury surface, $g(\Theta)$ is an entropic term accounting for ion size, $F(P, T)$ is a function of P and T, ψ_A is the true potential at an adsorbed layer due to surrounding change, given by

$$\psi_A = \psi_\beta + [4\pi\beta\iota\sigma_\beta/(\varepsilon_1\iota + \varepsilon_2\beta)] \tag{34}$$

and a_- is the activity of the anion in the aqueous phase. Using the Flory[245] and Huggins[365] volume fraction type statistics to describe the entropy of mixing of the adsorbed counterions with the water molecules in the inner region, $g(\Theta)$ can be described by the power series

$$g(\Theta) = s^2(\Theta) + \tfrac{1}{2}s^3\Theta_2 + \cdots \tag{35}$$

where s is the ratio of the area occupied by a partially hydrated ion to that occupied by a water molecule and $\Theta = v/N_s$; $1/N_s$ is the area occupied by a water molecule in the inner region and v is the density of adsorbed ions on the inner Helmholtz plane.

Figure 7 illustrates the agreement between the theoretical curves and the experimental data of Grahame and Parsons,[300] Wroblowa et al.,[718] Lawrence et al.,[424] and Grahame.[299]

2.1.4. The Concept of Zeta Potential, the Shear Plane, and Electrokinetic Phenomena

In the presence of an electric field the particles of a hydrophobic colloidal dispersion will move along the lines of force of the field, proving that the particles themselves are electrically charged even though the whole solution is electrically neutral; movement of the particles under the influence of an applied electric field is one example of an electrokinetic phenomenon and is known as electrophoresis.

In the process of movement through the dispersion medium it is clear that a plane of shear must occur between the particle and the medium;

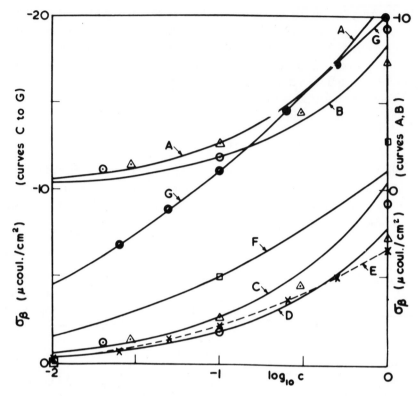

Fig. 7. Surface charge density of adsorbed anions σ_β (μC cm^{-2}) as function of $\log_{10}c$, where c is the anion concentration (mol dm^{-3}), for the ions Cl$^-$, Br$^-$, and I$^-$. Experimental values of σ_β (μC cm^{-2}) at $c = 0.1$ mol dm^{-3}: Cl$^-$ ion: (A) $\sigma_\beta = -2.7$[718]; (B) $\sigma_\beta = -1.9$[300]; $p = 1$; circles,[299] triangles[718] are experimental points. Cl$^-$ ion: (C, D) $p = 2$; same σ_β and experiments as for (A, B). Cl$^-$ ion: (E) constant ionic strength (1 mol dm^{-3}); $\sigma_\beta = -2.2$[424]; $p = 2$; crosses, experimental data.[424] Br$^-$ ion: (F) $\sigma_\beta = -5.1$[424]; $p = 2$; squares, experimental data.[424] I$^-$ ion: (G) $\sigma_\beta = -11.1$[299]; $p = 2$; double circles, experimental values.[299] (See Ref. 427.)

the precise location of this plane, if indeed it can be thought of as a plane rather than a region of definite thickness, is a problem which has not been resolved with complete satisfaction. Following the previous discussion, the most probable position of the plane might be expected to lie between the fixed Stern region of the double layer and the diffuse region, i.e., coincident with the outer Helmholtz plane; this has often been regarded as being the case and the potential at the shear plane, the zeta (ζ) potential, as being of the same magnitude as ψ_δ. For the shear plane coincident with the outer Helmholtz plane it can be shown that the electrophoretic velocity v_e of the particle is given by

$$v_e = \varepsilon\zeta E/4\pi\eta \tag{36}$$

where E is the strength of the applied field and η is the viscosity of the dispersion medium. The validity of this equation is subject to the following conditions:

(a) The double layer must be thin compared with the radius of the particle, i.e., $\varkappa a \gg 1$.

(b) The particle must be insulating, with negligible surface conductance.

Such a treatment, however, neglects the deformation of the external applied field which occurs in the close environment of the particle and Henry[338] showed that for a spherical particle subject to this effect the electrophoretic velocity is given by the equation

$$v_e = f_H \varepsilon\zeta E/\pi\eta \tag{37}$$

where f varies as a function of $\varkappa a$, being $\frac{1}{4}$ when $\varkappa a$ is very large and $\frac{1}{6}$ when $\varkappa a$ is small. The $\frac{1}{6}\pi$ form of the equation was also obtained by Hückel[364] from a consideration of the retardation effect of the diffuse layer upon the motion of the particle, and more recently Wiersema et al.[707] have reported a general solution of the problem which takes account of the effect of electrophoretic retardation, i.e., the drag on the particle exerted by solvent and counterions moving in the opposite direction, and the relaxation effect of the center of the ionic atmosphere lagging behind the particle (a further retardation effect). The authors define the function

$$\Phi = (6\pi\eta e/\varepsilon kT)v_e/E \tag{38}$$

and

$$q_0 = \varkappa a/\lambda \tag{39}$$

where $\lambda^2 = (z_+ + z_-)/2z_-$, z_+ and z_- being the valences of the small ions in the solution. Taking

$$\psi_0 = e\zeta/kT \tag{40}$$

and

$$m_\pm = (\varepsilon kT/6\pi\eta e^2)f_\pm \tag{41}$$

where the friction coefficients $f_{\pm}$ of the small ions are expressed as

$$f_{\pm} = Ne^2 z_{\pm}/\lambda_{\pm}^{\circ} \tag{42}$$

in terms of $\lambda_{\pm}^{\circ}$, the limiting equivalent conductances of the ions, the authors found that for symmetric electrolytes

$$\Phi = \gamma_0 f_1(\varkappa a) - \gamma_0{}^3[z^2 f_3(\varkappa a) + \tfrac{1}{2}(m_+ + m_-) f_4(\varkappa a)] \tag{43}$$

Thus the surface conductance, where it occurs outside the surface of shear, is taken into account by the transport equations $m_{\pm}$. With the same notation the Smoluchowski equation would become

$$\Phi = \tfrac{3}{2}\gamma_0 \tag{44}$$

and Hückel's equation would be

$$\Phi = \gamma_0 \tag{45}$$

and the approximation of Henry would have the form

$$\Phi = \gamma_0 f_1(\varkappa a) \tag{46}$$

Wiersema *et al.*[707] suggest that only surface conductance outside the shear surface is of importance in the determination of the ζ potential, since the ions inside the shear plane can be regarded as being at least as firmly bound as the water molecules, and hence effectively equate the ζ potential with ψ_δ. Dirkhin and Semenikhim[206] have questioned this assumption and suggested that it is necessary to take account of the transfer of ions across the outer Helmholtz plane, leading to the ζ potential and ψ_δ being unequal; by application of the phenomenon of diffusiophoresis* the authors derive somewhat complex equations which distinguish between ψ_δ and the ζ potential.

---

* Diffusiophoresis was first reported by Durkhin and Derjaguin[205] as being the motion of colloidal particles under the influence of a macrogradient of ionic concentration, i.e., a diffusion field in the absence of an external field caused by polarization of the double layer under the influence of a macrogradient of concentration. According to Durkhin,[206] the macrogradient of concentration in the bulk of the dispersion medium causes a drop in the concentration of ions in the outer Helmholtz plane of the double layer; when the double layer is thin, however, and since equilibrium exists between the ions, the double layer rearranges according to the distribution of the concentration of ions at the outer Helmholtz boundary and is polarized, i.e., a microgradient of concentration is induced in the vicinity of the particle.

2.2. Electrostatic Repulsive Forces between Colloidal Plates and Particles

Colloidal species together with their associated ions are electrically neutral; Coulombic repulsion between the particles does not therefore occur; an electrostatic repulsive force will arise, however, between the particles when the diffuse regions of their respective double layers overlap. It has been assumed in the early work of Verwey and Overbeek[680] and Derjaguin and Landau[170] and the later work of Glazman and co-workers [287–290] and Barboi[27–29] that the particle or plate potential ψ_0 is independent of the separation distance between the two surfaces; recently Frens et al.[265] have suggested from experimental studies that for the AgI/aqueous electrolyte interface at least, because of rapid Brownian collisions, this condition is not appropriate and Frens[264] considers that a better assumption is that the surface charge density σ_0 on the particle or plate is independent of the separation distance. Levine and Jones,[431] however, have shown that, at least in the case of larger separations, the interaction potential is the same whether the constant-potential or constant-charge condition is applied.

In considering the double layer interaction between two colloidal plates, Verwey and Overbeek[680] and Derjaguin and Landau[170] made use of the Gouy–Chapman theory for the diffuse region of the double layer (making the assumption of no Stern region in the electrical double layer) and the simplifying assumption that the potential at the plane midway between the plates is the sum of the potentials which would be produced by either plate in the absence of the other (Fig. 8). Thus the potential at a distance $x = d/2$, where d is the distance between the plates, is $\psi_{d/2}$ and since the potential curve passes through a minimum at this point, $(d\psi/dx)_{x=d/2} = 0$.

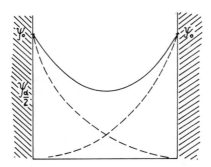

Fig. 8. The variation of potential as a function of the distance between two parallel plates spaced a distance d apart.

The first integration of the Poisson–Boltzmann equation, with the new boundary conditions for a uni-univalent electrolyte, yields the equation

$$d\psi/dx = -(8\pi kT/\varepsilon)^{1/2}[2\cosh(ze\psi/kT) - 2\cosh(ze\psi_{d/2}/kT)]^{1/2} \quad (47)$$

If $\phi = ze\psi/kT$, $\phi_0 = ze\psi_0/kT$, and $\phi_{d/2} = ze\psi_{d/2}/kT$, then

$$d\phi/dx = -(2\cosh\phi - 2\cosh\phi_{d/2}) \quad (48)$$

Integration of eqn. (48) between the limits of $x = 0$ and $x = d/2$, and $\psi = \psi_0$ to $\psi = \psi_{d/2}$ yields

$$\varkappa d = \sqrt{2} \int_{\phi_{d/2}}^{\phi_0} \frac{d\phi}{(\cosh\phi - \cosh\phi_{d/2})^{1/2}} \quad (49)$$

Verwey and Overbeek[680] have published numerical solutions of this integral giving $\psi_{d/2}$ for a variety of interplate distances and plate surface potentials ψ_0.

Since the relationship between surface charge σ and potential is given by the equation

$$\sigma = -\frac{\varepsilon}{4\pi}\frac{d\psi}{dx} = -\left(\frac{\varepsilon nkT}{2\pi}\right)^{1/2} 2\sinh\frac{\phi_0}{2} \quad (50)$$

then the charge of interacting double layers σ'_{dl} is given by the equation

$$\sigma'_{dl} = (n\varepsilon kT/2\pi)^{1/2}[\exp\phi + \exp(-\phi) - \exp\tfrac{1}{2}\phi_d - \exp(-\tfrac{1}{2}\phi_d)]^{1/2} \quad (51)$$

The free energy of the double layer G_{dl} can be obtained by a Debye–Hückel charging mechanism, thus

$$G_{dl} = -\int_{\psi=0}^{\psi=\psi_0} \sigma \, d\psi \quad (52)$$

Thus

$$G_{dl} = -(\varepsilon nkT/2\pi)^{1/2}\int_0^{\psi_0} 2\sinh\tfrac{1}{2}\phi_0 \, d\psi$$

$$= -(8nkT/\varkappa)(\cosh\tfrac{1}{2}\phi_0 - 1) \quad (53)$$

The repulsion between colloidal plates is not determined by the magnitude of the free energy as such but by the change in free energy as the two plates approach each other from an initial position of infinite separation; this change in free energy is equivalent to the work which must be done against forces arising from interaction between the double layers as they approach a separation distance d,

$$V_R = 2(G'_{dl} - G_{dl}) \tag{54}$$

where G'_{dl} is the free energy of the interacting double layer system

$$G'_{dl} = -(nkT/\varkappa)\{\varkappa d[\cosh(\tfrac{1}{2}\phi_d) - 1] + 2[(\sinh \tfrac{1}{2}\phi_0) - \sinh^2(\tfrac{1}{2}\phi_{d/2})]^{1/2}$$
$$+ \varkappa d(\sinh \tfrac{1}{2}\phi_d) - 4(\exp \tfrac{1}{2}\phi_{d/2}) \cdot E \tag{55}$$

where the elliptical integral E is defined as

$$E = \int_{\text{arcsin exp } \frac{1}{2}(-\phi - \phi_{d/2})}^{\pi/2} [1 - \exp(\phi - \phi_{d/2})]^{1/2} \, d \arcsin \exp \tfrac{1}{2}(-\phi - \phi_{d/2}) \tag{56}$$

Evaluation of the resulting complicated expression for V_R has been undertaken by Verwey and Overbeek[680] and Kruyt.[115]

The problem of the resulting force between two double layers has also been considered by Langmuir[422] and Derjaguin[157] and for large separations, i.e., little interaction between the plates, a simple solution is obtained. In this model the system is considered to be in equilibrium because at any point the force due to the hydrostatic pressure which exists between the plates in an aqueous medium is balanced by the force of the space charge:

$$dp + \varrho \, d\psi = 0 \tag{57}$$

Introducing the Poisson equation and restricting consideration to the perpendicular distance between the plates, we obtain

$$\frac{dp}{dx} - \frac{\varepsilon}{4\pi} \frac{d^2\psi}{dx^2} \frac{d\psi}{dx} = 0 \tag{58}$$

which can be integrated to give

$$p - (\varepsilon/8\pi)(d^2\psi/dx^2) = \text{const} \tag{59}$$

where the constant equals $P_{d/2}$, the hydrostatic pressure midway between the plates. The difference between $P_{d/2}$ and P_0, the pressure of the bulk solution, is the force P due to double layer interaction which is driving the plates apart; thus

$$P = P_{d/2} - P_0 = \int_{\psi=0}^{\psi=\psi_d} dp = -\int_{\psi=0}^{\psi=\psi_d} p \, d\psi \tag{60}$$

and using the Boltzmann substitution

$$p = -2nze \sinh\phi \tag{61}$$

then

$$P = 2nkT(\cosh\phi_d - 1) \tag{62}$$

For large separations of the plates the approximation

$$P = nkT(\phi_{d/2})^2 \tag{63}$$

is allowed, and since

$$V_R = -2\int_\infty^{d/2} P\, dd$$

V_R is then given by

$$V_R = \frac{64nkT}{\varkappa}(\exp -\varkappa d)\frac{\exp(ze\psi_0/2kT) - 1}{\exp(ze\psi_0/2kT) + 1} \tag{64}$$

2.2.1. Corrections to the Poisson–Boltzmann Equation for the Free Energy of Interaction of the Diffuse Double Layer Regions

Bell and Levine[43] have considered the situation which exists in relatively concentrated colloidal dispersions when overlap of electrical double layers occurs and the whole of the dispersion medium is comprised of the diffuse part of the double layers. In consequence there is no point in the dispersion medium where the ionic concentrations are unaffected by changes in either charges or mutual orientation of the particles. The electrostatic contribution to the double layer free energy is thus expressed in the form

$$G_e = E_e - TS_e \tag{65}$$

where E_e is the electrostatic energy of the system and S_e is an expression of the effects of ionic distributions. If the charging of the double layer is considered from the viewpoint of a Debye–Hückel mechanism and is carried out for a fixed volume distribution of ions, then the work done in charging G_e is identical with the electrostatic energy E_e. In the case of a nonuniform ion distribution, which characterizes the development of a double layer field in a concentrated dispersion, the ratio $-(G_e - E_e)/T$, i.e., the entropy term S_e, is subdivided by the authors into contributions from the dispersion medium S_e^D and from the interior of the colloidal particle S_e^I. Contributions to the latter may arise, in the case of emulsion droplets, for example, from a Boltzmann distribution of ions in the particle interior or, in the case of crystalline materials, from Schottky and Frenkel defects.[309] The contributions to S_e^D, which account for the volume effect of the hydrated ion, arise from the total number of water molecules per unit volume of the diffuse layer and the hydration shell which is regarded as being permanently attached to each ion.[233,269] This contribution is shown to lead to a decrease in the free energy change arising from electrolyte concentration changes in the dispersion medium compared to the case of a dilute sol.

The authors have extended their model[429] to allow for the additional effects due to the fluctuation potentials, cavity potentials, and the compressibility of the dispersion medium, together with the variation of permittivity with electrolyte concentration.[643] Although the calculations are not complete, the corrections indicate that a decrease in repulsion will occur, but, at least in the case of 1:1 electrolytes, due to partial cancellation of differing correction factors, the overall effect upon colloid stability may well be minimal.

Samfeld and co-workers[617] have applied the local thermodynamic approach to the repulsive energy between colloidal particles arising from interaction between overlapping diffuse double layers. The repulsive force per unit surface area of colloidal plate (π) immersed in a 1:1 electrolyte solution is given by the equation[616]

$$\pi = \frac{\Gamma_0 RT c_a' [(\gamma'_{+0}/\gamma_{+0})e^{-\phi} + (\gamma'_{-0}/\gamma_{-0})e^{\phi} - 2(\Gamma_0'/\Gamma_0)]}{1 + \Gamma_0' c_a' [(\gamma'_{+0}/\gamma_{+0})V_{+0}^* e^{-\phi} + (\gamma'_{-0}/\gamma_{-0})V_{-0}^* e^{\phi}]} \tag{66}$$

where $\gamma_{\pm 0}$ are the activity coefficients of the ions at zero field, c_a' is the molar concentration of the electrolyte, Γ_0 is the osmotic coefficient at zero field, $V_{\pm 0}^*$ are the standard molar volumes, and ϕ is the reduced electric potential $zF\psi/RT$.

For the case where activity and osmotic coefficients are assumed to be unity in the whole system, π becomes

$$\pi = 2\Gamma_0' c_a' RT(\cosh\phi - 1)\frac{1}{1 + \Gamma_a' c_a'(V_{+0}^* e^{-\phi} + V_{-0}^* e^{\phi})} \tag{67}$$

which, when compared with the classic DLVO equation for the repulsive force

$$\pi_{DLVO} = 2c_a' RT(\cosh\phi - 1) \tag{68}$$

suggests that the difference between the two equations is due to account being taken of the finite size of the ionic components and the nonideality of the system.

Comparison of eqn. (19) with the Poisson–Boltzmann equation

$$\ln(c_a/c_a') = -zF\psi/RT \qquad (69)$$

shows that the discrepancies between the results of Sanfeld and the Verwey–Overbeek model are related not only to the finite ion size ($V_{\pm 0}^*$) and the nonideality of the solution ($\gamma_{\pm 0}$) but also to the polarizability ($\gamma E/\partial C$) and to the electrostrictive effect ($P_0 - P$) given by eqn. (18).

The combination of these equations enables the repulsive pressure to be determined as a function of interparticle distance assuming (as shown in earlier discussion) that the permittivity ε is given by eqn. (13).

For spheres of radius a with a shell of radius δ separating the diffuse layer from the adsorbed layer, the repulsive energy per unit area of surface of two spheres V_R at a separation d is given by the relation

$$V_a = 2\pi a \int_{d_0/2}^{\infty} V_\pi(d)\, \delta d \qquad (70)$$

Table II contains the thermodynamic quantities for aqueous NaCl, LiCl, and HCl, and the parameters of the Verwey–Overbeek model. Based on

TABLE II. Thermodynamic Parameters for the Sanfeld Model of the Electrical Double Layer

	$V_{+0}^{*,a}$, cm³ mol⁻¹	V_{-0}^*, cm³ mol⁻¹	$\partial\varepsilon/\partial c_+$, cm³ mol⁻¹	$\partial\varepsilon/\partial c_-$, cm³ mol⁻¹	$\langle\gamma_0\rangle$	Γ_0'
HCl	0.1	18	-17×10^3	-1.5×10^3	1	1
NaCl	-1.6	18	-8×10^3	-1.5×10^3	1	1
LiCl	-1.1	18	-12×10^3	-1.5×10^3	1	1
LiCl Γ_0' [b]	-1.1	18	-12×10^3	-1.5×10^3	1	0.94
LiCl $\langle\gamma_0\rangle$ [c]	-1.1	18	-12×10^3	-1.5×10^3	1.02, 1.04, 1.1	1
Verwey-Overbeek	0	0	0		1	1

[a] For values of V_{S0}^*, and $\partial\varepsilon/\partial c_s$ see Hepler.[339]
[b] The value of $\Gamma_0' = 0.94$ is reported by Robinson and Stokes[604] for an aqueous solution of 10^{-4} mol cm⁻³.
[c] $\langle\gamma_0\rangle$ is the average value of γ_{+0}'/γ_{+0} or γ_{-0}'/γ_{-0} in the diffuse layer. The different values of $\langle\gamma_0\rangle$ for LiCl are chosen arbitrarily to show in which direction and to what extent activity coefficients influence the calculations.

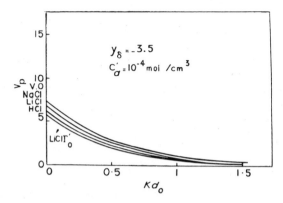

Fig. 9. Reduced repulsive potential energy V_p as a function of the reduced distance $\varkappa d_0$ for the Verwey–Overbeek model (VO), NaCl, LiCl, LiCl ($\Gamma_0' = 0.94$), and HCl. (See Ref. 617.)

these data, ($h = 3 \times 10^7$ es cgs units; $T = 298°K$; $V_{H_2O}^* = 18 \text{ cm}^3 \text{ mol}^{-1}$; and $\varepsilon_0 = 78.54$) the reduced repulsive energy V_p has been determined:

$$V_p = (\varkappa^2/4\pi a c_a' RT)V_a \qquad (71)$$

and Fig. 9 illustrates V_p as a function of the reduced distance $\varkappa d_0$ for the Verwey–Overbeek model, NaCl, LiCl, LiCl ($\Gamma_0' = 0.94$), and HCl at the same reduced potential ($\psi_\delta = 90 \text{ mV}$); the values of V_p obtained are smaller than those obtained by the Verwey–Overbeek model except at large inter-particle distances and follow the sequence

$$\text{Verwey–Overbeek} > \text{NaCl} > \text{LiCl} > \text{HCl}$$

The authors suggest that this type of lyotropic series is due to the polar-izability effect ($\partial\varepsilon/\partial C$) and the dielectric saturation effect ($\partial\varepsilon/\partial E$). A similar effect is observed when the spheres touch each other ($\varkappa d_0 = 0$), the discrep-ancy becoming larger with increasing electrolyte concentration (Fig. 10).

Figure 9 shows that in the case of LiCl ($\Gamma_0' = 0.94$) the osmotic coeffi-cient acts in the same direction as the polarization effect, but Fig. 10 shows that the differences due to the mean ionic activity coefficients are much larger and in the reverse order with respect to the Verwey–Overbeek model.

2.2.2. The Contribution of the Stern Region to the Electrostatic Repulsion Energy between Particles

In the Poisson–Boltzmann equation for the repulsion energy between colloidal particles repulsion is considered as arising from the overlapping

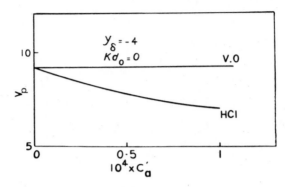

Fig. 10. Reduced repulsive potential energy V_p as a function of electrolyte concentration for the Verwey–Overbeek model (VO) and HCl when the diffuse layers are completely blended.

and interaction of diffuse double layers; corrections to the model arising from the volume of the ions in the double layer, their hydration and fluctuation potentials, the compressibility of the solvent medium, and other factors have been considered. The surface potential referred to in the previous section is due to the neglect of interactions involving the Stern region of the double layer, ψ_δ, and not the potential at the surface of the particle, ψ_0; any possible contribution to the repulsion energy between colloidal particles arising from the presence of a layer of specifically adsorbed ions within the Stern region of the double layer has been neglected.

Very few theoretical examinations have been published which consider this effect, although there are several reports by Lyklema and co-workers [64,682] on the flocculation of aqueous AgI dispersions from aqueous electrolyte in the presence of neutral molecules (e.g., alcohols) in which the authors attribute the flocculation behavior to competitive adsorption of alcohol molecules into the Stern region of the double layer. Specific adsorption of this kind is considered to be responsible for the lyotropic character of the relative efficiency of the alkali metal ions in promoting flocculation; it seems possible that a similar series would arise from considerations of the variation in effective molar volumes of the ions in the diffuse region of the double layer, but discussion of this is postponed to the appropriate section of this review.

One of the few published reports on the effect of specific adsorption in the Stern region upon the repulsion energy between particles is that of Deveraux and de Bruyn,[186] who have considered the effect of specific adsorption on the interaction between two colloidal plates on the basis of the Grahame model of the electrical double layer (Fig. 11). Points 0 and

x_5 locate the two solid surfaces; x_1 and x_4, the inner Helmholtz planes, corresponding to the layers of specifically adsorbed ions, and x_2 and x_3, the limits of the diffuse region. The permittivity is different for each region; $\varepsilon_{i,i+1}$ is the permittivity of the region $x_1 < x < x_{i+1}$. In the regions $0 < x < x_1$, $x_1 < x < x_2$, $x_3 < x < x_4$, and $x_4 < x < x_5$ the potential is assumed to vary linearly with distance; in the diffuse region the potential is described by the Poisson–Boltzmann equation.

For the diffuse region of the double layer Verwey and Overbeek[680] have shown that the free energy per unit area of a plate G_{dl} can be given by an alternative expression to eqn. (52):

$$G_{dl} = G(0) + \int_0^1 \int_{x_2}^{x_3} (\varrho' \psi' / \lambda) \, dx \, d\lambda \tag{72}$$

where λ is a parameter describing the state of the discharge process, being unity initially and finally zero when the system is discharged; ϱ' and ψ' are the space charge density and potential as functions of λ. The authors show that for the Gouy–Chapman double layer

$$G(0) = -[\varepsilon_{23}/8\pi(x_3 - x_2)](\psi_2 - \psi_3)^2 \tag{73}$$

and G reduces to

$$G_{dl} = \frac{2nkT}{\varkappa} \left[(\xi_3 - \xi_2) \left(\frac{L}{2} + 1 \right) - \int_{\phi_2}^{\phi_3} \frac{d\phi}{d\xi} \, d\phi \right] \tag{74}$$

where $\phi = ze\psi/kT$, $\xi = (8\pi n z^2 e^2/\varepsilon_{23}kT)^{1/2}x = \varkappa x$, and $C = -2 \cosh \bar{\phi}$, where $\bar{\phi}$ is the value of ϕ at the minimum of potential.

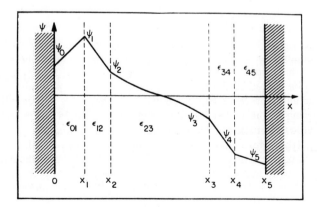

Fig. 11. Double layer interaction according to the Grahame model (see Ref. 186).

By application of Gauss's law to the Grahame model at $x = 0$, x_1, x_2, x_3, x_4, and x_5, $G_G(0)$ for this model is evaluated:

$$G^G(0) = \frac{- \varepsilon_{01}(\psi_0 - \psi_1)^2}{8\pi x_1}$$

$$- \frac{(\psi_1 - \psi_4)^2}{8\pi\{[(x_2 - x_1)/\varepsilon_{12}] + [(x_3 - x_2)/\varepsilon_{23}] + [(x_4 - x_3)/\varepsilon_{34}]\}}$$

$$- \frac{\varepsilon_{45}(\psi_4 - \psi_5)^2}{8\pi(x_5 - x_4)} \tag{75}$$

and G_{dl}^G is found to be

$$G_{dl}^G = \frac{2nkT}{\varkappa}\left[(\xi_3 - \xi_2)\left(\frac{L}{2} - 1\right) - \int_{\phi_2}^{\phi_3} \frac{d\phi}{d\xi}\, d\phi\right]$$

$$- \frac{\varepsilon_{01}(\psi_0 - \psi_1)^2}{8\pi x_1} - \frac{\varepsilon_{12}(\psi_1 - \psi_2)^2}{8\pi(x_2 - x_1)} - \frac{\varepsilon_{34}(\psi_3 - \psi_4)^2}{8\pi(x_4 - x_3)}$$

$$- \frac{\varepsilon_{45}(\psi_4 - \psi_5)^2}{8\pi(x_5 - x_4)} \tag{76}$$

The energy of interaction between the two double layers is shown by the authors to be

$$V^G = V_R + (8nkT/\varkappa)(\cosh \tfrac{1}{2}\phi_{2\infty} + \cosh \tfrac{1}{2}\phi_{3\infty} - \cosh \tfrac{1}{2}\phi_2 - \cosh \tfrac{1}{2}\phi_3)$$

$$+ [\varepsilon_{12}/8\pi(x_2 - x_1)][(\psi_1 - \psi_{2\infty})^2 - (\psi_1 - \psi_2)^2]$$

$$+ [\varepsilon_{34}/8\pi(x_4 - x_5)][(\psi_{3\infty} - \psi_4)^2 - (\psi_3 - \psi_4)^2] \tag{77}$$

where V_R is the interaction energy between the two plates according to the Gouy–Chapman model, and $\psi_{2\infty}$ and $\psi_{3\infty}$ are the potentials at the diffuse layer limits in the absence of interaction.

It is clear that repulsive interaction between the two plates will be increased by specific adsorption compared to the case when only diffuse layer interactions are considered. However, the authors do not give a quantitative assessment of the magnitude of the difference. In addition, the authors do not comment upon the relative magnitudes of ε_{01}, ε_{12}, ε_{23}, ε_{34}, and ε_{45}, but it would appear reasonable that the particular value of $\varepsilon_{i,i+1}$ might be determined from the model of Hasted[325] using the appropriate values of δ_+ and δ_- based upon the hydration of the ions comprising the particular layers.

3. THE ATTRACTIVE FORCES BETWEEN COLLOIDAL PARTICLES

3.1. London–van der Waals Attractive Forces

The addition of electrolyte to a hydrophobic dispersion results in the flocculation of the dispersion, caused by the reduction in the repulsive interaction between the particles; Fig. 12 (from the data of Kruyt[415] illustrates the marked suppression of the potential energy of repulsion which results from decreasing values of $\varkappa a$ between two spherical particles caused by the addition of a symmetric 1:1 electrolyte. Suppression of the double layer in this manner leads to instability in the colloidal dispersion,

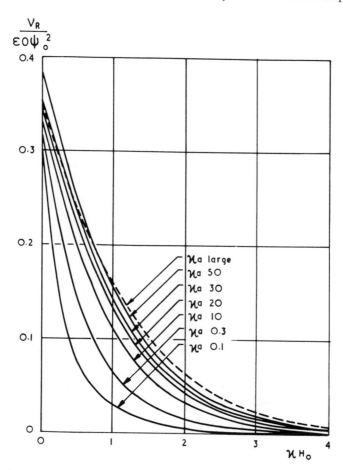

Fig. 12. Potential energy of repulsion between two spherical particles for different values of $\varkappa a$ (see Ref. 415).

revealing the presence of an attractive force which is independent of the chemical nature of the particle.

The only attractive forces sufficiently general in character to provide the necessary energy for flocculation are the London dispersion forces; such a force is one component of the wider group of attractive forces known as van der Waals forces, the other two being the Keesom force and the Debye force. The Keesom force originates in the interaction of adjacent dipoles and the Debye force from the induction of a dipole in an atom caused by the presence of a dipole in an adjacent atom. The London dispersion force, however, results from the charge fluctuations in an atom associated with the motion of the electrons; such fluctuations produce a transient dipole moment in the atom, and a phase difference between the transient dipoles of adjacent atoms results in an attractive force between them.

All three types of van der Waals forces show a similar dependence upon the interatomic distance, the energy of attraction varying inversely with the sixth power of the distance; however, in the case of aggregates of atoms, e.g., colloidal particles, the total interaction energy due to Keesom and Debye forces is not equal to the sum of the separate interaction energies but is much smaller. The London energy between two atoms, on the other hand, is, to a first approximation, independent of the interaction with other atoms; thus the total London interaction energy between two colloidal particles is found by simple addition of the individual energies.

The London dispersion interaction[446] between two atoms of polarizability α is given by

$$V_{\mathrm{L}} = -3h\nu\alpha^2/4r^6 \tag{78}$$

where $h\nu$ is the energy corresponding to the main dispersion frequency ν in the dispersion spectrum of the atom. According to Slater and Kirkwood[635] for a system of S oscillating electrons per atom

$$V_L = -(3h\nu\alpha^2/4r^6)\sqrt{S} \tag{79}$$

and since ν can be expressed in terms of the electronic charge e and the electron mass m_e, then

$$\nu = (1/2\pi)(e^2/m_e\alpha)^{1/2} \tag{80}$$

and

$$V_{\mathrm{L}} = -\frac{3\alpha^{3/2}he}{8\pi r^6}\left(\frac{S}{m_e}\right)^{1/2} \tag{81}$$

the polarizability α being obtained from optical data:

$$\alpha = \frac{3}{4\pi N} \frac{n^2 - 1}{n^2 + 2} \tag{82}$$

where n is the refractive index at long wavelengths.

For particle diameters or separation distances $\gtrsim 10$ nm a reduction in the attractive force between the particles results from the fact that the time of travel required for an electromagnetic wave to pass from one particle to another is longer than the time of revolution of the electrons—termed the retardation effect by Casimir and Polder[101]; thus V_L becomes

$$V_{\mathrm{L}} = -(3\alpha^2 he/4r^6) f(p) \tag{83}$$

where $p = 2\pi r/\lambda$ and $\lambda = c/\nu$. For $0 < p < 3$

$$f(p) = 1.01 - 0.14p \tag{84}$$

and for $p > 3$

$$f(p) = 2.45/p - 2.04/p^2 \tag{85}$$

Clayfield et al.[120] have recently reported an exact solution of the Casimir–Polder equations for the interaction of two spheres and of a sphere and plate; in the latter case the interaction is shown to be reduced by about 27% for a 3 nm sphere 100 nm from a plate, becoming larger with increasing distance and particle size.

3.2. Attractive Forces between Two Parallel Colloidal Plates

London dispersion forces between atoms are of only relatively short range but Kallman and Willstätter[391] showed that, because of the approximate additivity of dispersion forces, the attractive forces between colloidal particles are of much longer range and in fact are of the order of colloidal dimensions. From this consideration Verwey and Overbeek[680] have shown that the attractive potential between two infinite colloidal plates of thickness δ a distance d from each other in a vacuum is given by

$$V_A = -\frac{A}{48\pi} \left[\frac{1}{\frac{1}{2}d^2} + \frac{1}{(\frac{1}{2}d + \delta)^2} - \frac{2}{(\frac{1}{2}d + \frac{1}{2}\delta)^2} \right] \tag{86}$$

If $d \gg \delta$, then the following approximation results:

$$V_A = -A/12\pi d^2 \tag{87}$$

For the inverse situation, i.e., $d \ll \delta$,

$$V_A = -A\delta^2/2\pi d^4 \tag{88}$$

The constant A in eqns. (86)–(88) is the Hamaker constant, where

$$A = \tfrac{3}{4}\pi^2 q^2 h v \alpha^2 \tag{89}$$

Here q is the number of atoms contained in 1 cm^3 of the particle material.

In practice, since colloidal plates are surrounded by a dispersion medium such as water, and not a vacuum, the Hamaker constant must take account of the fact that each plate displaces an equal volume of water, and

$$A = A_{11} + A_{00} - 2A_{10} \tag{90}$$

where A_{11} refers to London dispersion force between plates, A_{00} to the force between water molecules, and A_{10} to the force between plates and water molecules.

Since characteristic frequencies do not vary appreciably for different substances,

$$A_{10} = (A_{11}A_{00})^{1/2} \tag{91}$$

then

$$A = A_{11} + A_{00} - 2(A_{11}A_{00})^{1/2} = (A_{11}^{1/2} - A_{00}^{1/2})^2 \tag{92}$$

A typical value of A for water ($A_{00} = 0.6 \times 10^{-12}$ erg) has been reported by Slater and Kirkwood.[635]

3.3. London Attractive Force between Spherical Particles

For two spheres of equal radius a with a distance r between their centers, in a vacuum, the attractive force between the particles has been given by Hamaker[320] as

$$V_A = -\frac{A}{6}\left(\frac{2a^2}{r^2 - 4a^2} + \frac{2a^2}{r^2} + \ln\frac{r^2 - 4a^2}{r^2}\right) \tag{93}$$

For small separations of the spheres the attractive force decays more slowly than in the case of flat plates and for a minimum distance between the plates d

$$V_A = -Aa/12d \tag{94}$$

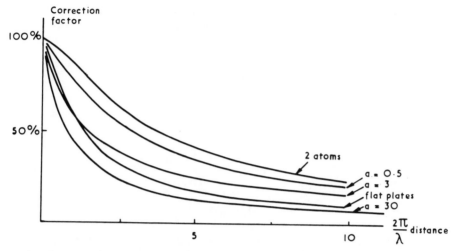

Fig. 13. Retardation correction factor for the London dispersion force between two equal spherical particles. The particle radii are respectively 0.6, 3, and 30 times $\lambda/2\pi$, where λ is the main London dispersion wavelength. The abscissa is the shortest distance between two spheres, expressed in multiples of $\lambda/2\pi$. (See Ref. 415.)

It should be noted that eqns. (93) and (94) are not corrected for retardation; this can be done in a similar manner to the case of flat plates and Fig. 13 (taken from the data of Kruyt[415]) illustrates the retardation correction factor as a function of the London characteristic wavelength for particles of different radii.

3.4. The Effect of an Adsorbed Layer on the London–van der Waals Dispersion Energy

Vold[684] has considered the effect of a layer of adsorbed molecules upon the interaction between two spherical particles; the classical Verwey–Overbeek view of interactions must be modified for particles which are covered by a layer of adsorbate, the adsorption energy of which is large enough to prevent contact between base particles. At distances of closest approach an additional repulsion energy, of range approximating to the length of the adsorbed molecule, results from their interference with each other's motion, with consequent decrease in entropy. For the limiting case when the homogeneous layer of adsorbate consists of a layer of water molecules the interaction energy is given as

$$V_A = -\frac{1}{12}(A_{11}^{1/2} - A_\infty^{1/2})^2 H\left(\frac{r + 2\delta}{2a}, 1\right) \tag{95}$$

where δ is now the thickness of the layer of bound water molecules, A_{11} refers to the material of the particle and A_{00} to the solvent, and $H([r + 2\delta]/2a, 1)$ is given by the relation

$$H\left(\frac{r + 2\delta}{2a}, 1\right) = \left[\left(\frac{r + 2\delta}{2a}\right)^2 + \left(\frac{r + 2\delta}{a}\right)\right]^{-1}$$
$$+ \left[\left(\frac{r + 2\delta}{2a}\right)^2 + \left(\frac{r + 2\delta}{a}\right) + 1\right]^{-1}$$
$$+ 2\left[\ln\left(\frac{r + 2\delta}{2a}\right)^2 + \left(\frac{r + 2\delta}{a}\right)\right]$$
$$- 2\ln\left[\left(\frac{r + 2\delta}{2a}\right)^2 + \left(\frac{r + 2\delta}{a}\right) + 1\right] \tag{96}$$

Taking an arbitrary distance of closest approach $r = 0.3$ nm between particles, and thicknesses of the bound solvent layer of 0, 0.5, 1.0, and 2 nm, the corresponding values of the Hamaker function are shown in Table III; for a coefficient with a value of 5×10^{-3} the data in Table III suggest that only small particles with thick solvate shells have insufficient interaction energy to prevent redispersion of the particles by thermal energy. As a particular example, if $\frac{1}{12}(A_{11}^{1/2} - A_{00}^{1/2})^2 \simeq kT$ and $10kT$ is regarded as being sufficient to prevent redispersion, particles of 10 nm are easily stabilized, particles of 20 nm only when the adsorbed layer is thicker than 0.5 nm, particles of 50 nm only when the adsorbed layer is thicker than 1.0 nm, and so on.

Vold has generalized eqn. (95) for the case of particles with any number of adsorbed layers, providing their composition is a function only of distance

TABLE III. Values of the Hamaker Function for Hydrated Spherical Particles Separated by 0.3 nm

Particle radius, nm	Thickness of hydration shell, nm			
	0	0.5	1.0	2.0
10.0	27.0	4.1	1.7	0.5
20.0	58.9	10.9	5.0	1.8
30.0	94	17.4	8.6	3.6
50.0	167	31.9	16.2	7.3
100.0	333	70	36.7	17.6

from the center of the adsorbing particle:

$$V = - \tfrac{1}{2} \sum_{i,j=1} \sum_{n} a_i a_j H_{ij} \tag{97}$$

where $i,j = 0$ refers to the medium and successive layers of adsorbed material are numbered from the outside inward as $1, 2, \ldots, n-1, n$, the unsolvated particle being numbered n. Here

$$a_i = (A_i^{1/2} - A_{i-1}^{1/2}) \tag{98}$$

$$H_{ij} = H_{ji} = H\left(\frac{\delta_{ij}}{2a_K} \frac{a_L}{a_K}\right) \tag{99}$$

where a_K is the radius of whichever shell, i or j, has the smaller radius, a_L is the radius of the other shell, and δ_{ij} is the separation between the surfaces of the two spheres.

One other case of particular interest to this review is that of interaction between spherical particles of radius a, having an adsorbed layer of oriented molecules or ions, e.g., an amphipathic ion, where the adsorbed layer is treated as two shells, an inner shell with a Hamaker constant A_I and thickness δ_I, and an outer shell, of Hamaker constant A_0 and thickness δ_0. At a distance of minimum separation $\delta_{\min}$ the dispersion interaction is

$$
\begin{aligned}
V_A = -\tfrac{1}{12}[&(A_0^{1/2} - A_M^{1/2})^2 H_0(x, y) + (A_I^{1/2} - A_0^{1/2})^2 H_I(x, y) \\
&+ (A_P^{1/2} - A_I^{1/2})^2 H_P(x, y) + 2(A_0^{1/2} - A_M^{1/2}) \\
&\times (A_I^{1/2} - A_0^{1/2}) H_{I0}(x, y) + 2(A_I^{1/2} - A_0^{1/2}) \\
&\times (A_P^{1/2} - A_I^{1/2}) H_{IP}(x, y) + 2(A_0^{1/2} - A_M^{1/2}) \\
&\times (A_P^{1/2} - A_I^{1/2}) H_{0P}(x, y)]
\end{aligned} \tag{100}
$$

where the subscripts M and P refer to the medium and the particle, respectively, and

for H_0: $x = (d - 2\delta_0 - 2\delta_I)/2(\delta_{\min} + \delta_0 + \delta_I),$ $y = 1$

for H_I: $x = (d + 2\delta_I)/2(\delta_{\min} + \delta_I),$ $y = 1$

for H_{I0}: $x = \dfrac{d - \delta_0 - 2\delta_I}{2(\delta_{\min} + \delta_I)},$ $y = \dfrac{\delta_{\min} + \delta_0 + r\delta_I}{a + \delta_I}$

for H_P: $x = d/2\delta_{\min},$ $y = 1$

for H_{IP}: $x = (d - \delta_I)/2\delta_{\min},$ $y = (\delta_{\min} + \delta_I)/\delta_{\min}$

for H_{0P}: $x = (d - \delta_0 - \delta_I)/2\delta_{\min},$ $y = (\delta_{\min} + \delta_0 + \delta_I)/\delta_{\min}$

Evidence for the discussion of specific cases is unfortunately lacking but the author draws the broad conclusion that unless the A values involved are much smaller than those known at present, then only for very small particles or for thick layers of adsorbed material can the London dispersion force be weakened to the extent that ordinary thermal motions can be expected to maintain the particles in a dispersed state; in addition, Vold suggests that maximum protection from a layer of given thickness of adsorbed material is obtained when the Hamaker constants for the particle, medium, and sorbate are in the order $A_M < A_P < A_S$ or $A_M > A_P > A_S$.

Osmond et al.[542] have recently attempted a more detailed assessment of the effect of the relative magnitudes of A_S, A_M, and A_P upon the dispersion interaction V_A and conclude that due to an algebraic error the sequences $A_M < A_P < A_S$ and $A_M > A_P > A_S$ are incorrect and for a maximum effect should be $A_S{}^* < A_M < A_P$ or $A_S{}^* > A_M > A_P$, where $A_S{}^*$ is given by the equation

$$A_S{}^* = \frac{A_M^{1/2}(H_P - H_{PS}) + A_P^{1/2}(H_S - H_{PS})}{H_P + H_S - 2H_{PS}} \tag{101}$$

Vincent[681] has extended the above models[542,681] to a consideration of the interaction between a sphere and a plate and between two colloidal plates; in addition, he has considered the case of interaction between two plates covered with adsorbate layers at a sufficiently close distance for intermingling of the adsorbate layers to occur. By assuming the density of the adsorbed layers to be uniform both before and after overlap (although higher in the latter case), a uniform osmotic pressure exists across the barrier which then acts as a "medium" for the residual particle interaction. The interaction between the overlapped adsorbed layers is incorporated into the nonideal osmotic term in the steric repulsion[518,546] between the adsorbed layers, which contains a term $\beta - \chi$, where β is an entropy of mixing parameter and χ a corresponding enthalpy term.

Vincent derives the relationship

$$\chi_d = (z/2\pi^2)(A_S^{1/2} - A_M^{1/2})^2 \tag{102}$$

where χ_d is the dispersion force contribution to the enthalpy of mixing, and z is the lattice coordination number according to the Flory theory.[246] A plot of χ_d vs. A_S/A_M is given in Fig. 14, which shows that steric repulsion is maximal when $A_S = A_M + \chi_d = 0$. As A_S/A_M increases or decreases, so steric repulsion is reduced and at a critical value, $\beta - \chi$ becomes negative and flocculation occurs.

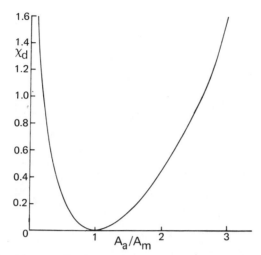

Fig. 14. The dispersion contribution $\varkappa_d$ to the Flory enthalpy of mixing parameter as a function of the ratio of the Hamaker constant of the adsorbate A_A to that of the medium A_M (lattice coordination number $z = 6$) (see Ref. 681).

Becher[41] has also determined the interaction energy between flat plates covered by an adsorbed layer and Fig. 15 illustrates V_A for plate thicknesses of 666.7 nm, $A_P = 8.1 \times 10^{-12}$ erg, $A_S = 6.4 \times 10^{-13}$ erg, $A_M = 6.1 \times 10^{-12}$ erg, and for an adsorbate layer of 0, 0.5, 2.5, and 5.0 nm thickness; the results show that at small separations the adsorbed layer reduces the van der Waals interaction by a factor of five, irrespective of the thickness of the adsorbed layer.

3.5. The Lifshitz Theory of Condensed Medium Interactions

In the macroscopic treatment of Lifshitz[435] no assumption as to the additivity of dispersion forces is made; in fact it is recognized that the atomic and molecular forces responsible for attraction between colloidal particles may be both nonadditive and temperature dependent. The fundamental assumption of the Verwey and Overbeek[680] treatment of the attractive force between colloidal particles is that the major, if not total, contribution of London dispersion forces, i.e., the electromagnetic correlations, are due to fluctuations which occur in a narrow frequency band of the ultraviolet region. In addition, since the original treatment of Hamaker applies to dilute gases, the necessary modification to allow for the presence of the dispersion medium is obtained through an electromagnetic analog of buoyancy together with a "static" dielectric correction factor. Parsegian

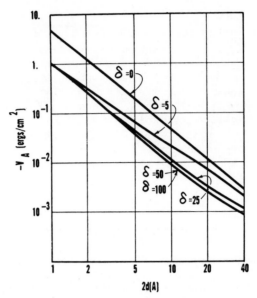

Fig. 15. The relation between the London dispersion force and interparticle distance ($2d$) between two parallel flat plates, for $R = 666.7$ nm, $A_P = 8.1 \times 10^{-12}$ erg, $A_S = 6.4 \times 10^{-13}$ erg, $A_M = 6.1 \times 10^{-12}$ erg, as a function of the thickness of the adsorbed layer δ. In this region the curves for $\delta = 5.0$ and 10.0 nm are indistinguishable. (See Ref. 41.)

and Ninham[562] suggest, however, that a procedure valid for dilute gases is not suitable for the study of condensed medium interactions, particularly those involving a highly polar substance such as water.

These authors apply the Lifshitz theory to a model of two planar parallel layers of nonpolar material of thickness b, material 2 separated by a region of thickness a, and material 3 bounded by two semiinfinite regions of material 1. The energy of interaction between the plates $V(a, b)$ is given by the expression

$$V(a, b) = \frac{h}{16\pi^2 a^2} \int_0^\infty d\xi \int_0^\infty x \, dx \ln[1 - \bar{\Delta}_{31}^2 \text{eff} \, e^{-x}] \tag{103}$$

where $d\xi = (2\pi kT/n) \, dn$, n being the material index, and

$$\bar{\Delta}_{31}\text{eff} = (\bar{\Delta}_{32} + \bar{\Delta}_{21} e^{-xb/a})/(1 + \bar{\Delta}_{32}\bar{\Delta}_{21} e^{-xb/a}) \tag{104}$$

and $\bar{\Delta}_{jk} = (\varepsilon_j - \varepsilon_k)/(\varepsilon_j + \varepsilon_k)$. If materials 1 and 3 are identical and $\varepsilon_1 \equiv \varepsilon_3 \sim \varepsilon_2$, this equation reduces to

$$V(a, b) = -\frac{A}{12\pi a^2} \left[1 - \frac{2a^2}{(a + b)^2} + \frac{a^2}{(a + 2b)^2} \right] \tag{105}$$

where $I_{jk,lm} = \int_0^\infty \bar{\Delta}_{jk}\bar{\Delta}_{lm}\,d\xi$ and $A = -12\pi a^2 h I_{21,21}/16\pi^2 a^2$, which is very similar in form to the result derived by Verwey and Overbeek for a similar configuration from pairwise summation procedures and A takes on the form of a Hamaker constant. The authors point out, however, that the actual interaction between two planar slabs cannot be described in terms of a constant A; this is achieved by defining a "Hamaker function" with the property that the free energy of interaction is given by

$$G(a, b) = A(a, b)/12\pi a^2 \tag{106}$$

In the limit of zero temperature and no retardation, essentially as for the model of Verwey and Overbeek, with pairwise summation of London

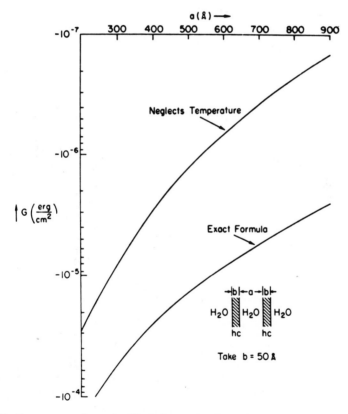

Fig. 16. Free energy of van der Waals interaction between two planar nonpolar slabs in water as a function of the distance between the slabs. Calculation by the complete formula [eqn. (106)], bottom line. Essentially the same curve results when retardation is neglected. Upper line: calculation using the low-temperature approximation [eqn. (107)]. (See Ref. 562.)

dispersion forces for the case of identical materials 1 and 3 the "Hamaker function" becomes

$$A_6(a, b) = -\frac{3hI_{21,21}}{4\pi} \left[1 - \frac{2a^2}{(a+b)^2} + \frac{a^2}{(a+2b)^2} \right] \qquad (107)$$

Figure 16 compares the free energy of interaction between two planar nonpolar hydrocarbon slabs in water calculated by the above method with the free energy for the restricted model, neglecting effects of temperature, i.e., the Verwey–Overbeek model. The data show that because of the highly polar nature of water, much of the van der Waals interaction comes from polarization at infrared and microwave frequencies rather than at ultraviolet frequencies[530]; in addition, the data call into question the basic assumption of the Verwey–Overbeek model, that pairwise summation of London dispersion forces reflects the total van der Waals interaction with sufficient accuracy. It also becomes clear from this analysis that, because of the low-frequency contributions, the van der Waals force contains a large temperature-dependent component.[561]

4. GENERAL PROPERTIES OF THE TOTAL INTERACTION

The general character of the total interaction between colloidal particles has been determined by Verwey and Overbeek[680] and Derjaguin and Landau[170] and is now known as the Derjaguin–Landau–Verwey–Overbeek or DLVO theory of colloid stability. The total interaction between colloidal particles is obtained by the superposition of the double layer and the van der Waals forces

$$V_T = V_{dl} + V_A$$

which in the simplest case involves the interaction between the diffuse double layers associated with colloidal flat plates of infinite extent; thus we have

$$V_T = \frac{64nkT}{\varkappa} [\exp(-\varkappa d)] \frac{\exp(ze\psi_\delta/2kT) - 1}{\exp(ze\psi_\delta/2kT) + 1} - \frac{A}{12\pi d^2} \qquad (108)$$

for the case where the thickness of the plates is much greater than the distance between them. The repulsion energy between the plates decays in an exponential manner with a range approximated by the thickness of the double layer associated with each plate and is finite for all values of the interparticle distance. The van der Waals attraction decreases with the inverse power of the distance between the plates, becoming very large and negative for small interparticle distances. In consequence of the dissimilar

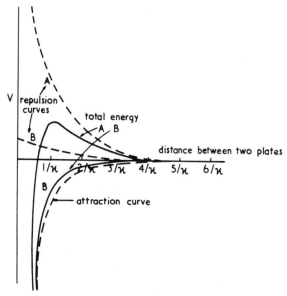

Fig. 17. The total interaction energy between two flat colloidal plates for one attraction curve and two repulsion curves of different height (see Ref. 415).

behavior of the two forces with distance between the particles, the attractive force predominates at small and large separations; within the intermediate range of distances repulsion may be predominant, but this is dependent upon the specific values of attraction and repulsion. Figure 17 illustrates the different curves of total interaction which arise; the more shallow repulsion curve results in a total interaction which decreases in a monotonic manner with distance, whereas the higher repulsion curve results in a total interaction energy which achieves a maximum at a distance which corresponds to the thickness of the diffuse double layer ($1/\varkappa$).

4.1. Variation of the Total Interaction with Changes in the Diffuse Double Layer Potential

If the electrolyte concentration is held constant and the value of the Hamaker constant is known, then the variation of the total interaction energy with variation in the diffuse double layer potential can be found. Figure 18, based upon the data of Kruyt,[415] (actually in this case for the interaction of two spherical particles), shows that a decrease in ψ_0 results in a decrease in the height of the repulsive energy barrier, although even at a ψ_0 value of 19.2 mV a repulsion barrier of the order of kT is still evident,

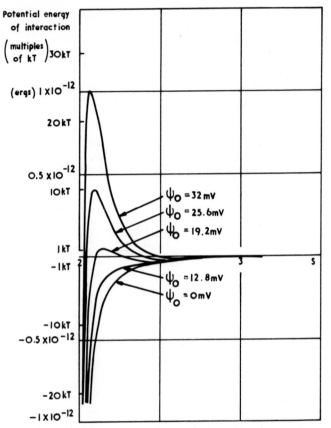

Fig. 18. The influence of the surface potential ψ_0 on the total potential energy of interaction of two spherical particles. $a = 10^{-5}$ cm; $A = 10^{-12}$ erg; $\nu = 10^6$ cm^{-1}. (See Ref. 415.)

together with a clearly defined secondary minimum at greater distances than the thickness of the double layer.

4.2. The Effect of Electrolyte Concentration and the Valence of the Counterions upon Total Interaction

If the Hamaker constant is known and a fixed value is assigned to ψ_δ, then the variation of the total interaction (a) as a function of the concentration of added electrolyte and (b) as a function of the valence of the counterion at constant electrolyte concentration can be examined. Figure 19, from the data of Sonntag and Strenge,[642] shows the interaction for a Hamaker constant of 5×10^{-13} erg and $\psi_\delta = 100$ mV of two parallel plates with variable concentration of a 1:1 electrolyte. Increasing concentration of the counterion reduces the height of the repulsive energy barrier until at a particular concentration of electrolyte the potential barrier V_T becomes zero; under this condition

$$\frac{64nkT}{\varkappa} [\exp(-\varkappa d)] \frac{\exp(ze\psi_\delta/2kT) - 1}{\exp(ze\psi_\delta/2kT) + 1} = \frac{A}{12\pi d^2} \qquad (109)$$

In addition, $dV_T/dd = 0$. Hence $\varkappa d = 2$, and by expressing the concentration as c mol dm^{-3} in place of n ions cm^{-3}, eqn. (109) becomes

$$c = 8 \times 10^{-22}(f^4/A^2z^6) \quad \text{mol dm}^{-3} \qquad (110)$$

where

$$f = \frac{\exp(ze\psi_\delta/2kT) - 1}{\exp(ze\psi_\delta/2kT) + 1}$$

and $T = 298°K$, $\varepsilon = 78.55$.

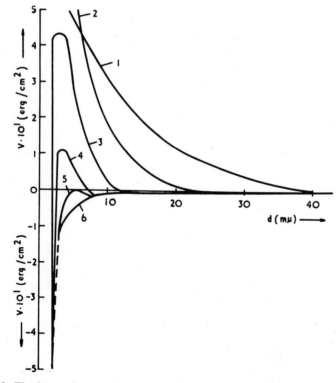

Fig. 19. The interaction as a function of interparticle distance for $A = 5 \times 10^{-13}$ erg, $\psi_\delta = 100$ mV, and variable ionic strength (mol dm^{-3}): (1) 10^{-3}; (2) 5×10^{-3}; (3) 2.5×10^{-2}; (4) 5×10^{-2}; (5) 7.5×10^{-2}; (6) 5×10^{-1}. (See Ref. 642.)

At sufficiently high values of f, which is usually the case for colloidal dispersions, f may be approximated by unity; in addition, since the flocculation concentration is inversely proportional to the sixth power of the valence of the counterion, eqn. (110) enables the relative flocculation concentrations of univalent, divalent, and trivalent electrolytes to be obtained as

$$100 : 1.6 : 0.13$$

which is in good agreement with the empirical relationship determined from experimental data and known as the Schultze–Hardy rule.[619]

4.3. The Effect of the Stern Layer on the Total Interaction between Colloidal Particles

The DLVO theory of colloid stability, as previously outlined, does not explain the specific influence of the nature of the counterion upon the interaction between metallic surfaces[174] nor the dependence of the concentration of flocculating electrolyte on the surface potential of the colloidal particle. Kruyt and Klompé[416] have found that for aqueous silver iodide sols the flocculating concentration of electrolyte increases with the activity of the counterion in the dispersion medium and therefore with the surface potential ψ_0, according to the Nernst equation. Divalent and trivalent ions, on the other hand, show a decrease with increasing activity of the ion.

Levine and Bell[428] have extended the DLVO theory of colloid stability to include the effects of discreteness of charge due to adsorption of ions into the inner and outer Helmholtz regions of the Stern layer; the double layer force per unit area between colloidal plates is now given by the expression

$$p_e(h') = 64nkT\,[\exp(-2\varkappa h')]t^4 \qquad (111)$$

where $t = \tanh(ze\psi_\delta/4kT)$, when the two parallel platelike particles of infinite thickness are separated by a distance $2h$. Since the electrical double layer repulsion pressure is a function of the charge distribution in the diffuse double layers, it follows that the separation distance involved is actually that between the two outer Helmholtz planes, i.e., $h' = h - \delta$, where δ is the thickness of the Stern region and ψ_δ is the potential of the outer Helmholtz plane. The corresponding interaction energy is given by

$$V_R' = (64nkT/\varkappa)[\exp(-2\varkappa h')]t^4 \qquad (112)$$

the van der Waals–London attraction energy between the plates being of the form

$$V_A' = -A/48\pi(h' - d)^2 \tag{113}$$

where d is an empirical value, of the order of a molecular radius, which allows for the lack of knowledge of the thickness of the inner region. From the DLVO conditions for critical coagulation

$$\varkappa(h' + d) = 1 \tag{114}$$

and

$$\varkappa d \exp(-2\varkappa d) = \frac{52}{z^2} \frac{\varepsilon d}{A} \left(\frac{kT}{e}\right)^2 t^2 \tag{115}$$

Making use of a simplified form of eqn. (31), the authors show that

$$r - \ln r + p \ln(1 - pbr) = D \mid z \mid + \frac{ze\psi_\delta}{kT}$$

$$+ \frac{Et^2}{z^2} \left(\sinh \frac{ze\psi_\delta}{2kT}\right) \exp(2\varkappa d)$$

$$- 4 \ln \mid t \mid + \frac{Gt^2 \exp(2\varkappa d)}{1 + \varkappa a} \tag{116}$$

and

$$\frac{e\psi_\delta}{kT} = \frac{e\psi_0}{kT} + \frac{d}{\varXi} \frac{Et^2}{z^3} \left(\sinh \frac{ze\psi_\delta}{2kT}\right) \exp(2\varkappa d) - \frac{\delta r}{\varXi gz} \tag{117}$$

where

$$E = \frac{52}{KA} \frac{\varXi}{\delta} \frac{\varepsilon^2}{2\pi} \left(\frac{kT}{e}\right)^2, \qquad G = \frac{26kT}{A}$$

$\varXi$ is the thickness of the inner Helmholtz plane, K is the integral capacity of this region, C is a factor depending on ε, ε_1, ε_P, β, $\varXi$, and the electrolyte concentration [see eqn. (31)], and the constant $D(z)$ is specific to the adsorbed cation and the nature of the plate wall. The electrolyte concentration at coagulation is now given by the expression

$$n = 107.5[\varepsilon^3(kT)^5 t^4 \exp(4\varkappa d)/(ze)^6 A^2] \tag{118}$$

which can be expressed in the equivalent form

$$n^* = 107.5t^4 \exp(4\varkappa d) \tag{119}$$

where the "reduced concentration" n^* is defined by

$$n^* = (ze)^6 A^2 n/\varepsilon^3(kT)^5$$

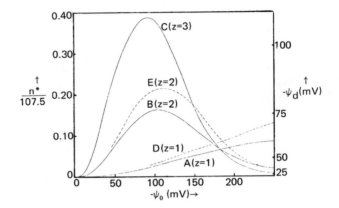

Fig. 20. A plot of t^4 versus ψ_0, i.e., of the coagulating concentration n versus the surface potential. $n^*/107.5 = (ze)^6\, An^2/[107.5\; \varepsilon^3(kT)^5] = t^4 e^{4\varkappa d}$. For full curves (A, B, C) $\delta = 0$; for broken curves (D, E) $\delta = 2A$.

A plot of $n^*/107.5$ versus surface potential ψ_0 is shown in Fig. 20 for three valences $z = 1$, 2, and 3; when $d = 0$ it is seen from eqn. (118) that the Schultze–Hardy rule is obeyed if t has the same value for the three valences at a given ψ_0, i.e., at the point of common intersection of the three curves, but at other values the coagulation concentration passes through a maximum value which depends upon the surface potential ψ_0. Table IV gives the values determined for the characteristics of the coagulating electrolyte concentration at the maximum and at the Schultze–Hardy point.

TABLE IV. Characteristics of Plots of Coagulating Electrolyte Concentration versus Surface Potential[a]

				At max in ψ_δ				At Shultze–Hardy point			
Z	Curve	p	D(Z)	r	σ_β	$-\psi_\delta$	$-\psi_0$	r	σ_β	$-\psi_\delta$	$-\psi_0$
1	A	2	2.96	1.09	1.89	63	280	0.159	0.275	56.6	184
2	B	4	3.09	1.10	0.95	39	105	5.48	4.75	28.3	184
3	C	7	5.54	1.14	0.657	27	90	9.32	5.39	18.9	184

[a] σ_β in μC cm^{-2}, ψ_0 and ψ_δ in mV. Parameter values: $K = 22.5\ \mu$F cm^{-2}, $A = \tfrac{4}{3} \times 10^{-12}$ erg cm^2, $N_s = 5 \times 10^{14}$, $Z\sigma_\beta{}^\circ = 1.73$ ($\sigma_\beta{}^\circ$ in μC cm^{-2}), $i/\delta = \tfrac{1}{3}$, $g = \tfrac{3}{2}$, $d = 0$, $\gamma = 1$, $E = 4{,}62$, $bz^2 = 0.0433$.

5. EXPERIMENTAL EVIDENCE FOR THE MODIFICATION OF THE DLVO THEORY OF HYDROPHOBIC COLLOID STABILITY BY HYDRATION INTERACTIONS

5.1. Primary Hydration of the Colloidal Particles

The theoretical studies reviewed in earlier sections, certainly in the case of the classic DLVO theory, do not allow for hydration of the particle itself, "primary hydration," as a stabilizing effect against particle flocculation. Subsequent refinements of the DLVO theory allow for the possibility of some kind of hydration structure as influencing the parameters which govern colloid stability, but whether this factor is a significant quantity appears to be a very open question.

The problem is further complicated if the colloidal particle carries on its surface an adsorbed layer of a material different to that of the particle itself; can any hydration structure of this layer be regarded as "primary hydration" or should this be more reasonably termed "secondary hydration"? If this difference is indeed real, does it have an appreciable effect upon the stability of the colloidal system?

Evidence for the importance of the difference between the two hydration states comes from the nmr studies of water in disperse systems.[123,382] In the earlier paper Johnson et al.[382] report studies of the spin–spin relaxation in polyvinyl acetate dispersions of mean particle diameters 80 and 130 nm, stabilized with sodium octadecyl sulfate. Figure 21 illustrates the dependence of the spin–spin relaxation rate T_2^{-1} on the particle–particle separation distance as a function of temperature for the 80-nm particles; the data show that a solvation energy barrier exists which is constant at low particle concentrations and disappears at higher temperatures. At lower temperatures and higher concentrations, however, an appreciable barrier to interparticle contact is observed due to partially immobilized water molecules surrounding the particles. Similar behavior is observed with the smaller particles but the relaxation rate starts to rise rapidly at a much shorter interparticle distances than in the case of the larger particles. Thus it seems that the total amount of "structured" water per unit surface area is larger, the larger the particles; in addition, the dependence of the extent of water structuring on particle diameter, as measured by the ratio of the relaxation times for the two PVA particles at a fixed separation and temperature, obeys a lower power law than that given by the London dispersion interaction between particles.

The nmr evidence, which is also supported by studies of flocculation

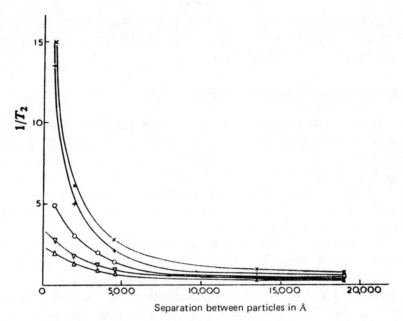

Separation between particles in Å

Fig. 21. The dependence of the spin–spin relaxation rate T_2^{-1} of the water protons on the interparticle separation for 80-nm polyvinyl acetate particles at ($\times$) 2, ($+$) 12, ($\bigcirc$) 23, ($\bigtriangledown$) 33, and ($\triangle$) 45°C (see Ref. 382).

rates,[382] suggests that the particles are surrounded by an extended region of ordered water $\sim$30 Å thick, which is of significant importance in stabilizing the dispersions. It should be noted, however, that no attempt was made to remove the surfactant, sodium octadecyl sulfate, which was initially present as a stabilizer in the preparation of the dispersion; it is therefore at least possible for the extended region of ordered water to be associated with the adsorbed layer of surfactant on the particle surface, i.e., secondary hydration, rather than with the polyvinyl acetate surface—primary hydration.

The later work of Clifford and co-workers[123] suggests that this is indeed the case; nmr relaxation data were obtained on polystyrene dispersions of mean particle size 89 and 520 nm, which were initially prepared using sodium decanoate as stabilizing agent, but were subsequently exhaustively dialyzed to remove the surfactant.

If it is assumed that water affected by the particle surface exchanges with bulk water and bulk water is denoted as state A and the modified "surface water" as state B, it has been shown[713] that

$$\frac{1}{T_2} = \frac{1}{T_{2A}} + \frac{p_B}{p_A}\left(\frac{1}{T_{2B} + s_B}\right) \tag{120}$$

where T_2 is the observed relaxation time, T_{2A} and T_{2B} are the relaxation times of the two states, p_A and p_B are the populations of the two states, and s_B is the average residence time of a proton in state B. There should therefore be a linear relationship between the observed relaxation rate $1/T_2$ and $V/(1 - V)$, where V is the volume fraction of the solid; Figure 22 illustrates the spin–spin relaxation time for two distinct concentration regions for the 89-nm particles over a range of temperatures. In the non-flocculated region ($<20\%$) the authors suggest that the observations can be explained by the presence of a few tightly bound water molecules on the particle surface—too few in fact to form a single monolayer.

It appears therefore to be the case that particles whose surface is devoid of adsorbed surfactant have little hydration associated with them; particles

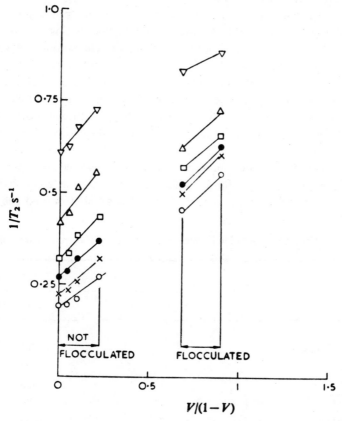

Fig. 22. The dependence of the spin–spin relaxation rate of the water protons T_2^{-1} on $V/(1 - V)$, where V is the volume fraction of solid, for suspensions of 89-nm-diameter polystyrene particles; ($\bigcirc$) 58; ($\times$) 47; ($\bullet$) 37; ($\triangle$) 20; ($\triangledown$) 10°C (see Ref. 123).

with an adsorbed surfactant layer, on the other hand, appear to have associated with them a hydration layer of appreciable thickness. This layer is obviously stabilized by the presence of the surfactant molecules and, although the polar head groups of the surfactant molecules will be specifically hydrated, a hydration layer of such a thickness could only be achieved by hydrophobic hydration of the nonpolar hydrocarbon chain of the surfactant molecule. It should be mentioned, however, that although nmr studies suggest little hydration of a "bare" particle surface, such studies only observe the most tightly bound water molecules (see Chapter 6) and water associated in a more dynamic sense with the particle would not appear to be markedly different from bulk water by this technique; hydration (in a dynamic sense) of the bare particle might well therefore be appreciably larger than found in this case.

It is particularly relevant in a consideration of "primary hydration" to recall the early work of Freundlich[268] on the flocculation of hydrophobic sulfur sols prepared by the addition of an alcoholic solution of sulfur to water or by the addition of acid to sodium thiosulfate solution; Table V contains the critical coagulation concentrations (CCC) of alkali metal salts for the two sulfur sols. The sol prepared by addition of an alcoholic solution of sulfur to water shows very little sensitivity to the nature of the flocculating cation; the sol prepared from sodium thiosulfate shows a very marked sensitivity to the nature of the cation and a lyotropic series very similar to that observed with proteins and nucleic acids is observed; it is particularly remarkable that $LiSO_4$ is far less effective as flocculating agent than LiCl; possibly a correlation exists here between this behavior and the strong

TABLE V. Critical Coagulation Concentration (CCC) of Electrolytes for Sulfur Sols

Electrolyte	CCC, m dm^{-3}	
	Alcoholic sol	Thiosulfate sol
$Li_2SO_4/2$	35	1500
LiCl	34	750
NaCl	33	190
KCl	32	85
RbCl	31	80
CsCl	30	95

capacity for hydrating biopolymers exhibited by the SO_4^{2-} ion. Freundlich suggests that the thiosulfate sulfur sol binds detectable quantities of pentathionic acid $H_2S_5O_6$ which is responsible for the binding of a greater quantity of water than is observed in the case of the alcoholic sulfur sol. It may be that other factors such as particle size, which was apparently not determined, will influence the extent of hydration, but the correlation of the CCC with the hydration characteristics of the alkali metal ions (Volume 3) strongly implicates hydration as a factor in the stability of the "thiosulfate" sol.

Freundlich[268] also reports studies of the flocculating capability of mixed electrolytes upon As_2S_3 sols, and found that the addition of LiCl at amounts less than that required for flocculation has a very marked effect upon the amount of the second electrolyte required to produce flocculation; in particular the Li^+ ion appears to stabilize the sol against the action of Mg^{2+} ions, a much higher concentration of which is required to produce flocculation than with sols in the absence of Li^+. Such behavior again correlates with the hydration behavior of Li^+ and Mg^{2+} ions in their interactions with biopolymers. The Na^+ ion has a much less marked effect and the K^+ ion appears to be without effect.

The evidence reviewed in Volume 3 of this treatise has shown that the Li^+ ion may be regarded as a strong promoter of order in liquid water, the Na^+ ion less so, and the K^+, Cs^+, and Rb^+ ions as structure breakers; of the alkaline earths, the Mg^{2+} ion is a particularly strong promoter of water structure. The very clear correlation between the hydration characteristics of the ions and their flocculating capability with respect to certain sols points to the involvement of hydration structure in the stabilization of those sols. Accepting the importance of hydration, can this factor be reasonably described as primary hydration? Lack of further evidence makes it difficult to be firm in this matter, but there are certainly suggestions[684] that for small colloidal particles close approach of heavily hydrated Li^+ ions to the particle surface would produce a hydration layer of nanometer dimensions; such a layer would result in a reduction of interparticle attraction and hence increased stability. The presence of such a hydrated layer of Li^+ ions would also make it particularly difficult for a Mg^{2+} ion with a compatible hydration shell to approach close to the particle surface.

Such an interpretation would also be in agreement with the model of Devillez et al.,[188] since their suggestion is that, because of the effects of ion hydration, the ions would tend to concentrate in the surface layer in the order

$$H^+ < Li^+ < Na^+$$

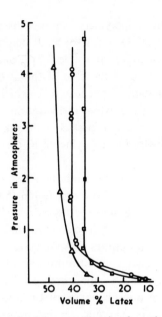

Fig. 23. Pressure as a function of volume concentration for polystyrene latex dispersions at pH 8.0 and 25°C in various 10^{-1} mol dm^{-3} electrolyte solutions. Particle radius 352 nm. ($\triangle$) LiCl; ($\bigcirc$) RbCl; ($\square$) CsCl. (See Ref. 30.)

These authors also suggest that the effect of molar volume of the counterions is to reduce their concentration in the diffuse layer and hence to increase the repulsive energy, but that for small ions this term can be neglected; this may be an oversimplification, since in reality it is not the volume of the counterion which is of importance but the volume of the hydrated counterion; the greater the hydration volume, the lower the counterion concentration in the diffuse layer will be and hence the greater the colloid stability.

This appears to be supported by the work of Barclay et al.[30] In an ingenious experiment based essentially upon exerting a known pressure upon the disperse system and observing the change in pressure which results, these workers determined pressure–volume fraction curves for polystyrene latices from which the surfactant had been removed. Figure 23 illustrates the data for particles of 352 nm radius in the presence of 10^{-1} mol dm^{-3} solutions of various electrolytes—the lowest electrolyte concentration at which flocculation could be detected. A marked difference is seen between the volume fractions at which the steep pressure rise commences for the different counterions; the order is

$$V_{Li^+} > V_{Rb^+} > V_{Cs^+}$$

At higher pressures a solid disk of the flocculated dispersion is obtained and electron micrographs show that after compression the latex flocculated in CsCl shows a disordered array while that flocculated in LiCl shows an ordered array. According to the sedimentation studies of Kruyt,[415] a disordered array indicates the presence of a secondary minimum in the total interaction–distance relationship. Since the secondary minimum is usually only observed at relatively high electrolyte concentrations, the data suggest preferential concentration of the ions within the double layer in the order

$$Cs^+ > Rb^+ > Li^+$$

which is in agreement with the model of Devillez et al.[188] The earlier rapid rise in the pressure of the dispersion under compression in the presence of the Cs^+ and Rb^+ is proposed by the authors as being due to the early presence of secondary minimum flocculated aggregates.

The data of Lyklema[456] also support an interpretation along these lines; by utilizing the technique of potentiometric titration of dispersed AgI in the presence of a potential-determining electrolyte over a temperature range 5–85°C, he determined the relative adsorption and hence surface charge of AgI over a range of ψ_0 values for solutions of $LiNO_3$, KNO_3, and $RbNO_3$. Figure 24 illustrates the variation of surface charge with

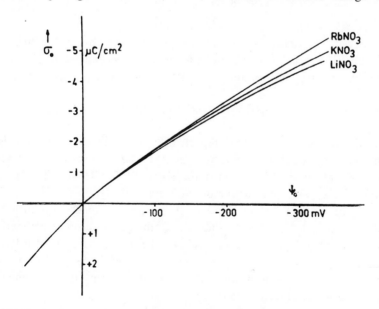

Fig. 24. Surface charge versus potential for an AgI dispersion at 23°C; ionic strength 10^{-1} for $LiNO_3$, KNO_3, and $RbNO_3$ (see Ref. 456).

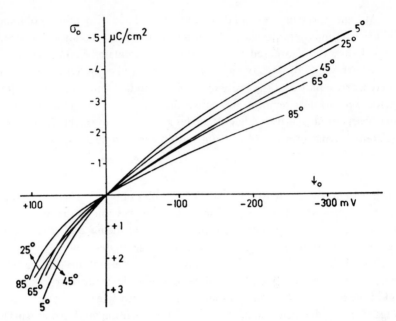

Fig. 25. The temperature dependence of the double layer charge for an AgI dispersion in 10^{-3} mol dm^{-3} K biphthalate $+ 0.099$ mol dm^{-3} KNO$_3$; all curves are plotted with respect to their own zero point of charge (see Ref. 456).

potential at a constant ionic strength of 0.1 mol dm^{-3} and the influence of the counterion; Fig. 25 illustrates the temperature dependence of the double layer charge for KNO$_3$ solution. The data show that at a fixed negative ψ_0

$$\sigma_0(\text{Rb}^+) > \sigma_0(\text{K}^+) > \sigma_0(\text{Li}^+)$$

Thus the lyotropic series first reported by Freundlich for other sols is reflected in the double layer charge of AgI and confirmed by the CCC values of AgI reported by Klompé and Kruyt[416] for RbNO$_3$, KNO$_3$, and LiNO$_3$ of 126, 135, and 165 mmol dm^{-3}, respectively.

The decrease in surface charge with increasing temperature illustrated by Fig. 25 would appear to be due to a gradual desorption of counterions from the surface layer of the particle and the double layer in turn having a more diffuse character. This is confirmed by reference to Fig. 26: The surface charge at 85°C as a function of the ionic strength of the counterion is identical for all three counterions.

Lyklema reports evidence from the $\sigma_0(\psi_0)$ studies for a phase transition in the surface hydration layer of AgI at approximately 50°C which he

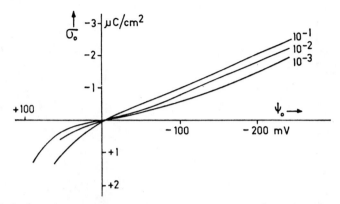

Fig. 26. Surface charge versus potential for an AgI dispersion at 85°C; ionic strength 10^{-1}, 10^{-2}, 10^{-3}; cation: Li^+, K^+, or Rb^+ (see Ref. 456).

interprets as breakdown of the structured water layer followed by loss of specificity in the flocculation capability of the counterion; Fig. 27 illustrates the ratio of the flocculation concentrations of $LiNO_3/KNO_3$ and $RbNO_3/KNO_3$ and is strong confirmatory evidence for this interpretation.

A similar lyotropic sequence of $Cs^+ > K^+ > Li^+$ was obtained by Abendroth[1] in studies of the adsorption of H^+ and OH^- ions on pyrogenic silica in solutions of LiCl, KCl, and CsCl over the pH range 1.8–9.0; Fig. 28 shows that at pH values below $\sim$5 the silica dispersion (CAB-O-MIL M7) has an essentially neutral surface but at higher pH values OH^- adsorption becomes easier in the presence of water-structure-breaking cations such as

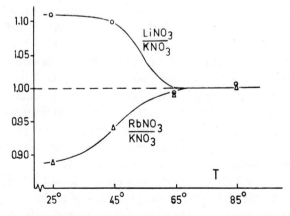

Fig. 27. The ratios of flocculation concentrations against temperature; negative AgI sols in $LiNO_3/KNO_3$ and $RbNO_3/KNO_3$; $\psi_0 = 220\,mV$ (see Ref. 456).

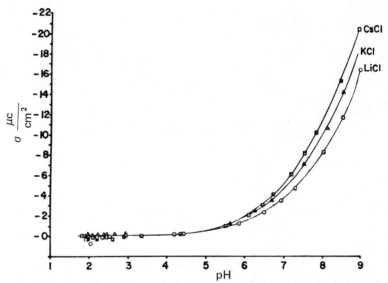

Fig. 28. The surface charge density of a silica dispersion (CAB-O-MIL M-7) as a function of pH in $10^{-1}\,mol\,dm^{-3}$ solutions of CsCl, KCl, and LiCl (see Ref. 1).

Cs^+. Differential capacity studies as a function of pH confirm that counterion adsorption decreases in the sequence $Cs^+ > K^+ > Li^+$, which is in agreement with the previously reviewed data.

Bérubé and de Bruyn,[61] from potentiometric titrations of a rutile (TiO_2) dispersion, have determined differential capacity curves for this material in the presence of $LiNO_3$, $NaNO_3$, and $CsNO_3$ and found a reverse sequence of adsorption, $Li^+ > Na^+ > Cs^+$. They suggest that the preferential adsorption of the water-structuring ions is due to interaction of the ionic hydration shell with the hydration layer of the oxide surface which, from infrared[434,720] and nmr[478] studies is known to be hydrated by a strongly *chemisorbed* layer of water. It is difficult to understand why interaction of the hydration regions can be expected to lead to preferential adsorption of the structure-promoting species; from consideration of the entropy contribution that would arise from the restrictions that the hydration regions would impose on each other a repulsive energy term might be expected. This is confirmed by the mixed electrolyte studies of Freundlich on the flocculation of an As_2S_3 sol, since the Mg^{2+} ion appears to be repelled from the particle surface in the presence of the Li^+ ion, both ions of course being regarded as structure promoters of water It seems more probable that the structure of the chemisorbed layer of water. on the rutile surface is incompatible with the hydration sphere of the Li^+ ion (see Volume

3) and a net decrease of solvent order results, giving rise to preferential adsorption of the ion. In the case of the less preferentially adsorbed Cs^+ ion it is possible that the electrostrictively oriented water molecules on the surface of the Cs^+ ion are geometrically compatible structure with the specifically adsorbed water layer on the oxide surface; hence reduced adsorption is the result.

Discussion of the effects of ions upon the stability of hydrophobic colloids has so far been restricted to considerations of the effects of cations upon negatively charged particles. Similar lyotropic series are observed in the effects of anions upon the stability of both negatively and positively charged dispersions.

Bérubé and de Bruyn[61] in their report of adsorption of ions at the negatively charged TiO_2 surface found that specific adsorption of anions decreased in the order

$$Cl^- \simeq ClO_4^- \simeq NO_3^- > I^-$$

Comparison of this sequence with the anionic hydration effects reviewed in Volume 3 reveals that the above sequence corresponds also to decreasing solvent structure promotion capability by the ions. The authors argue that, as in the case of the cations, preferential adsorption of the structure-promoting ions occurs because of constructive interaction between the hydration shell of the TiO_2 surface and that of the ion.

The alternative model previously outlined, of incompatible hydration layers, would also explain the lyotropic sequence observed for the anions, the disruption of hydration structure leading to preferential adsorption of the Cl^- ion while the more strongly structure breaking iodide ion, which can be expected to be hydrated by an electrostrictively oriented water layer, suffers a net repulsion by interacting with the hydration layer of the rutile surface. It is clear that an explanation of this kind implies that the hydration layer of the oxide surface must be of a fundamentally different nature to that of hydrophobic surfaces, such as AgI, As_2S_3, S, Au, etc. It is to be expected therefore that the specific adsorption of anions on a negatively charged AgI surface should be of a reverse order to that observed with TiO_2; Bérubé and de Bruyn quote (without reference) evidence that this is indeed the case.

The implication of a fundamentally different surface hydration layer on the surface of oxides is supported once again by the data of Freundlich[266] for the CCC values of anions with a *positively* charged Fe_2O_3 sol and the data of Gann[278] and Ishizaka[368] for positively charged Al_2O_3 sols (see

TABLE VI. Critical Coagulation Concentrations (CCC) of Electrolytes for Positively Charged Hydrophobic Sols

Electrolyte	CCC for the Fe_2O_3 sol, mmol dm^{-3}	CCC for the Al_2O_3 sol, mmol dm^{-3}	
		Gann[278]	Ishizaka[368]
NaCl	9.25	77	43.5
KCl	9.0	80	46
$BaCl_2/2$	9.65	—	—
KBr	12.5	150	—
KI	16	300	—
KCNS	—	42	67
KNO_3	12	—	60
$BaNO_3/2$	14	—	—
K_2SO_4	0.205	0.28	0.30
Tl_2SO_4	0.22	—	—
$MgSO_4$	0.22	—	—
$K_3Fe(CN)_6$	—	0.10	0.080
$K_4Fe(CN)_6$	—	0.08	0.053

Table VI). The sequence of specific adsorption of the anion is in each case

$$Fe(CN)_6^{4-} \gg Fe(CN)_6^{3-} > SO_4^{2-} \gg Cl^- > Br^- \simeq NO_3^- \simeq CNS^- > I^-$$

The adsorption sequence of anions with respect to an oxide surface appears to be unchanged, irrespective of whether that surface is positively or negatively charged; this remarkable fact implies that it is the oxygen atoms on the surface which play a very dominant role (presumably by hydrogen bonding) in orienting the surface layer of water molecules—such a layer will most likely be of a structure incompatible with that existing in normal water; it appears from studies reported elsewhere in this review that this type of hydration is best classified as hydrophilic hydration. Nappa[521] also observed a similar lyotropic series for the flocculation effectiveness of anions with respect to a polyvinyl acetate dispersion stabilized by poly-(ethylene *oxide*) chains, which appears to be further confirmation of the importance of the oxygen atom; this paper will be discussed in greater detail in the following section.

5.2. Hydration of the Adsorbed Layer at the Particle Surface—"Secondary Hydration"

The theoretical studies of Vold[684] suggest that, at least in the case of small colloidal particles, an adsorbed layer of a material different to that of the particle will result in a considerable diminution of the London dispersion force attractive term, possibly to its complete suppression. Considerable evidence is available[117,409,466,467] that the adsorption of a layer of surfactant on the particles results in a considerable increase in the stability of the dispersion toward flocculation by electrolyte.

The point of particular interest to this review is the possible involvement of hydration structures within the adsorbed layer in the stabilization of the hydrophobic dispersion; the results of Ottewill and Walker[546] are of importance within this context. These workers studied the influence of the nonionic surfactant n-dodecyl hexaoxyethylene monoether, $[C_{12}H_{25}(CH_2CH_2O)_6OH]$–$C_{12}E_6$, upon the stability of polystyrene dispersions (40–60 nm diameter). The changes in stability were measured as a function of added concentrations of lanthanum nitrate and nitric acid; at $C_{12}E_6$ concentrations above the critical micelle concentration of the surfactant the particles were stabilized even under conditions of zero charge. The data obtained suggested that at optimum conditions the particles were covered with a monolayer of $C_{12}E_6$ some 5 nm thick and that the monolayer contained approximately 0.26 g $C_{12}E_6$/ml and 0.74 g H_2O/ml; it thus appears that hydration of the adsorbed surfactant layer must play an important role in the stabilization of the dispersion against flocculation.

Mackor[466] and Mackor and van der Waals[467] have suggested that stabilization in the presence of adsorbed layers arises from changes in configurational entropy as the particles approach one another; Ottewill and Walker suggest, on the other hand, that the most effective stabilization occurs when a complete monolayer of surfactant, which must be vertically oriented since the thickness correlates well with the length of the $C_{12}E_6$ molecule, covers the surface of the particles. Such a configuration must reduce the extent of configurational changes which can occur. The calorimetric studies of Corkill et al.[135] place a value of -41.2 kJ mol^{-1} on the heat of solution of $C_{12}E_6$ in water; this high energy of solvation implies that ethylene oxide–water contacts are favored over ethylene oxide–ethylene oxide contacts, presumably, as outlined earlier in this discussion, by interaction between the oxygen atom and a water molecule.

Napper[520] has also reported studies of the flocculation behavior of polyvinyl acetate dispersions stabilized by adsorbed layers of poly(ethylene oxide) (PEO) by determination of the critical flocculation temperature, i.e., the temperature at which flocculation occurred. By comparison with studies of the flocculation of polyvinyl acetate stabilized with poly(12-hydroxystearic acid) in hexane the author shows that the enthalpy and entropy of dilution parameters of the Flory equation (72) for PEO in water are both negative, while the reverse occurs for poly(12-hydroxystearic acid) (PSO) in hexane. Napper suggests from this evidence that the PSO-stabilized dispersion owes its stability to an entropic mechanism arising from interaction between polymer chains; the PEO-stabilized dispersions, however, are envisaged as owing their stability primarily to a positive enthalpic repulsion term arising from the increased degree of freedom of water molecules released from the chains on interpenetration.

Gargallo and co-workers[279] have also observed a lyotropic series in measurements of the zeta potentials of paraffin droplets stabilized with stearic acid in the presence of different electrolytes; Table VII correlates measured values of the ζ potential at an electrolyte concentration of 10^{-4} mol dm^{-3} with the Jones–Dole B coefficient,[386] the solvation entropy of the ions, and the estimated thickness of the shear plane at the particle surface. Table VIII shows the values of surface potential ψ_0 of a spread monolayer of stearic acid on a series of aqueous electrolyte substrates; it is clear that no variation in ψ_0 can be detected which corresponds to the observed variations in ζ potential. The data in Table VII make it clear that the observed variations in the ζ potential are correlated with the detailed hydration behavior of the ions. The authors suggest that entrapment of the hydration structure of the ions within the double layer of the particles results in an increase in the distance between the shear plane and

TABLE VII. ζ-Potential Measurements of Stearic Acid-Stabilized Paraffin Dispersion in the Presence of Various Chlorides

Ion	ζ, mV ($c = 10^{-4}$ mol dm^{-3})	B,[a] 25°	S_{solv},[b] J deg^{-1} mol^{-1}	d_{sp},[c] nm ($c = 10^{-4}$ mol dm^{-3}
Li$^+$	-41.3	$+0.1495$	$+14.9$	34
K$^+$	-60.6	-0.0070	$+102.9$	21
Rb$^+$	-80.0	-0.030	$+124.7$	13.4
Ba^{2+}	-46.2	$+0.220$	$+12.6$	—

[a] The Jones–Dole B coefficient of viscosity.[386]
[b] The solvation entropy of the ions.[582]
[c] The estimated thickness of the shear plane at the particle surface.[549]

TABLE VIII. Surface Potential Values of a Spread Monolayer of Stearic Acid on Electrolyte Substrates

Electrolyte	ψ_0, mV	
	$c = 10^{-5}$ m dm^{-3}	$c = 10^{-4}$ m dm^{-3}
LiCl	-115 ± 5	-120 ± 5
KCl	-120 ± 5	-125 ± 5
RbCl	-120 ± 5	-122 ± 5

the particle and a consequent lowering of the electrophoretic mobility, and hence the ζ potential of the particle. It is particularly striking that the divalent Ba^{2+} ion, which according to the DLVO theory of colloid stability should be much more effective in suppressing the ζ potential, is only as effective as the Li$^+$ ion; the B values in Table VII indicate that the Ba^{2+} ion is also hydrated to a similar degree, confirming the importance of the hydration structure in stabilizing the particle.

Gargallo *et al.* do not make it clear in their report how the hydration structure interacts with the particle surface, but it seems most probable that the "hydrophobic hydration" (see Chapter 1, Volume 4) associated with the hydrocarbon tails of the stearic acid is compatible with that surrounding the structure-promoting ions, thus preventing their closer approach to the particle surface, resulting in the movement of the shear plane away from the particle surface.

Napper[521] has extended his investigations into the stabilizing effects of poly(ethylene oxide) chains upon poly(vinylacetate) dispersions to include the effects of electrolytes in the dispersion medium; as in previous work the flocculating capability is determined from the critical flocculation temperature of the dispersion, but in this investigation a 2 mol dm^{-3} solution of the electrolyte was employed. The CFT values of the various electrolytes are listed in Table IX together with the θ temperature, i.e., the temperature at which the electrolyte converts water into a θ solvent for the poly(ethylene oxide) chains; a clear correlation exists between the CFT values and the θ temperatures, strongly implicating solvent structure in the stability of the dispersion. A lyotropic series of decreasing flocculating power is obtained

$$SO_4^{2-} \gg Cl^- > NO_3^- > CNS^- > Br^- > I^-$$

As we have seen in earlier discussion, this sequence is in close agreement

TABLE IX. Critical Flocculation Temperatures (CFT) and θ Temperatures of a Poly(vinyl acetate) Dispersion in the Presence of Various Electrolytes at 2 mol dm^{-3} Concentration

Electrolyte	CFT, °K	θ, °K
LiCl	359 ± 2	363 ± 5
NH$_4$Cl	349	349 ± 3
NaCl	332	333
KCl	329	330
RbCl	329	329
CsCl	332	333
MgCl$_2$	361 ± 3	353 ± 5
CaCl$_2$	358	355
SrCl$_2$	350	346
BaCl$_2$	357	358
K$_2$SO$_4$ ^{a}	305 ± 2	307 ± 3
KNO$_2$	337	338
KBr	346	351
KCNS	344	—
KI	358	—

^{a} 0.405 mol dm^{-3}.

with those found for other surfaces in which oxygen atoms are an important component; an explanation in terms of earlier discussion can therefore be offered—incompatibility between the hydration structure originating from the oxygen atom and that of the structure-forming anions, resulting in destruction of solvent structure and preferential adsorption (the SO$_4^{2-}$ ion is a particular exception and will be discussed later). The electrostrictive hydration of the "structure-breaking" ions can be expected to be compatible with the specific hydration of the oxygen atoms, resulting in a repulsive energy term.

The author proposes that the explanation of the series lies in what is essentially the reverse behavior of that previously outlined; the specific hydration of the oxygen atoms is regarded as compatible with that surrounding the structure-making ions. By a somewhat convoluted argument the author invokes the three hydration regions of the Frank and Wen model of ionic hydration [256] to explain the sequence observed, i.e., that the solvent structure-promoting ions are found to be the most effective flocculating agents. A further lyotropic series for the cations is also observed; their flocculating effectiveness is as follows:

$$Rb^+ = K^+ = Na^+ = Cs^+ > NH_4^+ = Sr^{2+} > Li^+ = Ca^{2+} = Ba^{2+} = Mg^{2+}$$

In this sequence the reverse situation exists compared with the anions; the most structure-breaking ions are now the most effective flocculants and the structure-making ions are the weakest in this respect. The author relies for an explanation of this reverse behavior upon involvement of the three hydration regions of the ions to different extents from those observed with the anions and rejects any explanation based upon changes in hydration of the stabilizing moieties.

An alternative explanation can be put forward for this inverse behavior on the basis of the different type of hydration thought to exist at hydrocarbon surfaces compared to that around the oxygen atoms. Nandi and Robinson[517] have shown from studies on proteins and polypeptides that the contributions to the stability of the biopolymer from the hydration structures associated with the nonpolar and polar regions of the molecule are different but additive; they also found that the effects of added electrolytes are additive. In particular, anions appear to affect the hydration of the polar regions, while cations interfere with the hydration of the nonpolar residues, possibly due to the compatibility of hydrophobic hydration structure and that surrounding "structure making" ions. If this is the case with ethylene oxide, then the anions will be expected to interact with the hydrophilic hydration regions surrounding the ether oxygen atoms and the cations with the hydrophobic hydration sphere surrounding the methylene groups. This would result in a reversal between structure-making and structure-breaking ions in the two series and the results observed by Napper would be expected.

As in the case of biopolymers generally, the sulfate ion appears to be unique in its effect upon the stability of hydrophobic colloids whether stabilized with an adsorbed surfactant layer or not. The data of Bull and Breeze[92] have shown that proteins are preferentially hydrated in the presence of sulfate ion but that the ion itself is apparently unhydrated. If we apply this finding to the observed effect on hydrophobic colloid stability, the extremely poor flocculating power of Li_2SO_4, even compared to LiCl, for a negatively charged sulfur sol[268] may be due to induced hydration on the surface of the sol particles. The extremely effective flocculating power of the sulfate ion for a positively charged Fe_2O_3 or Al_2O_3 sol, on the other hand, is to be expected on the basis of Schulze–Hardy rules for an unhydrated doubly charged ion.

Several Russian workers[54,371] have also reported the much enhanced effect of the sulfate ion compared to Cl$^-$ and the equivalence of K$^+$ and Na$^+$ in promoting flocculation of latices stabilized by nonionic surfactants containing appreciable numbers of ethylene oxide groups. The data are broadly interpreted by these workers in terms of a hydration stabilization structure.

5.3. The Importance of Hydration As a Factor in the Stability of Hydrophobic Colloids, in the Presence of Short-Chain Monohydric Alcohols

In the previous section we have seen that adsorption of a strongly surface active layer on the particle results in an appreciable contribution from hydration effects to the improvement in stability of the dispersion; it now remains to examine what contributions from solvent interactions are to be expected to the stability of a hydrophobic dispersion in a dispersion medium containing short-chain monohydric alcohols.

Bijsterbosch and Lyklema[64] have studied the surface charge of an AgI dispersion as a function of the potential of a cell which consisted of an Ag/AgI electrode, the solution containing the dispersion, the non-electrolyte under investigation, and a reference electrode; the influence of urea, n-propyl, n-butyl, iso-butyl, sec-butyl, tert-butyl, and n-amyl alcohol were investigated; all solutions carried a swamping excess of 0.1 mol dm^{-3} KCl. Fig. 29 illustrates the surface charge versus cell potential for various concentrations of n-butyl alcohol; similar curves were obtained for all other alcohols but urea gave a totally different plot (Fig. 30). Figure 31 shows the shift in the zero point of charge (zpc) of AgI from that in pure water as a function of the alcohol concentration. The authors state that adsorption of the alcohol molecules at the particle surface, with displacement of a hydration layer, is responsible for the behavior seen in Figs. 28 and 31. The sigmoidal curves of Fig. 31 are extrapolated to high alcohol contents and the linear region of the plot is used to determine the surface coverage by alcohol molecules from a Langmuir adsorption isotherm[185] approach and subsequently the free energy of adsorption of the alcohol on the particle surface. Other applications of the Langmuir adsorption isotherm at high alcohol concentrations are utilized to confirm the value of the free energy of adsorption of the various alcohols (Table X). The magnitudes of the values of the free energy of adsorption confirm the postulate of the authors that adsorption of the alcohol molecules occurs with the hydrocarbon groups adjacent to the AgI surface and the hydroxyl group protruding into the dispersion medium.

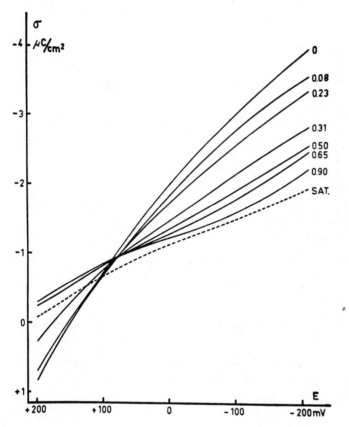

Fig. 29. Surface charge as function of potential for an AgI dispersion in various concentrations of *n*-butyl alcohol. Concentration of added KNO_3 is 0.10 mol dm^{-3} (see Ref. 64).

The authors do not attempt to explain the sigmoidal nature of the shift in the zpc curves and imply that the mechanism determined at high alcohol concentrations adequately explains the behavior over the whole alcohol concentration range studied. In view of our extensive knowledge on the nature of alcohol water mixtures (Volume 2 and Chapter 1, Volume 4) this is unlikely to be the case; conversion of the concentration scale of Fig. 31 to mole fraction reveals that a close correlation exists between the inflection points observed for the various alcohols and that alcohol concentration at which maximum structuring of the solvent is thought to occur. Thus, although the explanation favored by the authors may well be true at high alcohol concentrations, where the "normal" water structure is thought to be totally disrupted by the solute, at lower concentrations the

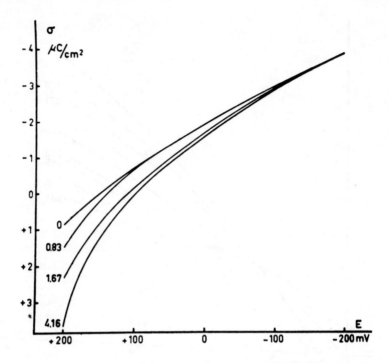

Fig. 30. Surface charge as a function of potential for an AgI dispersion in various concentrations (mol dm^{-3}) of urea. Concentration of added KNO_3 is 0.1 mol dm^{-3}. (See Ref. 64.)

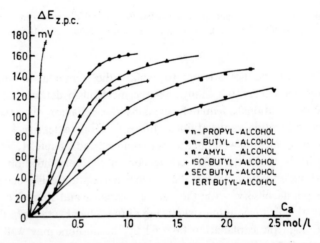

Fig. 31. The shift of zero point charge of an AgI dispersion as a function of added alcohol concentration in the presence of 0.001 mol dm^{-3} KNO_3) (see Ref. 64).

TABLE X. Standard Gibbs Free Energy of Adsorption ΔG° of Various Alcohols from Aqueous Solution onto AgI as Determined by Different Models

Alcohol	Average value of ΔG°, kJ mol^{-1}
n-Propyl	-9.2
n-Butyl	-13.2
iso-Butyl	-13.4
sec-Butyl	-11.3
tert-Butyl	-9.2

rapid movement of the zpc indicates the buildup of relatively thick layers of structured alcohol–water solvent around the particles. The effects of urea are also of interest from the point of view of solvent structure; the authors suggest that the data are explicable on the basis of adsorption of urea at positively charged silver ion sites in the AgI surface; they do not, however, explain why at urea concentrations >4 mol dm^{-3}, i.e., where urea is known to be an extremely effective breaker of solvent structure, the data suggest that the surface charge of AgI is rapidly approaching infinity! A totally destructured solvent may well permit the rapid desorption of structure-breaking I$^-$ ions away from the surface, leaving a massive excess of Ag$^+$ ions and hence the large positive surface charge.

Ottewill and Vincent[545] have studied the adsorption of MeOH, EtOH, PrOH, and BuOH on a polystyrene dispersion (mean diameter 220 nm) from which stabilizing surfactant had been removed by extensive dialysis; Fig. 32 shows the composite adsorption isotherms for ethanol and propanol onto the latex particles and Fig. 33 illustrates the adsorption of n-butanol. Discussion by the authors is centered mainly on the butanol adsorption isotherm with similar reasoning applying, by implication, to the adsorption of ethanol and propanol. The authors suggest that the initial region of the isotherm (up to mole fraction $X_A = 0.0025$) is the result of adsorption of butanol molecules with hydrocarbon groups nearest the surface at hydrophobic sites on the polystyrene surface. For $0.0025 < X_A < 0.006$, where adsorption rises steadily, it is considered to be due to alcohol molecules adsorbing with the OH group nearest the surface at hydrophilic sites; for $0.006 < X_A < 0.015$, multilayer adsorption of butanol molecules is proposed.

Comparison of the ethanol and propanol adsorption isotherms with partial molar volume data[257] and sound absorption maxima[96] found in

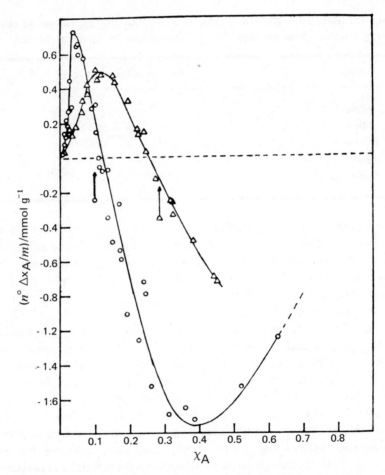

Fig. 32. Composite isotherms for the adsorption of ethanol and propanol onto a polystyrene latex surface at 25°C: (△) ethanol; (○) n-propanol; arrows indicate mol fraction at which receding contact angle becomes zero (see Ref. 545).

alcohol–water mixtures shows that the maximum in the adsorption isotherm again occurs at the alcohol concentration that corresponds to
maximum structuring in the solvent medium; unfortunately, Ottewill and
Vincent did not continue the adsorption experiments with butanol to the
limit of solubility, but there seems little doubt that a similar maximum
would have been observed. The existence of such a maximum and the
negative adsorption at higher alcohol concentrations strongly suggest that
an alternative explanation of the data is required. Initial adsorption of
alcohol molecules at the polystyrene surface is very probably as envisaged

by the authors, with the hydrocarbon groups adsorbing at hydrophobic sites on the polystyrene surface; it is difficult to see why the subsequent adsorption of alcohol molecules at hydrophilic sites via the OH group should occur at all, since presumably such sites are already heavily hydrated and it is unlikely that an alcohol molecule adsorbing via its OH group would succeed in displacing a water molecule. It seems more probable that a complex array of hydrated alcohol can build up around the particle by hydrophobic interactions using the initially adsorbed alcohol molecules as nucleation sites. At alcohol concentrations in excess of the adsorption maximum collapse of this structure would lead to the release of alcohol molecules from the environs of the particle with apparent desorption (see Chapter 1, Volume 4). The difference in adsorption behavior between polystyrene and Graphon surfaces, shown by Ottewill and Vincent, can be explained by the early, much stronger adsorption of butanol on the

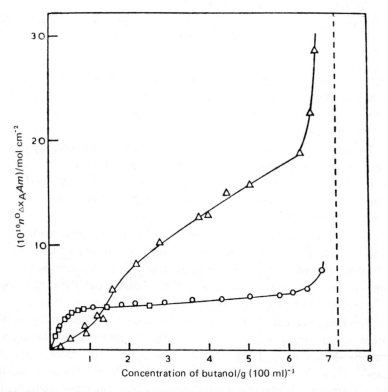

Fig. 33. Isotherms for the adsorption of *n*-butanol onto a polystyrene latex surface ($\triangle$) and on a Graphon surface ($\bigcirc$); dashed line indicates solubility limit of *n*-butanol in water (see Ref. 545).

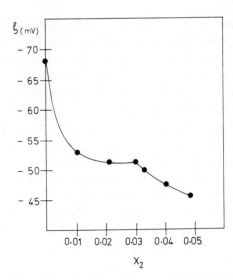

Fig. 34. The apparent ξ potential of a polystyrene dispersion after removal of surfactant as a function of tert-butanol concentration.

Graphon surface, giving a much more closely packed layer in which the alcohol molecules are too close to each other to allow the optimum buildup of the solvent structure which occurs on the polystyrene surface where the alcohol molecules are more widely spaced.

Kostin and co-workers[414] have also reported that the stability of butyl rubber dispersions containing potassium oleate increases markedly in MeOH, EtOH, n-PrOH, n-butyl, and n-amyl alcohol–water mixtures at the alcohol concentration that corresponds to maximum structuring of the solvent medium.

Eagland and Wardlaw* have determined electrophoretic mobilities of a 140-nm polystyrene dispersion, from which the surfactant had been removed by the mixed bed ion exchange technique of Vanderhoff,[674] in a range of tert-butanol–water mixtures; Fig. 34 illustrates the data, plotted as "effective ζ potential" versus X_A.

The rapid decrease in the effective ζ potential with increasing alcohol concentration occurs over the region in which structuring of the aqueous phase by the presence of the alcohols is thought to be predominant. The plateau region observed appears to coincide with the alcohol concentration that corresponds to maximum structuring of the dispersion medium. The

---

* Unpublished results.

further decrease in the effective ζ potential at higher alcohol concentrations corresponds to the region in which collapse of the solvent structure caused by an excess of alcohol molecules is thought to occur. Following the reasoning of Gargallo et al.,[279] at alcohol concentrations below approx $X_4 = 0.04$ the particle appears to be surrounded by a solvation layer which is at its maximum thickness at the alcohol concentration and which corresponds to maximum structuring of the solvent. This in turn will move the shear plane outward from the particle surface, resulting in an apparent lowering of the ζ potential at the shear plane. At higher alcohol concentrations the further decrease of the effective ζ potential suggests that complete monolayer or even multilayer adsorption of alcohol molecules onto the particle surface is occurring which can be expected to extend the shear plane further from the particle surface. Such an interpretation agrees well with that proposed for the data of Ottewill and Vincent.[545]

Vincent et al.[682] have recently reported on the effect of n-butanol on the stability and surface charge of aqueous AgI dispersions in the presence of various uni-univalent and di-univalent electrolytes. The authors report that no simple relationship exists between surface charge on AgI and stability of the dispersion; with increasing butanol concentration the surface charge decreases continuously, whereas the stability passes through a maximum (Figs. 35 and 36). All the systems examined show that maximum flocculation values of electrolytes are obtained when the butanol concentration is $\sim$0.54 mol dm^{-3} ($X_A \sim 0.01$ mol fraction), the alcohol concentration at which maximum solvent structuring is thought to occur. It is particularly interesting that the lyotropic series for the 1:1 electrolytes undergoes an inversion with regard to dispersion stability on passing through the maximum flocculation value, a behavior which is not reflected in the surface charge data. The authors interpret the data as indicating competitive adsorption between cations and alcohol molecules at the particle surface which affects the stability of the dispersion simultaneously in three ways:

1. Ions in the Stern layer are replaced by neutral alcohol molecules, resulting in a lowering of ψ_δ and increased instability.

2. If at essentially constant surface charge, counterions are forced from the Stern layer to the diffuse layer, ψ_δ would increase, leading to sol stability.

3. Due to increased thickness of the Stern layer, the reference plane of the repulsive energy is moved out slightly compared to the reference plane of the van der Waals attraction energy, a trend which is expected to increase the stability of the dispersion.

Such an interpretation would certainly be in agreement with the behavior to be expected from interaction of the structure-making ions and the alcohol–water solvent structure. Figure 35 shows that in the region where solvent structure is increasing the lyotropic series is of the form

$$Li^+ < K^+ \simeq Na^+ < Rb^+$$

in promoting flocculation; the strongest structure-forming ion is the weakest flocculating ion due, on the basis of the model explained earlier, to repulsive interaction between the structured solvent layer surrounding the particle and that surrounding the ion. The structure-breaking ions, e.g., Rb^+, are of course least affected by this interaction. It is to be expected therefore that the counterions are forced into the diffuse layer by the increased thickness of the structured layer surrounding the particles, in agreement with proposals 2 and 3 of Vincent *et al.* At higher alcohol concentrations collapse of the solvent structure occurs, which effectively reverses proposals 2 and 3, and the Li^+ ion is not capable of adsorbing into the Stern layer of the particle to a greater extent than the larger ions. A similar sequence is apparent with the divalent ions (Fig. 36), but at higher alcohol concentra-

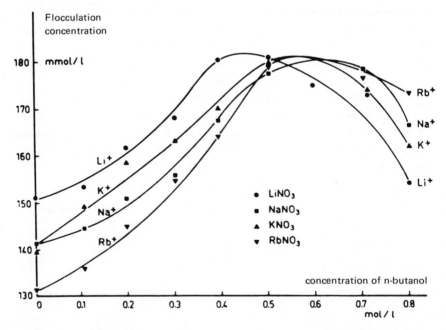

Fig. 35. The stability of an AgI dispersion in the presence of *n*-butanol. Flocculating cations: Li^+, Na^+, K^+, Rb^+. (See Ref. 682.)

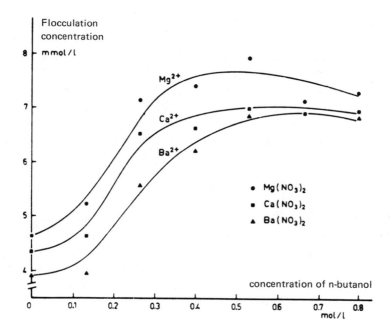

Fig. 36. The stability of an AgI dispersion in the presence of n-butanol. Flocculating cations: Mg^{2+}, Ca^{2+}, Ba^{2+}. (See Ref. 682.)

tions inversion of the lyotropic series does not occur; this may well be due to the very strong structure-promoting capabilities of the alkaline earth cations, which may still retain a structured solvent sheath under conditions when the 1:1 electrolytes can no longer do so.

6. CONCLUSIONS

The theoretical models discussed in Sections 2–4 outline the possible involvement of hydration and solvent structure in the modification of the DLVO theory of colloid stability; it is by no means clear, however, from these studies that such factors can be expected to have a major influence on the stability of hydrophobic dispersions. The data in Section 5 suggest that hydration structures in general play a significant role in the stabilization of hydrophobic colloids, particularly in the cases where enhancement of the "natural solvent structure" can be made to occur.

The evidence from studies of a wide variety of surfaces which contain oxygen atoms is that hydration of such surfaces is unique and not compatible with the "native" structure of water. Presumably hydrogen bonding between the water molecules and the oxygen atoms occurs, resulting in yet another layer of oxygen atoms and resultant incompatibility. Such a hydration layer might best be described as "hydrophilic hydration"[259,658] and resembles the hydration states of carbohydrates.

It is perhaps pertinent to conclude this review with a statement by Freundlich, made in 1922 in his book "Kolloid und Kapillare Chemie": "It is unmistakable that the flocculation values of the ions of the alkali metals arrange themselves in the lyotropic series

$$Li > Na > K > NH_4$$

with many negative hydroxide sols this is very striking. Thus the flocculation value ranges for a negative V_2O_5 sol from 130 mmol liter^{-1} for LiCl to 7.7 for RbCl. Whether here the parallelism between adsorbability and lytotropic series makes itself felt, or whether the hydration of the ions asserts itself in a way not hitherto considered by theory, remains undecided. I am almost inclined to believe the latter alternative."

CHAPTER 2

Properties of Water in Capillaries and Thin Films

J. Clifford

Unilever Research Port Sunlight Laboratory
Wirral, Cheshire, England

1. INTRODUCTION

One of the most controversial aspects of the study of water is research into its properties when present in capillaries and thin films, and the interpretation of these properties in terms of the effect of surfaces on water structure. The surface of a liquid can be considered to be a "fault" in its structure and, in principle, there can be no doubt that the presence of a boundary must influence to some extent the properties of the liquid near it. It has often been supposed that for a highly structured liquid like water the effect may be larger and more dependent on the nature of the surface than for a less structured liquid.

In addition, it has been suggested that the properties of water in the presence of surfaces depend not only directly on the interaction of the surface with water but also on the distance between adjacent surfaces, i.e., on the extent of the water domain.[125] This concept is based on the idea that the structure which determines the properties of bulk water can only exist if a certain minimum number of water molecules can take part in it. The magnitude, range, and permanence of these effects on properties have been much debated, and water in thin films, pores, or capillaries has been investigated to determine their nature and importance.

Among the first systematic experimental investigations were those by Hardy[321] in 1912. In the 1920's and 1930's the investigations of Bangham, [23,25,26] Griffiths,[304] Bastow and Bowden,[38] and Frumkin[274] contributed much to this field but it has been Derjaguin and his co-workers who have been most active in investigating the modification of water properties by interfaces.[158] Over the last forty years they have carried out very many quantitative studies of the physical properties of liquids near surfaces.

In recent years the availability of spectroscopic and similar techniques has resulted in studies of the molecular properties of liquids at interfaces and in thin films, though these methods have often been hampered by the need to develop special devices to make possible the investigation of the small quantities involved. They have the advantage of providing information directly about molecular rotations, vibrations, etc., and are free from some of the assumptions involved in deriving molecular properties from thermodynamic or bulk rheological measurements. It seems likely that the main advances in the study of water in thin films in the next few years will come through the use of such methods.

To an increasing extent, the practical importance of investigating the effect of surfaces on water has been realized. The solution of many industrial problems, connected, for example, with lubrication, corrosion, flotation, foaming, emulsification, colloid stability, wetting, precipitation of obnoxious materials from effluents, etc., depends in part at least on the understanding of the properties of thin liquid layers. Postulated changes in the properties of liquid near a surface have frequently been used as a convenient but unverifiable explanation of differences between experimental results and theoretical treatments, particularly in aqueous disperse systems, and the need for a more productive approach is obvious.

Biological systems in particular involve chemical and physical interactions in water in or near interfaces and if the structure and properties of water are much influenced by the presence of surfaces, this must greatly affect life processes. It has been postulated, for example,[439] that all the water in living cells is in a state profoundly different from normal water and the ion selectivity of cells has been explained on this basis.

Another system in which thin water films may be of great practical importance is ice. The existence of a thin liquid film on ice was first suggested by Faraday[234] and the concept has recently been used to explain meteorological phenomena.

Another recent development has been the controversy about anomalous water.[235] It has been suggested that water condensed on to silica or glass surfaces takes up a form which has a different structure and very different properties from ordinary water and that it retains this form when removed from the surface. Although its unusual properties are now generally attributed to the presence of impurities, the study of anomalous water has greatly stimulated the development of techniques for examining surface water and thinking about its structure.

The study of these phenomena involves measurements near the limits of detection and the use of specially adapted methods, giving results whose interpretation often includes many assumptions and extrapolations. Also, as is often the case with surface studies, these investigations require enormous care in the elimination of the effect of impurities on the observations. Small and apparently unimportant differences in experimental technique used in different laboratories can give widely differing results. Consequently, without far more detailed information about experimental methods than is usually provided in scientific papers, assessment of the validity of work in this field is difficult. Also, even where experimental results can be considered reliable, their interpretation in terms of special water structure at interfaces involves allowances for other effects by means of force calculations of the kind used in theories of colloid stability. Although the general correctness of these is well established, it cannot at present be said that they yield results of sufficient precision to establish differences between water structure in bulk and water structure near surfaces with any certainty. A great deal of ingenuity has been devoted to overcoming these difficulties and significant advances have been made, but the properties of water in thin films remain to some extent an area for controversy and faith.

2. THERMODYNAMIC PROPERTIES OF WATER IN THIN FILMS

2.1. Disjoining Pressures

2.1.1. Principles

The classical thermodynamic discussion of the properties of liquid–liquid, liquid–vapor, and liquid–solid interfaces is carried out entirely in terms of such overall properties as surface free energies, surface excess concentrations, etc. and cannot in itself give any information about the depth of surface interactions. The simple classical capillarity theory leads to the familiar equations for capillary rise, contact angle, etc., and, in particular, to the Kelvin equation[665] for the vapor pressure of liquids with a curved interface, i.e., liquids in capillaries or in drops:

$$\gamma\left(\frac{1}{r_1} + \frac{1}{r_2}\right) = \frac{RT}{V_m} \ln \frac{P}{P_s} \tag{1}$$

where γ is the specific surface free energy, r_1 and r_2 are the principal radii of curvature (positive when the surface is convex), V_m is the molar volume of the liquid, P is the vapor pressure over the interface, and P_s is the corresponding saturation vapor pressure over a flat surface. (Even on this basis the behavior of water in capillaries and porous solids is likely to be quantitatively different from that of most other liquids, because of the large surface free energy.)

However, it was realized that this approach is inadequate if one or more of the dimensions considered becomes comparable with the range of action of the surfaces forces.[155] In this case the specific surface free energy is not independent of the form of the liquid, i.e., the thickness of the film. To make possible the thermodynamic treatment of such a situation Derjaguin[155] introduced the concept of disjoining pressure.

The disjoining pressure is the force at equilibrium required to remove a small increment of thickness of the thin liquid layer. In effect it is the change of free energy with thickness and is given by the expression[173]

$$\gamma = \gamma_0 + \int_h^\infty \Pi\, dh \tag{2}$$

when γ is the specific surface free energy of the thin liquid film, γ_0 is the specific surface free energy of an infinitely thick liquid film, h is the thickness of the film, and Π is the disjoining pressure. A later, more general definition to allow for differences between the properties of the liquid in thin film and those of the bulk liquid is due to Derjaguin and Shcherbakov.[175] This is

$$RT \ln(P/P_s) = V_m(\delta w/\delta h) \tag{3}$$

where V_m is the molar volume of the liquid, P is the vapor pressure in equilibrium with the thin film, and P_s is the standard vapor pressure of the liquid at the temperature T. The term $-\delta w/\delta h$ is the disjoining pressure Π. If Π is positive, the thin film is stable. This expression should apply to any thickness of film from the first monolayer upward.

These expressions refer only to a flat surface. If a curved liquid–air interface is involved, this will possess a capillary pressure given by the Kelvin equation and this, in effect, will be the equivalent of the disjoining pressure when relatively wide capillaries are involved and surface effects are negligible. The total effect on the vapor pressure is given by[557]

$$\gamma\left(\frac{1}{r_1} + \frac{1}{r_2}\right) - \Pi = \frac{\delta w}{\delta h} = \frac{RT}{V_m} \ln \frac{P}{P_s} \tag{4}$$

Thus a disjoining pressure might be measured directly by studying the effect of hydrostatic pressure on film thickness or indirectly by measuring vapor pressures over stable thin films of known thickness.

Depending on the liquid and on the surface considered, Π can be positive or negative and can change sign as h is varied. For liquid–solid systems with a finite contact angle Π must be negative for certain film thicknesses. As the film is thinned a region of instability is reached and it separates into thicker and thinner regions—i.e., a thin film and a drop.

For a liquid that does not form a finite contact angle with the solid, the difference between the classical and the disjoining approaches is shown in Fig. 1.[596] Figure 1(A) shows a classical meniscus for a liquid that wets the solid. The height of rise a is given by

$$a = (2\gamma)^{1/2}/\varrho g \tag{5}$$

where γ is the surface free energy, ϱ is the density, and g is the gravity constant. No liquid phase is present above the meniscus although an adsorbed monolayer could exist. Figure 1(B) shows a meniscus with a liquid film; in this case

$$\gamma = \gamma_0 + \int_h^\infty \Pi \, dh \tag{6}$$

At any height the hydrostatic and surface forces balance, and there will be a corresponding decrease in the vapor pressure at that height which will be given by eqn. (3).

The existence of a disjoining pressure does not necessarily imply modification of the liquid structure by the surface. The disjoining pressure can be due to many effects. It is usual to consider that these effects are additive[628]

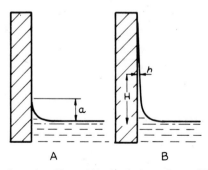

Fig. 1. Formation of wetting films: (A) classical meniscus, (B) meniscus with wetting film.

and to split the disjoining pressure into a number of components[557]:

$$\Pi = \Pi_{LL} + \Pi_{SL} + \Pi_{el} + \Pi_I$$

Π_{LL} and Π_{SL} are disjoining pressures due to van der Waals interactions. Π_{LL} is the disjoining pressure of the liquid film in the absence of solid, i.e., the free energy change due to van der Waals forces in splitting a thin film of liquid from the bulk, and Π_{SL} is the disjoining pressure due to the effect of the solid on the liquid film. Π_{el} is an electrostatic term to account for double layer effects, and Π_I is the change in free energy due to specific interactions between the liquid layer and the surface, leading to molecular orientation or hydration. If one is concerned with investigating the changes in liquid structure near the surface, then a comparison of measured disjoining pressures with calculated values of Π_{LL}, Π_{SL}, and Π_{el} is involved, to obtain Π_I. Here, too, even if Π_I is insignificant, the behavior of water in thin films would be expected to be different from that of most other liquids because of its different bulk properties and the effect of these on Π_{LL}, Π_{SL}, and Π_{el}. For example, for simple nonpolar liquids the disjoining pressure might be expected to arise almost entirely from London dispersion forces, i.e., $\Pi = \Pi_{SL} + \Pi_{LL}$, and to fall steadily to zero as the film thickness is increased. It has been shown that this in fact happens.[182] For water even in the absence of a Π_I term the different variation with film thickness of Π_{LL}, Π_{SL}, and Π_{el} would be expected to lead to a much more complex behavior.

2.1.2. The Measurement of Disjoining Pressures

Thus to investigate the structure of a liquid near an interface it is necessary to measure the disjoining pressure, to calculate contributions to it from van der Waals forces and from double layer forces, and to obtain the Π_I term, as a quantitative measure of alterations in liquid structure, as a difference. This is extremely difficult and it cannot be said that much definite information about the value or nature of Π_I has been obtained. Nevertheless, the work done in establishing the reality of disjoining pressures in thin liquid films—particularly in aqueous systems—in measuring them, and in calculating van der Waals and electrostatic contributions to them, represents a most important achievement in the field of surface chemistry.

Disjoining pressures have been measured in three types of system. (a) Free liquid films, i.e., films with two liquid–vapor interfaces, (b) films with a solid substrate, i.e., liquid films between a solid and a vapor or liquid or between two solids, and (c) emulsion films, i.e., films between liquid–liquid interfaces, e.g., an aqueous film between two oil droplets.

(a) *Free Liquid Films.** Pure liquids do not form free films. For water, Derjaguin *et al.*[176] showed that when contamination by surface-active material was prevented, air bubbles coalesced instantly. Consequently, all studies of free films have been concerned with soap films where the surface properties are determined by an adsorbed layer of surfactant. The techniques and results of the whole study of soap films have been reviewed by Sheludko[628] and Kitchener.[402] Here we are only concerned with those aspects likely to provide information about the state of water in thin films.

Essentially the experiment consists in measuring film thickness and hydrostatic pressure or vapor pressure. The film can be formed by the bringing together of two gas bubbles,[176,597] by forming a double concave drop in the end of a glass tube and removing liquid by suction,[630] or on a fine wire frame drawn out of the liquid.[459] Its thickness is usually determined by optical methods, either interferometry[177] or reflectance, or measurements of the polarization of reflected light,[181] but conductrometric[623] and capacitance[639] methods have also been used. The hydrostatic pressure[177,597] applied to a film or the vapor pressure in equilibrium with it[181] can be measured, and also the rate at which the film drains under the influence of gravity or the hydrostatic pressure.[626] The change in equilibrium thickness with addition of electrolyte under constant pressure can also be measured.[597]

The principal uncertainty in this work has always been in the measurement of film thickness, since this involves assumptions about the nature of the liquid in the film. In the interferometric method, for example, it is necessary to assume a model for the film of two layers of adsorbed surfactant molecules separated by a layer of water, and to assume values for the refractive indices of the layers.

(b) *Films on Substrates.* All these techniques have also been used for liquid films on solid substrates. The apparatus used by Kitchener[597] in a particularly careful study is shown in Fig. 2. A gas bubble generated electrolytically inside the cell and collected in a cup is pressed against a silica surface and the thickness of the film formed is studied by a light reflectance method. It is found that traces of impurity, i.e., dust or organic contaminants, will prevent accurate measurements, so that great care must be taken to eliminate them. The technique was adapted from that of Derjaguin

* The subject of liquid films is treated in detail in Chapter 3.

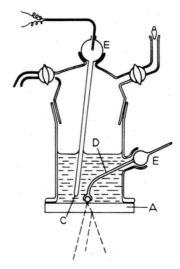

Fig. 2. Cell for disjoining pressure measurements: (A) polished silica plate; (B) aqueous solution; (C) electrolysis apparatus; (D) silica bubble holder; (E) ball and socket joints.

and Kusakov,[169] who studied films on glass surfaces. The thickness of the film for a given bubble pressure is measured. An alternative arrangement in which films thickness and vapor pressure are measured has also been used,[181] and very similar techniques have been employed to study emulsion films, i.e., thin films of aqueous solutions between oil droplets.[639]

2.1.3. The Calculation of Disjoining Pressures

The contributions to the disjoining pressure by van der Waals and electrostatic forces have been calculated often enough, but no attempt has been made to estimate Π_I theoretically, since the concept of structuring of water at an interface remains vague and qualitative. The theoretical considerations involved in the calculation of disjoining pressures are essentially the same as those of the DLVO theory for colloid stability (see Chapter 1) and the application of this to thin films is discussed at length by Sheludko.[628] Calculations of the van der Waals contribution have been based both on the microscopic[320] and the macroscopic[209,532] approach to the estimation of London dispersion forces. In general Π_{LL} will always be negative and Π_{SL} always positive. However, as Read and Kitchener have pointed out,[596] it is not reasonable to calculate Π_{LL} for water on the basis of dispersion forces alone. The effect of hydrogen bonding into a three-dimensional network which determines the cohesion of the bulk liquid will make Π_{LL} more negative if the average hydrogen bonding in a thin film is less than that in bulk water.

The electrostatic contribution Π_{el} has been derived by Langmuir,[422] Frumkin,[273] Verwey and Overbeek,[680] Derjaguin and Landau,[171] and Sheludko[628] for interacting double layers in free films, and modified by Read and Kitchener[597] for films between solid substrates where the two surfaces will have different potentials. Normally Π_{el} will be positive unless the surfaces are of opposite charge.

2.1.4. The Results of Disjoining Pressure Measurements

(a) *Free Films.* Derjaguin and Titievskaya[176] showed that the thinning of films is related to the applied hydrostatic pressure down to thicknesses of about 300 Å, after which some films burst and others attained an equilibrium thickness. The equilibrium reached is only metastable and the probability of forming such a film is determined by the rheological properties of the adsorbed monolayer and its ability to resist fluctuations.

Sheludko and co-workers have measured the rate of thinning of films. [623–625,629] They showed that the flow of liquid out of a film was zero at the adsorbed monolayer, which in effect behaves as a rigid plane, and that it is determined by the viscosity of the liquid and the pressure causing the thinning process. Great difficulties were experienced in obtaining homogeneous films, as would be expected, because of the slowness with which certain types of equilibration processes take place.[584] At thicknesses greater than 0.5 μm, where the Π_{LL} term is very small, and in the presence of 0.1 M KCl to eliminate Π_{el}, thinning took place under the applied hydrostatic pressure, as would be predicted by the assumption of classical capillarity theory and the normal bulk water viscosity, and was independent of the nature of the surface-active material in the adsorbed layers. Below a thickness of about 1000 Å the films drain faster than predicted, because of the negative disjoining pressure, Π_{LL}, and Sheludko was able to obtain Π_{LL} as a function of h.

In systems in which the double layer effect is not fully suppressed by electrolyte the thinning is slowed down and an equilibrium thickness can be reached. Sheludko and Exerowa[630,631] obtained for a film containing 0.01 N KCl a complete curve of disjoining pressure against thickness, from 1500 Å down to an equilibrium thickness of 250 Å, on the basis of measurements of rates of thinning. From measurements in which the double layer effect was suppressed, the Hamaker constant of water was obtained and found to agree with theory. These measurements indicate that the disjoining pressure in free films of thickness $> 250\,\text{Å}$ is due to van der Waals and electrostatic double layer forces, and that the viscosity of water in such films has its normal bulk value. Essentially similar conclusions can be drawn from other work on the thinning of free films in nonaqueous systems.[633]

Studies of films at equilibrium also indicate that for thick films (700–1500 Å[630] and 1900–5000 Å[387])[631] the electrostatic disjoining pressure is equal to the hydrostatic pressure. From equilibrium studies on thinner films in which van der Waals forces are important, Sheludko was able to derive a Hamaker constant for water similar to what he found from his kinetic measurements. Studies by Lyklema and Mysels[457] on large films extended this work down to a thickness of about 100 Å with essentially similar results.

Thus for free films it would seem that for thicknesses of more than about 100 Å properties are determined by electrostatic double layer forces and van der Waals forces and that the water in them is in the same state as bulk water.

Free films are only metastable and if the thickness is reduced below a certain value, fluctuation processes cause either rupture or a sudden reduction in thickness of part of the film to form a very thin "black" film.[387] (Which alternative occurs depends on the rheological properties of the adsorbed layer.) This film can be stable at thicknesses smaller than the critical thickness which is determined by the interplay of the dispersion and electrostatic forces. Some investigations distinguish between two types —"first" and "second" black films—by their variation in equilibrium thickness with respect to ionic strength.[387] Only the "second" type of black film is not being stabilized by electrostatic repulsion forces.

The second type of black film is very thin. Accurate measurements of thickness are difficult in this region but in general they seem to vary between 40 and 100 Å.[134,178,387,627,704] There is no direct evidence concerning the nature of the water in them. Some electrical conductivity measurements indicate that in these films the conductivity is lower than in the bulk solution,[579] but others, in a different system, indicate that it is higher, but their stability is not easily accounted for without assuming that a Π_I term is important. Certainly the possibility exists that the water in these films consists of two 10–15 Å thick hydration layers and not of bulk water. Also it is possible that the transformation from bulk water plus thin hydration layers, to thick hydration layers occur abruptly as the thickness is reduced. It is clear, however, that the critical water thickness where the change occurs is less than 100 Å.

Emulsion films have been little studied, but appear to behave in much the same way as free films,[638] with electrostatic and van der Waals forces controlling their behavior at thicknesses greater than 250 Å, again with possible water structure effects in black films.

(b) *Films on a Solid Substrate.* The study of films of water on a solid substrate is difficult because of the long times required for equilibration, and the enormous effect of contamination by dust and grease, which can only be avoided by very careful experimental techniques.[597] Early measurements by Derjaguin and Kusakov[169] were performed with films of water on lead glass. Very thick stable films were observed. For example, for a hydrostatic pressure of 1000 dyn cm^{-2} films about 700 Å thick were obtained in water containing 10^{-3} N NaCl. Even for 10^{-1} N salt solutions in which the double layer effect would be suppressed, films greater than 500 Å thick were obtained. Other workers, however, could not obtain films at all[216] or only in the presence of a wetting agent.[222] Because of the uncertain nature of glass surfaces, and the possibility of chemical attack of water on them, later work has been mainly on quartz or mica surfaces.

Bangham and Mosallam[24] found a stable film of adsorbed water on mica, about 20 Å thick at relatively low vapor pressures. Earlier work by Bangham[23,25] in which films were formed by a receding liquid on a freshly cleaned mica surface, and appeared to be much thicker, has been questioned.[557] Derjaguin and Zorin[182] studied 100-Å-thick films on glass, and Zorin and Churaev[732] found that stable regions of film 400 Å thick and 100 Å thick could coexist with abrupt steps between them. Derjaguin[175] has explained this in terms of two-phase films. The thicker film is the equilibration situation for bulk liquid on a boundary phase, while the thinner film is the boundary film itself, with its structure and properties, different from those of the bulk liquid, determined by surface forces. It is also possible that the system is the equivalent of a soap film with black areas. The stability of the thicker film will be determined by the balance of electrostatic and van der Waals forces, and the stability of the thin film by hydration and solvation effects, i.e., Π_I forces due to the effect of the surface on water structure.

Other liquids show different behavior.[182] Nonpolar liquids such as CCl_4 and C_6H_6 on glass[182] gave disjoining pressures which reduced steadily to zero as film thickness increased, as would be expected provided that $\Pi_{LL} + \Pi_{SL} > 0$. For polar liquids, however, the disjoining pressure becomes negative for thicknesses above about 50 Å, i.e., a drop of liquid could coexist with an equilibration film of that thickness. Even more complex behavior is found with benzene on mercury,[182,632] which has been explained both as being due to the variation of $\Pi_{LL} + \Pi_{SL}$ with thickness, and as being caused by the coexistence of a surface-oriented phase and a bulk phase.

Some very reliable work by Read and Kitchener[596,597] on water films on silica has clearly indicated the importance of avoiding contamination in investigations of liquid films on solid surfaces and has indicated that, for films thicker than 250 Å, the major factor causing stability is the electrostatic effect and that there is no evidence for any change in water structure in films of this thickness or greater.

A totally different method of measuring disjoining pressures is due to Barclay and Ottewill,[31] who measured the pressure created by dispersed plates of sodium montmorillite. They found reasonable agreement with predicted behavior on the basis of electrostatic and dispersion interactions, down to a thickness of 50 Å, but at lower thicknesses the forces are much greater and are attributed to solvation effects.

Thus, disjoining pressure measurements on water on solid substrates indicate in general that for films thicker than 100 Å the behavior is controlled by van der Waals and electrostatic forces. For thinner films an additional effect due to surface structuring of the water may play some part in determining film stability. The range of this effect seems to depend on the nature of the surface.

2.2. Other Thermodynamic Measurements

2.2.1. Heats of Immersion

The heats of immersion of solids in liquids can be used to study the effect of surfaces on the properties of water near them.[724] The heat of immersion is measured as a function of surface coverage, i.e., the number of layers of water, either from adsorption isotherms at different temperatures, or by means of calorimetric measurements of heats of wetting on solids on which various amounts of water vapor have already been adsorbed from the vapor phase.

For this purpose it is essential not only that the surface area of the dry solid must be known, but also that it should not vary with the amount of water adsorbed. For example, for solids where most of the surface area is in small pores, the heat of immersion falls off rapidly to a figure far below the enthalpy of the liquid surface, which should be the limiting value. In effect, the filling of pores reduces the accessibility of the surface to the wetting liquid. This effect can be allowed for from knowledge of pore structure[314] but in general nonporous systems are required. For most surfaces the enthalpy change—which is negative, heat being evolved—is high for small coverings but falls to a value of 118.5 ergs cm^{-2}, the value of the surface enthalpy of liquid water. This value is sometimes reached after only one monolayer, e.g., for water on asbestos,[689] but in other cases a thicker layer is needed, though not often more than three monolayers. For nonpolar surfaces, e.g., Graphon,[661] the heat of immersion is small and increases to 118.5 ergs cm^{-2} as coverage is increased. Water is adsorbed in patches around isolated hydrophilic centers and the apparent surface area increases as adsorption is increased.

Very complex results are found for some systems, e.g., clays,[728] where swelling, ion exchange, and other changes can occur and in such systems interpretation of the results is difficult, and reliable information about the nature of water in thin films cannot be obtained.

Figure 3 shows a typical graph of heat of immersion against pre-coverage of water for a hydrophilic heterogeneous surface, α-Fe$_2$O$_3$.[482] Initially there are high values as water is chemisorbed to give a hydroxylated surface. Subsequent water is physisorbed, the heat of immersion approaching the value of 118.5 ergs cm^{-2} after about two monolayers. This result, of course, describes only the enthalpy change involved when the solid is immersed, and indicates that after two monolayers it is simply that due to the change in total water surface. However, structural changes in the water may not lead to enthalpy changes large enough to be detected and may show up only in the entropy of the system. The entropy of immersion can be calculated[388] and this has been done for a few systems.[108] In general the

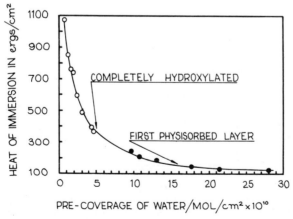

Fig. 3. Heat of immersion of α-Fe$_2$O$_3$ in water as a function of precoverage.

water in the first few layers would seem to be more structured than bulk water. This technique is only useful for well-defined solids and only for a very few layers of water molecules.

Essentially similar information for water in porous systems can be obtained from differential sorption heats of capillary condensation. It has been shown that small differences between these heats and the condensation heat for water can be adequately explained by classical theories in terms of the shape of the concave meniscus involved.[369]

An alternative approach is to measure the temperature variation of heats of immersion of solids in water.[38] A study by Tyler *et al.*[671] of the wetting of some well-characterized silicas, in addition to showing that the presence of micropores makes the interpretation of heat of immersion results very uncertain, also indicated that for unannealed silicas the heat of immersion was dependent on the temperature at which the immersion was carried out. This was interpreted as showing that the heats of immersion contain terms which reflect disturbances of the structure of liquid layers near the solid, but no estimate of the magnitude or range of the effect could be made.

2.2.2. Densities

The formation of surface layers can also be studied by observing volume changes by means of high-precision density measurements of suspensions of solid particles. The results can be expressed in terms of a surface concentration Γ, if they are compared with results for some reference liquid for which the surface concentration is defined as equal to zero. Surface concentrations (with reference to cyclohexane defined as zero) for

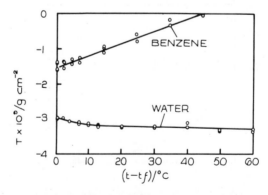

Fig. 4. Surface concentration of benzene and water as a function of the experimental temperature minus the freezing point of the liquid.

water and benzene on graphite[17] are shown in Fig. 4. For both liquids Γ is negative, indicating a less dense layer near the solid. It is impossible to derive the thickness of the affected zone from such measurements, since only an overall effect can be observed. However, it has been calculated that for water the effect was compatible with a zone 30 Å thick in which the density is 1% less than in the bulk liquid. For water near nylon, Dacron, glass fibers, and glass powder van Gils has reported positive surface excesses 1000 times larger.[675]

2.2.3. Phase Changes

The relationship between the melting point of a substance and the dimensions of a pore or capillary containing it has been studied on the basis of the Kelvin equation

$$\Delta T = \frac{RT}{\Delta H} \frac{1}{r} \qquad (7)$$

where ΔH is the latent heat of fusion, ΔT is the depression of the melting point below its bulk volume, and r is the radius of the capillary. This assumes that the vapor pressure of the liquid is lowered below its bulk value only by the curvature of the liquid interface, that the vapor pressure of the solid is that of the bulk solid, and that the properties of the liquid are not otherwise affected by its location.

The study of the freezing of liquids in porous solids is of limited use since it can only express the deviation from the Kelvin equation in terms of an overall effect and can make no statement about the thickness of modified layers. Also, porous solids do not consist of cylindrical capillaries of uniform size and are, in general, of complicated internal geometry. However, it is possible to compare results obtained for different liquids and different solids. In a number of investigations the radius of the pores was obtained from vapor pressure measurements which also assume the validity of the Kelvin equation. Consequently all that can be said on the basis of this work is that the relationship between ΔT, ΔH, and r is independent of the chemical nature of the adsorbed liquid or the adsorbed solid. It is found that water behaves like other liquids but the validity of the Kelvin equation is not proved. A more sophisticated approach by Bakaev et al.,[22] however, using various silicas, measuring capillary radius independently of the Kelvin equation and using thermal diffusion techniques for measuring the amount of water formed on melting, confirms a linear relationship between $1/r$ and ΔT (see Fig. 5), although the accuracy of

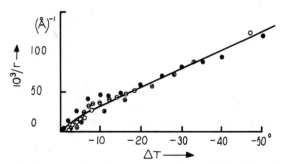

Fig. 5. The relation between the reciprocal pore radius and the diminution of the melting point of water. The different points refer to measurements on different silica samples.

measurement is not sufficiently high to preclude minor deviations from a straight line, which may occur between $r = 100$ and $20\,\text{Å}$. Within the experimental accuracy all the silica samples examined behave similarly. However, more recent work[570] in which nuclear magnetic resonance techniques are used to follow the freezing and melting of water in porous silicas indicates that a part of the water—identified as that in a monolayer on the silica surface and in small pores—never freezes at all within the temperature range used by Bakaev, and consequently his results must be considered as applying to only part of the adsorbed water.

The results obtained by Plooster and Gitlin,[581] who measured the heat capacities of water in commercial silica powders, indicated that for these systems the freezing behavior was abnormal for film thicknesses of up to about $40\,\text{Å}$. The temperature and heat of transition of the water to ice phase change is reduced substantially to below that of bulk water. For layers $100\,\text{Å}$ thick the heat capacity behavior becomes that of the bulk liquid.

3. NONEQUILIBRIUM PROPERTIES OF WATER IN THIN FILMS

3.1. Viscosity

The viscosity of a liquid is a property which is considered to be sensitive to structural changes in the liquid. Consequently, there have been many investigations of velocity of flow of liquids in thin films and capillaries as a means of studying the effect of surfaces. Much of the early work, reviewed by Henniker,[337] consisted of measurements on the flow of liquids through porous partitions. Increased viscosity for some liquids, including water in small pores, is claimed, but since the geometry of the pores was unknown, these results are not conclusive. Similarly, measurements of the rheology of suspensions of small particles are also indecisive, since the small differences between the observed viscosity and that calculated on the basis, for example, of the Einstein formula, can easily be accounted for by electroviscous effects, within the accuracy of the measurements and calculations. A very elegant method devised by Derjaguin et al.[167] not only avoids these disadvantages but also makes possible the observation, not only of an overall change in viscosity, but also of the variation of viscosity with distance from the surface. A uniform current of air is blown over a thin layer of liquid on a solid substrate and the viscosity is calculated from optical measurements of the film thickness profile. Pure nonpolar liquids show constant viscosity down to 10 Å from the surface. If surfactants are added in low concentrations, these appear to affect the viscosity near the surface—presumably by adsorption at the liquid–solid interface. Polar liquids retain bulk viscosities down to a critical distance from the surface but below this distance their viscosity changes, either increasing or decreasing—the effect being particularly noticeable for high-molecular-weight ethers. Polymeric liquids show changes at much greater distances from the surface than do low-molecular-weight liquids.

This method has been criticized,[328] but in general it would appear to be a valid technique for investigating flow of pure liquids near surfaces. Unfortunately, it has not been successfully applied to volatile liquids such as water. However, it indicates, at least, that experiments which show very thick, highly viscous layers (∼1 μm) of pure liquids are probably in error. Hayward and Isdale[328] have shown conclusively, by using very rigorous precautions to exclude foreign particles from their apparatus, that much of the early work indicating thick liquid layers was affected by the presence of dirt particles.

Water viscosity in pores and capillaries and thin films has been investigated by other techniques. Bulkley[90] investigated flow through glass and platinum capillaries and found no evidence for increased rigidity for thickness of 300 Å. Bastow and Bowden used a system with parallel glass plates and by two techniques[38,39] found no increased viscosity for 1000-Å-thick layers. On the other hand, Griffith,[304] Derjaguin,[156] and Fuks[275] found increased viscosity in water layers of thicknesses varying from 1000 to 10,000 Å. Many of these increased viscosities, however, could be due to the presence of dust particles. These cannot be removed from liquids by simple filtration[328] and in general must be considered to be present unless very elaborate precautions are taken to ensure their absence.

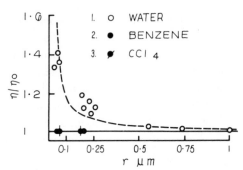

Fig. 6. Dependence of relative viscosity of water, CCl$_4$, and benzene on radius of quartz capillaries ($t \approx 20°$C).

More recent work has been reported by Churaev et al.[112] This is free from many of the objections to early work. The viscosity of liquids is measured in quartz capillaries with radii from 1 to 0.05 μm. The radii are determined by capillary pressure (by measuring the pressure of nitrogen needed to form and detach a bubble from the end of the capillary), and the viscosity by the motion of the liquid under a gas pressure differential. The results for benzene, carbon tetrachloride, and water are given in Fig. 6. The viscosity relative to the bulk viscosity is plotted as a function of capillary radius. It will be seen that while the viscosity of the nonpolar liquids is independent of capillary radius, that of water increases very much in small capillaries, being 1.5 times greater than the bulk value in a capillary of radius 0.05 μm. This could be accounted for by an 80-Å-thick layer of water on the capillary surface which is not sheared by the forces causing liquid flow. The effect decreases with increasing temperature and at 70°C the viscosity of water in these capillaries equals the bulk viscosity.

Studies of the viscosity of thin water films by shearing between surfaces of optically smooth rubber and glass[602,603] indicates that down to 70 Å the viscosity of the water retains its bulk value. The shearing forces here, though, are much greater than in the capillary flow experiments.

3.2. Ionic Mobilities

As an alternative to measuring the viscosity directly, ionic mobilities can be measured. To obtain a system whose geometry and surface characteristics are known, Anderson and Quinn[26] made use of a recently discovered method for making exactly defined capillary holes in mica. The track-etch process was used to make narrow pores with radii down to 30 Å in sheets of mica about 7 μm thick. The pore radii were measured by a Knudsen gas flow technique. Then the dc resistance of the membrane in 0.1 M KCl is measured, and the fluid viscosity derived from the ionic conductivity. It was found that there was no difference between the viscosity of the liquid in the pores and the bulk viscosity even for pores as small as 30 Å radius. The experiment was carried out at different temperatures over the range -2 to $+48$°C. The activation energies for conductivity for 220- and 380-Å pores were equal to those in the bulk liquid. For 56-Å pores a 7% increase in activation energy is observed, an effect which could be entirely accounted for by ascribing nonbulk properties to one layer of water molecules next to the pore walls.

These experiments seem to be less vulnerable to questioning than almost any other in this field, and certainly show that in the mica–electrolyte solution systems long-range water structure effects do not exist. However, it might be said that the presence of a relatively high concentration of ions disrupts the effect of the surface on the water. Also the measurement is of an effect in which there is no bulk movement of water relative to the walls of the capillary, and it is possible that the effect of surfaces on the flow of water past them depends on shear rate, e.g., it might be dilatent or thixotropic—in which case no effect might be observed for a nonflowing solution. Thus, although the results of Anderson and Quinn are of great importance, they do not completely eliminate the possibility of long-range water structure at interfaces.

However, Beck and Schultz,[42] using a similar system with pore radii from 45 to 300 Å, found that the flow of water through the capillaries under pressure was well accounted for by the Poiseuille equation taking the viscosity of water as equal to its bulk value, though the diffusion of small, uncharged molecules through the capillaries was slower than in bulk water.

3.3. Other Structure-Sensitive Properties

Other properties which have been studied with the aim of determining the nature of water in thin films have been thermal conductivity, double refraction, and dielectric behavior.*

---

* Drost-Hansen in his review article on the structure of water near solid interfaces[202] has included a great many other types of measurement but many have only been made on systems so complicated (e.g., clays and fibrous polymers) that many explanations other than surface modifications of water structure are available to account for minor deviations from expected behavior.

3.3.1. Thermal Conductivity

Metsik and Timoschenko[500] prepared systems containing about 1000 mica sheets compressed in stacks with water between them. They measured the thermal conductivity parallel to the layers as a function of the thickness of the water layers. They found that the thermal conductivity was greatly increased in these films. For 1000-Å films the thermal conductivity was found to be an order of magnitude higher than that of bulk water and three times that of ice. This is a surprising result, the range and magnitude of the surface effect being much larger than observed by other techniques.

3.3.2. Double Refraction

Derjaguin and Greene-Kelly[166] studied oriented layers of sodium montmorillite by optical techniques. Double refraction of layers of water molecules between clay platelets was observed and interpreted in terms of the orientation of water molecules by the clay surface. This effect seemed to be independent of water layer thickness up to 200 Å.

3.3.3. Dielectric Properties

The dielectric properties of a substance are a sensitive indicator of its physical state and consequently very many studies have been made of the dielectric properties of water adsorbed on solids or held between sheets of mica. However, the interpretation of the measurements is not unambiguous, since changes in capacitance of a complex system with at least two components and three phases are being measured, and it is difficult to separate those effects that are due to the adsorbed layer of liquid, particularly with complex systems such as clays or fibrous cellulose. Nevertheless the water–silica system especially has been widely investigated. For example, Kuro-saki[394] interpreted his results by postulating that in a silica gel, water first forms an immobile monolayer, then a liquidlike layer, and then several icelike layers. The presence of icelike rather than liquidlike layers is deduced from a maximum loss factor in the region of 10 kHz. No information is available about the dimensions of the pores in the sample. Similarly, from dielectric relaxation measurements of water adsorbed on γ-alumina,[196] it was found that relaxation rates were lower than those of bulk water and on the basis of the temperature variation of relaxation rates it was concluded that the adsorbed water is in a state intermediate between water and ice. Unfortunately, there is a lack of information about the thickness of the layers involved, though it is suggested that 100-Å-diameter pores are average for the adsorbent.

A similar study of a better-defined system α-Fe_2O_3 indicates[481] that at first water interacts chemically with the surface to give hydroxyl groups. The first layer physisorbed subsequently is rigidly bound. The dielectric constant of the system then rises slowly with water layer thickness. Relaxation times remain relatively long—closer to ice than water over the range of the coverage studied (up to seven monolayers). Here, too, the pore structure of the substrate is not reported.

Dielectric measurements on water held between sheets of mica[5,559] indicate that here also there is a significant slowing down of dielectric relaxation rates relative to bulk water.

4. MOLECULAR PROPERTIES OF WATER IN THIN FILMS

In most of the experiments discussed so far the properties of water in capillaries or thin films have been derived indirectly from measurements the results of which contain contributions from all the components of a complex disperse system. Consequently there must always be some uncertainty about interpretation in terms of water structure changes. Modern spectroscopic techniques, however, can be adapted to give information which is related only to the water component, and in principle it should be possible to use such techniques as nuclear magnetic resonance spectroscopy, infrared spectroscopy, and neutron scattering to give *direct* information about the state of water molecules in thin films.

4.1. Infrared Spectroscopy

The application of infrared spectroscopy to the study of chemisorbed water and to water physisorbed on surfaces to the extent of about a monolayer has been extremely successful in providing information about the vibrational and rotational states of adsorbed water molecules, and has shown conclusively that in these circumstances the water present does not resemble liquid water in its behavior.[399] However, applications of infrared spectroscopy to the study of the thicker layers of water, which are the subject of this chapter, have been few and have not added a great deal of understanding about the state of water in films. In general for layers thicker than a few angstroms the spectra are found to be very similar to that of liquid water[667] with broad overlapping bands; interpretation of differences of detail has been hindered by lack of understanding of the infrared spectrum of bulk water itself.

The absorption of infrared radiation at 2.93 μm has been used[134] to measure the thickness of thin water layers (in black soap films) on the basis of the assumption that the water extinction coefficient at this frequency is the same for water in thin films as for bulk water. Thicknesses obtained in this way are similar to those found by other methods. There is no doubt that in calcium fluoride cells, 0.4–1.7 μm thick the infrared adsorption behavior of water is indistinguishable from that of the bulk liquid within experimental error.[667]

An infrared study of water adsorbed on potassium bromide pressed disks has been reported,[666] but in the absence of information about pore sizes and surface area in these systems the results were difficult to interpret. However, it was concluded that adsorbed water, with properties determined by the surface, constitutes a layer at the most only a few molecules thick, and that water in excess of this is present in small clusters in pores, and that its properties are determined by cluster size rather than by surface effects.

An investigation of the infrared spectra of water adsorbed in zeolites as a function of water content has been made by Frohnsdorf and Kingston.[272] The adsorbent consisted of a system of cells, nearly spherical, with radii of 11.4 Å, bounded by alumina silicate rings. Each cell contains up to 21 water molecules. In this system surface hydroxyl groups are not formed, which greatly eases the task of interpretation of the spectra. For low water contents the results are considered to show that isolated water molecules exist, hydrogen-bonded to surface oxygens. At higher concentrations the spectra approach that of bulk water, though the extent of hydrogen bonding, even in the largest groups (21 molecules), appears to be less than that for bulk water at the same temperature.

4.2. Neutron Scattering Spectroscopy

Neutron scattering spectroscopy is a method well suited to the study of water in thin films, since most elements have a very low scattering cross section compared to hydrogen and in consequence the other components of disperse systems are essentially transparent to the neutron beam. The water alone is observed and the frequency spectra of molecular modes and the rates of intermolecular and diffusion motions of water molecules can be determined with great accuracy.

Early work on adsorbed water involved studies of the motions of isolated water molecules in natrolite,[73] but more recently White and his colleagues have been studying the slow neutron scattering of water in a number of well-defined disperse systems. Diffusional motions of water molecules lead to energy changes in the scattered neutrons which are shown in a broadening of the energy spectrum of the scattered beam (Fig. 7).[367] The results are expressed most simply in terms of self-diffusion coefficients measured over a very brief time scale, 10^{-12}–10^{-11} sec. The short times and distances involved eliminate macroscopic effects such as tortuosity factors. Work on clay platelets with layers of water one to two water molecules thick indicated diffusion coefficients increasing with layer thickness, but about an order of magnitude less than that of liquid water.[538] The use of inelastic neutron scattering indicated that the librational modes of the molecules in the thin layers was similar to that of molecules in bulk water. For thicker layers diffusion coefficients varied logarithmically with water layer thickness[539] (Fig. 8). This was explained by using the Stokes–Einstein and Eyring theories to calculate diffusion coefficients in the layer, employing the free energy of evaporation calculated from the Kelvin equation for the free energy of water in thin layers with curved interfaces. Good agreement with the observed results was obtained. This may be fortuitous but it is quite clear from the results that the thickness of surface layers of high viscosity must be quite small. The diffusion coefficient is half its bulk value in 15-Å-thick layers and does not seem to depend greatly on the surface charge density or on the nature of the cations in the clay—in fact, only on the layer thickness. Evidence from inelastic scattering at high energies indicates

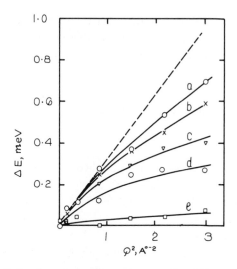

Fig. 7. Quasielastic line broadening for neutrons scattered from water in Li vermiculite samples at a temperature of 23°. Weight water/weight clay is equal to (a) bulk water, (b) 2.77, (c) 2.55, (d) 0.51, (e) 0.21. Q is the momentum transfer.

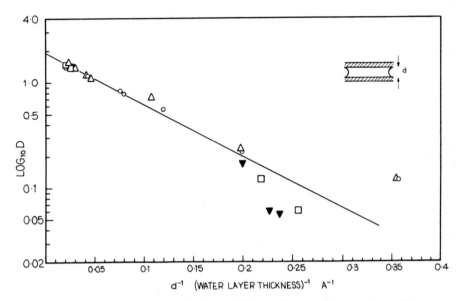

Fig. 8. The variation of the self-diffusion coefficient of water with water layer thickness in clays. Logarithm of self-diffusion constant for water (*D*) for (△) sodium montmorillonite, (○) lithium montmorillonite, (□) lithium vermiculite, and (▼) sodium vermiculite as a function of inverse water layer thickness. The solid line shows a theoretical curve derived from the Kelvin equation using a model for the clay water film shown by the inset.

that the characteristic modes of liquid water began to appear in layers more than three water molecules thick. In general, these results show that in these clays, water in layers more than 15 Å thick is very similar to liquid water as far as diffusive motion is concerned.

4.3. Nuclear Magnetic Resonance Spectroscopy

4.3.1. *Interpretation*

Nuclear magnetic resonance was applied to the study of water in disperse systems[644] very soon after the discovery of the technique, and a great deal of work has been done in this field. Most of it, however, has consisted of studies of chemisorbed water or water physisorbed in amounts of one monolayer or less. This has provided much understanding of the interaction between water molecules and surface groups. Such work has recently been reviewed.[555,600,712] In this section, however, only results concerned with relatively thick layers of water in pores or on surfaces will be considered.

Nuclear magnetic resonance studies of the effect of surfaces on water have usually involved the measurement of proton relaxation times. These have been used to provide information about molecular mobility in terms of a correlation time τ_c for molecular rotation or translation.

The interpretation of results on disperse systems is complicated. As Resing[600] points out, the heterogeneity of such systems manifests itself in the nmr signal in many ways. Diamagnetic susceptibility differences may broaden peaks and lead to apparent relaxation rates much lower than the real ones. If this is avoided by using pulse techniques correctly, then valid values for spin lattice (T_1) and spin–spin (T_2) relaxation times may be obtained, but these cannot usually be interpreted in terms of one correlation time for the protons. Exchange at different rates between protons in different environments, each with its own pattern of molecular mobility, will take place and the relaxation time measured will be a complex average for the system. It is possible to obtain useful information only by measuring several parameters on well-defined simple systems, at a number of water coverages, and over a temperature range. Most studies of this kind have been carried out on water in porous adsorbents. The importance of the pore size distribution of the adsorbent and of the presence of paramagnetic impurities was first realized by Winkler,[711] who studied systems containing micropores and macropores with slow exchange of water between them.

4.3.2. Charcoal Pores

Resing[601] measured the relaxation time of water adsorbed in pores of about 27 Å diameter in charcoal. The results are affected by the presence of paramagnetic centers in the adsorbent but it was possible to show that the water did not possess one characteristic correlation time, as in bulk water, but a whole spectrum of correlation times. For water adsorbed in charcoal pores the central value for the distribution is very similar to the correlation time in bulk water at the same temperature. Resing has also investigated other systems containing water in pores, including zeolites, bacterial cell walls, and proteins. In all cases distributions of correlation times were found, but with median correlation times different from bulk water. The results are shown in Fig. 9 as a relationship between a median jump frequency and temperature. It should be emphasized that in all these cases the water is in small pores and its state is such that it does not freeze. In general its behavior resembles that of liquid water more closely than that of ice.

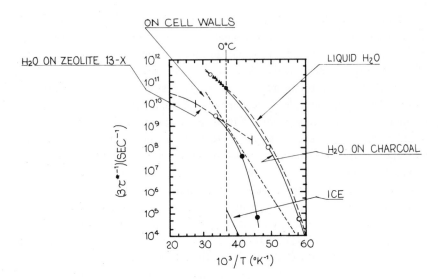

Fig. 9. Jump frequency $3\tau^{*-1}$ versus inverse temperature for water, ice, and adsorbed water.

4.3.3. Clays

Many nmr studies have been carried out on the layers of water between clay or mica platelets. Early work is reviewed by Graham.[296] The results have varied very much with the exact nature of the system investigated but it is clear that, in general, molecular motion is very restricted in thin layers (<10 Å) of water, relative to bulk water. For these systems proton doublet splittings due to the specific orientations of H_2O molecules over long periods are observed.[714] For thicker layers such splittings cannot be detected. From the doublet splitting of the deuteron resonance in clay–D_2O systems it could be deduced that only a few layers of water were oriented by the aluminosilicate surface.

4.3.4. Silicas

The behavior of water adsorbed in amounts corresponding to 0.5, 3.1, and 15 monolayers on nonporous silica surfaces has been studied.[122] The water adsorbed at the two lower coverages behaved quite differently from bulk water. It did not freeze and its molecular mobility was shown to depend largely on hydrogen bond interactions between the adsorbed water molecules and the –OH groups on the surface. When a pore system is created by compression of the silica in the manner used to form a pressed disk for infrared studies there is no change in the nmr relaxation behavior for silicas with coverage of 0.5 monolayer.

The samples with 15 monolayers of water behave quite differently. Most of the water will freeze and its behavior can be adequately described in terms of an average relaxation time governed by rapid exchange between protons in surface OH groups, protons in one or two layers of bound water, and protons in the rest of the water, which seems not to differ from bulk water as far as molecular mobility is concerned. When pores are created in the system by compression the state of the water is very much affected, as is shown in Fig. 10. Also, comparison of the behavior of porous and nonporous silicas with similar water coverages, shown in Fig. 11,

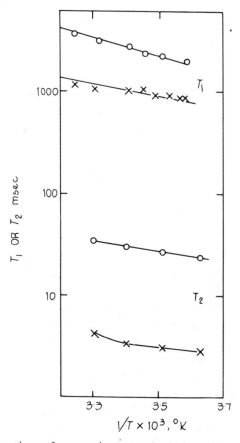

Fig. 10. Relaxation times of protons in water adsorbed on silicas to the extent of 15 monolayers: ($\bigcirc$) silica RA/2, ($\times$) compressed silica RA/2.

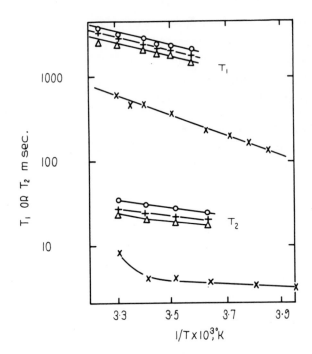

Fig. 11. Relaxation times of protons in water adsorbed on silica to the extent of 15 monolayers: ($\bigcirc$, $+$, $\triangle$) nonporous silicas, ($\times$) a porous silica gel.

indicates that it is the existence of pores rather than simply the effects of the surface that is important.

4.3.5. Polymer Latices

The state of water layers near polymer particles has been investigated. A preliminary study[382] was carried out on polyvinyl alcohol sols where flocculation rates were lower than those predicted by theory and which were in consequence thought likely to possess a layer of modified water. Studies of the variation of water relaxation times with particle concentration and with temperature (Figs. 12 and 13) indicated that a 30-Å water layer with a viscosity of about 1000 times that of bulk water existed around the particles and that when particles approached each other (to give intervening water layers less than 5000 Å thick) the effects became even more marked. The change of relaxation rate with temperature was also complex, showing abrupt changes at certain temperatures. However, it was found that these particles were not smooth spheres, but possessed very porous surfaces and attached polymer chains. The results clearly refer to water in this kind of porous system and not to water near simple surfaces.

Later work on systems containing smooth, nonporous polymer spheres[123] gave entirely different results and indicated that apart from the binding of a few water molecules on active groups on the polymer surface, there was no observable effect on water molecular mobility. However, when the systems are flocculated so as to form "pores" near the point of contact of the spheres, up-field chemical shifts and relaxation rate changes related to a distinct slowing down of the average molecular motion of the water were observed. It is clear that in these systems it is the confining of water within pores or between polymer chains rather than a specific interaction with the surface which effects its mobility.

4.3.6. Lamellar Phases

Nmr has also been used to investigate the mobility of water in thin films in systems of lamellar mesomorphic phases.[123] In these systems water layers of thickness up to 50 Å can be obtained. In general the environment of the water will be similar to that in a black soap film. In a decanol–sodium caprylate–water system the variation of the water proton spin–lattice relaxation rate with water content is shown in Fig. 14. It was interpreted as showing that for water layer thicknesses above 22 Å the variation

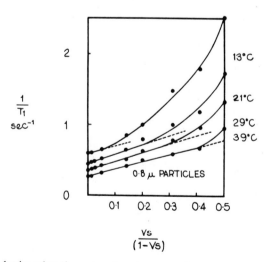

Fig. 12. Spin–lattice relaxation rates of water protons in an aqueous dispersion of 0.8 μm polyvinyl acetate particles as a function of $V_s/(1 - V_s)$, where V_s is the volume fraction of the solid.

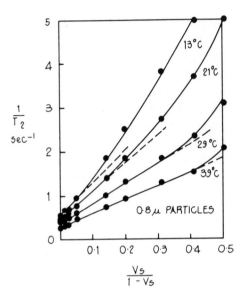

Fig. 13. Spin relaxation rates of water protons in an aqueous dispersion of 0.8 μm polyvinyl acetate particles as a function of $V_s/(1 - V_s)$, where V_s is the volume fraction of the solid.

of relaxation rate with thickness could be accounted for in terms of the exchange of water tightly bound to surface COO^- groups with the rest of the water in the layers, which could be assumed to possess bulk properties. For water layer thicknesses of less than 17 Å, however, the relaxation rate changes with thickness as the mobility of the water molecules drops rapidly. In 8.3-Å layers the average correlation time for molecular rotation is about three orders of magnitude larger than that for water molecules in the bulk liquids.

Here, too, it appears that it is the layer thickness rather than any merely specific surface effect which is important.

4.3.7. Chemical Shifts

In general water in thin films and in pores in adsorbents is sufficiently mobile to give narrow lines whose chemical shifts can be measured. Proton resonance spectra[125] of porous and nonporous silica particles of similar total surface area saturated with water (Fig. 15) indicate that water outside the pores has a chemical shift identical, within experimental error, with that of bulk water, when allowance is made for bulk susceptibility effects. Similarly, its spin–lattice relaxation time is only slightly lower than that of normal water, the linewidth being due to magnetic susceptibility differences between silica and water. Water in pores, however, shows a substantial (14 ppm) shift upfield and possesses a much shorter T_1 than bulk water. Part of the difference in this case could be due to exchange with Si–O–H protons, but similar behavior has been observed in a number of other

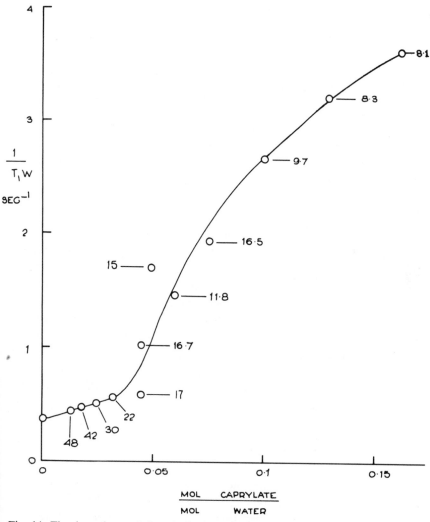

Fig. 14. The dependence of the spin–lattice relaxation rate of water protons $1/T_1 W$ on the ratio mol caprylate/mol water in the lamellar D phase of the sodium caprylate + decanol + water system. The numbers adjacent to the points refer to the water spacings (Å) calculated from X-ray data.

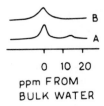

Fig. 15. Proton resonance spectra of dispersions of silica in water: (A) silica gel GC 50 (porous), (B) neosyl (nonporous).

porous systems, e.g., proteins[127] and polymer latices,[382] and water in small pores seems to exhibit, in general, chemical shifts to high field. This could indicate that these is less hydrogen bonding in the water in small pores than in bulk water even though the molecular mobility is lowered. This is entirely possible since the bulk water is bonded to other mobile water molecules while in small pores much of the binding would be to a stationary surface.

4.3.8. Diffusion Coefficient Measurements

Self-diffusion coefficients can be measured by pulsed nmr[316] and this technique has been applied to the study of water in disperse systems. Recent reviews[556,647] have indicated that there is still a great need for the development of the technique but preliminary results have been obtained on a number of disperse systems, including biological cells.[660] In general these indicate that self-diffusion coefficients of water in both biological and nonbiological capillaries with dimension of about 1 μm, though lower, are of the same order of magnitude as self-diffusion coefficients of bulk water. So far the systems examined have not been sufficiently well defined to establish whether tortuosity effects can account for the differences. However, the existence of thick, viscous water layers is clearly not possible.

5. TEMPERATURE ANOMALIES

Another approach to the study of water structure at interfaces is the investigation of thermal anomalies in the properties of the vicinal water. This method is associated particularly with Drost-Hansen.[199,202] It is recognized that the determination of the absolute values of such parameters as viscosity near surfaces is fraught with difficulties and uncertainties, and consequently the relative values at different temperatures are measured and peculiarities in the temperature variation of the parameter are noted. Thermal anomalies are interpreted as arising from structural transitions in the zone affected by the surface, the implication being that the structures involved are stable over well-defined ranges, between which higher-order phase transitions take place. A very odd aspect of the results obtained is that temperatures at which the anomalies are seen appear to be independent of the nature of the surface, which can vary from an ionic crystal to a liquid hydrocarbon. The main disadvantage of the temperature anomaly approach is that it provides no information at all concerning the amount of water involved in the postulated surface structures. The phenomena might be due equally well to two or three layers or to two or three hundred layers of water molecules.

It is now generally accepted that the properties of bulk water do not show thermal anomalies,[113,412,612,621] i.e., they vary smoothly with temperature over the range usually studied. However, evidence exists that some surface properties vary with temperature in an unusual manner.

For phenomena associated with the pure water–air interface the evidence for anomalies is open to question. Thus, some investigators[575,663] have observed inflections in the surface tension–temperature curve (Fig. 16) while others[283] claim that no such inflections exist. The evidence for temperature anomalies in the surface tension of aqueous solutions of organic molecules is much stronger. For example, they have been observed for the interfacial tension between hexane and alcohol–water mixtures.[258]

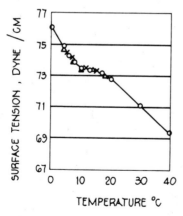

Fig. 16. Surface tension of water, measured by Timmerman and Bodson.[663] Circles represent data obtained by the capillary rise method. Triangles represent data obtained by the drop weight method. Crosses represent data obtained by the bubble pressure method.

For water–solid systems, too, the temperature variation of interfacial phenomena appears to change abruptly at certain temperatures. Disjoining pressures, for example, have been measured as a function of temperature. [572,573] The variation with temperature of the disjoining pressure of a 100-Å layer of water between two quartz plates has been measured and results of the kind shown in Fig. 17 are observed. The variation of the viscosity of water with temperature in a similar system has been investigated[573] and inflections at about the same temperatures have been observed.

The thermal anomaly approach has also been applied to spectroscopic studies. The intensity of the infrared absorption band for water at $2100\ \mathrm{cm}^{-1}$ in a 12-μm calcium fluoride cell plotted against temperature shows an anomaly near 30°.[613] However, the origin of this has been disputed on the grounds that the band is composite and regular shifts of band positions with temperature could account for the observed variation.

Nmr studies of relaxation times of water in colloidal suspensions showed anomalies near 3° and 31°.[382] These are now known to be associated with the existence of pores and of long-chain polymers extending from the particle surface. In systems free from these complications anomalies were not observed.[123] Most of the phenomena in fact have been observed in complex systems; for example, the properties of cellulose,[200] clay,[592] and a vast variety of biological preparations[203] show anomalous temperature dependence.

In general it seems likely that in many systems temperature anomalies do occur in the properties of water near interfaces, and since water is the

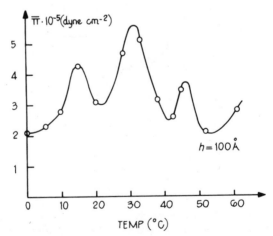

Fig. 17. Disjoining pressure of water between quartz plates: separation between plates 100 Å.

only component common to all these systems, it can be inferred that these anomalies are associated with water near a surface. They are more evident in complex systems, and those where surfaces are not physically well defined, than in simple systems with relatively clear interfaces. They might be said to be a property of water in systems with rough practical interfaces rather than the smooth, plane interfaces sought by surface chemists. Clearly they are found, above all, near interfaces where the superficial surface area is very small compared to the surface area available to the water molecule, i.e., in surfaces with irregularities, attached polymer chains, small pores, etc. Unless great care is taken in the preparation of the surface, nearly all surfaces are of this type and this kind of interface always occurs in biological systems. Consequently temperature anomaly phenomena, while providing no information in themselves about the behavior of water near simple defined interfaces, are of enormous significance for understanding the behavior and role of water in the "witches brew"[255] which comprises biological systems.

6. ANOMALOUS WATER

6.1. History

"Poly Water Drains Away." The title of a recent article in _Nature_[15] probably sums up the present views of a majority of scientists, including most of those who have investigated its properties. It has been suggested[331] that the story of anomalous water really begins in the seventeenth and eighteenth centuries with the "water into earth" controversy, involving such scientists as von Helmont, Boyle, and Scheele. Certainly in 1934 Wilsdon[710] discovered that the properties of water in glass capillaries, and particularly its vapor pressure, depended on whether the water is introduced from the vapor or the liquid phase. However, the systematic study of the anomalous properties of water condensed into glass or quartz capillaries begins with the rediscovery of the phenomenon by Fedyakin in 1961–62.[235] It was postulated that the molecular arrangement of water was being controlled by the surface during the course of condensation in a silica or glass capillary with the result that the structure formed was different from that of bulk water. From 1962–1967 its existence was largely ignored except in the institution where it was discovered, where its physical properties were studied, a task involving great ingenuity and some uncertainty in interpretation of results. This was due to the minuteness of the amounts of anomalous water which could be prepared and the consequent need to develop new techniques for its investigation.

From 1967 the importance of the scientific, industrial/commercial, and social consequences that might follow from the existence and large-scale manufacture of this new form of water were widely recognized. Over the next four or five years much wealth and scientific resources were expended in research into anomalous water—perhaps more than on all other studies of water in thin films and capillaries over the last century. Experimental investigations into anomalous water were carried out in laboratories all over the world and possible structures worked out by a variety of theoretical approaches and the results, discussions, and controversies were widely reported in the scientific and the popular press.

The general conclusions to be drawn from all this work is that the existence of a new form of water is by no means established.

In the course of work on the physical properties of water in glass capillaries[235] Fedyakin noted that, in general, only very narrow capillaries (with a radius <1 μm) conferred unusual properties on condensed water. However, under certain circumstances the water in much wider capillaries (with radii of 2–50 μm) behaved in a manner different from bulk water. If the water was condensed in these capillaries from the vapor state either by transfer from normal water in the same capillary or from an atmosphere containing water vapor at less than 100% relative humidity, then its properties resembled those of the water in the narrow capillaries rather than those of bulk water. Thus, measurements of thermal expansion over the temperature range -20 to $+30°$ showed a maximum density at $-20°$ instead of 4° and a larger coefficient of expansion than bulk water, the exact value depending on the particular sample investigated. That anomalous water had a lower vapor pressure than bulk water followed from the method of preparation, and from the growth of anomalous water columns at the expense of bulk water columns in sealed capillaries. Its viscosity was found to be several times higher than that of ordinary water.

Over the next few years micromethods were developed by Derjaguin et al. for measuring the physical properties of the small quantities of anomalous water that were obtainable (all attempts to prepare it in amounts larger than a few micrograms were unsuccessful), and improved methods of preparation were developed to counteract suggestions that the phenomena observed were due to the effect of impurities.

In its latest form[393] the apparatus used consisted of a Pyrex vessel of the form shown in Fig. 18, notable for the absence of joints or stopcocks. Contamination is guarded against by the provision of cold traps. The

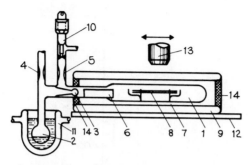

Fig. 18. Apparatus for preparing anomalous water: (1) Chamber, (2) side-arm, (3) diaphragm, (4) tube, (5) tube, (6) striker, (7) capillary holder, (8) capillaries, (9) heater, (10) tap, (11) condenser, (12) comparator stage, (13) microscope objective, (14) plugs.

capillaries, drawn from pure quartz tubing, are supported in a molybdenum cradle. The apparatus is held at 450° under high vacuum for 48 hr and the experiment performed by breaking a glass seal by remote control and admitting into the chamber the vapor from water maintained at a lower temperature than the capillaries. Some of the capillaries are then found to contain anomalous water.

Some physical properties of the condensate were measured, with the following results.

6.2. Physical Properties

6.2.1. *Viscosity*

The viscosity of anomalous water was estimated by applying a gas pressure difference across a capillary containing a thread of the substance and measuring its rate of motion.[165] The results indicated that, at first, the viscosity was up to 20 times higher than that of water but that after a few measurements it dropped to the normal value for bulk water. High rates of movement resulted in a much more rapid falloff of viscosity. After standing for several hours it recovered its initial value. (Its behavior is similar to that of a thixotropic gel.)

6.2.2. *Thermal Expansion*

The relationship between the temperature and the volume of anomalous water is shown in Fig. 19. This could be measured by following the change in length of a thread of the material.[163,235] The temperature of maximum

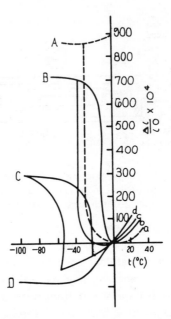

Fig. 19. Thermal expansion of anomalous water. Curve a shows the behavior of normal water, curves b, c, and d the behavior of water containing different concentrations of the anomalous component ($C_b < C_c < C_d$).

density is displaced to below 0° and the coefficient of expansion is larger than that of water and almost constant from 0 to 30°.

6.2.3. Freezing

The freezing–melting behavior was studied.[163,164,172] This is complex and varies very much from sample to sample. (In fact it is now known to depend on the concentration of an anomalous component in a normal water solvent.) This behavior has been described in detail by Everett *et al.*[225,227] They have established that much supercooling takes place but that eventually normal ice separates out and remains in equilibrium with a liquid phase over a wide temperature range. As they point out, this is only possible if the system contains at least two components which cannot be interconverted by a reversible chemical reaction. They show that the freezing–melting behavior is that of a two-component system exhibiting a eutectic and indicate the similarity of anomalous water to gelatin gels.[227,454] One of the components of anomalous water is clearly normal water. The other, which can be concentrated by removing water as ice, is usually called "polywater." On the basis of the volume change during freezing Derjaguin estimated a molecular weight of about 180 for the polywater component.[180]

6.2.4. Density

Another way of concentrating anomalous water to give polywater was arrived at during measurements of the density of the material. Drops of anomalous water suspended in a density gradient column made from tetrachloroethylene and a hydrocarbon, increased in density with time as ordinary water, which is slightly soluble in the column liquid, was removed. Eventually material with a density of about 1.4 g cm^{-3} was obtained[172] (the exact value must be regarded as doubtful[472]).

6.2.5. Surface Tension

The surface tension of anomalous water has also been measured.[184] The results vary from sample to sample and the high, and time-dependent, viscosity makes measurements rather uncertain, but carefully prepared samples have a higher surface tension than does ordinary water. The surface is richer than the bulk in the normal water component, a behavior characteristic of many solutions of inorganic substances. Indeed it has been calculated that the negative adsorption of polywater at the air–water interface is much the same as that of simple inorganic salts.[227]

6.2.6. Refractive Index

A special method was developed for refractive index measurements and a value of about 1.5 for polywater is obtained.[183] If this value is considered with the density, a molar refraction (for 1 mol of H_2O) identical with that of water is obtained. (However, the density may be in error.)[472]

6.2.7. Conductivity

The electrical conductivity has not yet been measured for samples prepared in quartz capillaries in the most recent apparatus but measurements on material obtained in glass capillaries[236] indicated a value of about 10^{-3} ohm^{-1} cm^{-1}, a figure which corresponds to a solution of about 0.01 N NaOH in water. Derjaguin has argued that salt solutions of concentration high enough to produce the observed vapor pressure lowering would show higher electrical conductivities than those observed, but as Howell[361] points out, this is not true if a hydrosol is the solute.

6.2.8. Vapor Pressure

Vapor pressure studies have been carried out on columns of anomalous water by measuring the variation in their length with changes in relative humidity.[179] The lowered vapor pressure of anomalous water is a consequence of the presence of an involatile component, i.e., polywater, as predicted by Raoult's law. It was found that the product of the length of the liquid column and the relative lowering of the vapor pressure $\Delta p/P_0$ was not constant, as would be predicted by Raoult's law, but varied with $\Delta p/P_0$.

This was attributed to negative deviations from Raoult's law. By evaporating off all the ordinary water and measuring the length of the polywater column, a molecular weight of about 200 was calculated for the polywater.

6.2.9. Volatility

Distillation of anomalous water through a heated zone indicated that polywater is volatile above about 200° and stable up to 500°.[161,163] Polywater prepared by the removal of water from samples of anomalous water has been studied by heating it in a sealed tube and measuring its rate of evaporation. Derjaguin et al. have used such measurements to calculate a value for the latent heat of evaporation of 25 J mol^{-1} compared with 40 J mol^{-1} for normal water.[161] However, the validity of these calculations has been questioned[228] and it has been shown that an alternative approach gives a latent heat of evaporation similar to that of bulk water.

6.3. Methods of Preparation

The investigation of anomalous water in laboratories other than those where it was discovered began about 1968. Results varied very much but in general it was found possible to prepare a material similar to the original anomalous water. The proportion of capillaries capable of producing the substance, however, varied in an unpredictable manner. Also, although samples of anomalous water were found to have roughly the same physical properties as those observed by the original discoverers, important differences in detail were noted. Particularly significant were the observations of Everett and his co-workers on the thermodynamic properties of anomalous water[225-228] from which it became clear that it consists of a number of crystalline and gel-like solutes (together forming polywater), in solution of otherwise normal water. (Howell and Lancaster[362] have reproduced the vapor pressure behavior with mixtures of silica gel and water.)

Among the most important of the new contributions to anomalous water studies were investigations of the method of preparation, the use of modern spectroscopic and similar techniques to determine the chemical identity of the anomalous material, and theoretical studies of possible structures for polywater.

Studies of methods of preparation indicated the importance of preventing contamination. As soon as capillaries are drawn they become coated with substances present in the atmosphere surrounding them. Also the silica surface is in a state of strain and in places may not correspond to the quartz conformation but to some more reactive type of silica. Other possible sources of contamination are[33] (i) gross salt contamination, resulting from surface migration; (ii) residues in the capillary, resulting from the cleaning process; (iii) leaching out of compounds from the silica walls; and (iv) surface migration and vapor transport of apparatus impurities (greases, etc.).

When meticulous care was taken to eliminate impurities, the success ratio, i.e., the proportion of capillaries found to contain anomalous water, was very much reduced and some investigators were either unable to obtain anomalous water in pure quartz capillaries under stringent conditions or else only able to prepare it by using a 100% relative humidity atmosphere.[362] Others obtained material indistinguishable from bulk water.[33] It can be inferred that many samples, though not necessarily all, have been contaminated and that the presence of impurities can give rise to material which behaves like anomalous water.

6.4. Analysis

The use of microanalytical methods gave different results with different samples. Organic impurities were present in early Russian samples[145,729] and in some others,[425] originating presumably from grease or pump oil in the apparatus or from human sources (desquamated skin material or breath aerosol). Most analyses were hindered by the minute amounts available. Many impurities arising from the method of cleaning of capillary tubes have been found. Microanalysis of some samples has shown the presence of a great variety of inorganic contaminants. Sodium, potassium, boron, sulfate, carbon, chlorine, nitrogen, and silica have been found in various samples, often in high concentrations.[33,610] Sodium sulfate and sodium borate in particular have been identified.[589] Silica has sometimes been found both by electron probe microanalysis and by microchemical methods and it has been shown that early work, in which silica was not discovered, may be in error because of the difficulty of detecting silica in thin samples by electron probe methods.[37]

Spectroscopic methods have, in general, not given reproducible or unequivocal evidence concerning the nature of anomalous water. Infrared spectra of anomalous water showed the normal water bands and additional bands at about 1100, 1600, and 1400 cm^{-1}. The spectra differed considerably from sample to sample and changed with time. Lippincot[442] interpreted the infrared absorption at 1600 cm^{-1} as being the OH stretching band normally observed at 3400 cm^{-1} in water and at 3756 and 3691 cm^{-1} in water vapor. On the basis of the shift he suggested that polywater contained a symmetric O–H–O group, isoelectronic with the FHF$^-$ unit in bifluoride. Subsequently, however, Rousseau[610] showed that the band at 1100 cm^{-1} was due to the SO$_4^{2-}$ ion and that the bands at 1600 and 1400 cm^{-1} could be due to carboxyl and bicarbonate groups. It has also been shown that many features of the spectrum are similar to those observed for silica–water systems,[225] and mixtures of water and inorganic salts have been prepared with spectra matching those of anomalous water.[36,37] The infrared spectra of poly H$_2$O and poly D$_2$O show the same anomalous absorption peaks.[145] This indicates that the peaks are unlikely to be due to modification of hydrogen bonding in proton-containing groups, but rather to impurities.

Raman spectra[442] have been found to be nonreproducible[8] and ambiguous.[610] Nmr spectra have in all cases given a peak whose chemical shift and relaxation behavior are identical with ordinary water.[709] Concentrated samples have, in addition, given an extra broad peak 300 Hz downfield from the normal water peak.[508,558,576] A similar peak is found in the proton spectra of acids,[558] silica solutions,[508] and, for that matter, of empty capillary tubes[589] (perhaps due to protons adsorbed on the silica surface).

Mass spectrometry has indicated the presence of organic and inorganic impurities in many samples.[610] Analysis of early Russian samples indicated the presence in large amounts of lipids and phospholipids, presumably from human sources.[729] In all cases the peaks due to normal water are found, and no peak with a mass that could be due to a water polymer has ever been detected.[575,610] Silica compounds, e.g., H$_{11}$Si$_3$O$_7^+$ have been found as major components of polywater prepared by the most careful methods.[153] It has been suggested[540] that the observed peaks could arise from decomposition products of silicone oils used in the apparatus but this has been disputed.[227] Polywater samples (anomalous water concentrated by evaporation) have been examined by ESCA[609] and large amounts of inorganic salts found.

6.5. Theoretical Studies

If it is assumed that polywater is some form of liquid containing only hydrogen and oxygen in 2:1 molar ratio and differing structurally from normal water, then, in principle, it should be possible to predict a structure for it, either on purely theoretical grounds by quantum mechanical calculations of stability, or on the basis of spectroscopic evidence. The original structures proposed[111,183,219] were based on the assumption that only physical interactions between normal water molecules were involved, giving a more closely packed array of H_2O molecules than occurs in ordinary water. However, these models were seen to be clearly incompatible with many of the properties of the substance, particularly its thermal stability, and were rejected in favor of models involving changed chemical bonds.

There have been many of these—tetrahedral $(H_2O)_4$ molecules,[70] two- and three-dimensional networks,[441] linear chains,[536] ring structures,[499] and a variety of other polymeric forms.[392] The whole array of possibilities has been critically reviewed by Kamb,[392] who concludes that no adequate structural basis for anomalous water as a H_2O polymer has been advanced. Similarly O'Konski and Levine[537] conclude that there are no plausible structures for polymers of water more stable than the ordinary hydrogen-bonded species. The validity of the spectroscopic evidence and therefore the assumption of the existence of a very strong symmetric O–H–O bond on which many structures are based is also very doubtful. In general, it is becoming clear that significant and surprising changes in hydrogen bond stereochemistry concepts would be required to explain anomalous water in terms of a new H_2O polymer. However, it also seems that present theoretical techniques are inadequate to predict the probability that a postulated structure can exist as stable entity together with ordinary water. Thus, although no entirely satisfactory model for anomalous water has been yet produced, there is no reason to suppose a theoretical basis could not be found for any H_2O polymer species which is found to exist by experiment.

6.6. Silica–Water Interactions

Clearly most of the samples of anomalous water prepared so far owe their anomalous properties to the presence of impurities, i.e., substances other than water or silica. In fact a strong case has been made out for saying that *all* anomalous water owes its properties to impurities.[33] However, it is still possible that some of the preparations do not contain such impurities and that for these the polywater, which in solution in water confers on it anomalous properties, is either a silica compound produced by the chemical effect of water vapor on a silica surface, or indeed a polymer of water produced by the effect of a silica surface on water molecules condensed on it.

In the absence of consistent analytical data it is not possible to decide between these alternatives, although it can be said that some samples undoubtedly contain silica. The interaction between water and silica has been studied by Prigogine and Fripiat,[583] who have put forward a theory for the formation of anomalous water based on the chemical attack on strained regions of the silica surface by isolated H_2O molecules. It has been shown that the degree of dissociation in the first monolayer of water adsorbed on silica is 1%—10^6 times greater than in bulk water,[270] and that silica can be directly dissolved in acidic water by the electrophilic attack of protons. Gingold, too, has shown that leaching of the capillaries can take place.[280]* Adsorption of water molecules onto silica is followed by solution of the surface to give a silica gel, which is then diluted by condensation of further water molecules. This process seems at least credible and in accordance with existing knowledge of silica–water systems. The low-molecular-weight silica acids found could undoubtedly distil at 200° and would be decomposed at high temperatures with the elimination of water.

6.7. Conclusions

From the published work it can be concluded that:

1. When water is condensed from the vapor into small (2–50 μm) capillaries of quartz or glass a substance is sometimes obtained, whose physical properties are distinctly different from those of pure bulk water. These physical properties are by no means entirely reproducible from sample to sample but in general the material obtained has a lower vapor pressure, a higher specific gravity, a higher viscosity, and a very different thermal expansion behavior from water. This substance is not obtained when water is drawn into capillaries.

---

* O'Brien[535] provides an alternative possible explanation for the fact that anomalous water is only found by condensation from the vapor and not by drawing water into capillaries. On the basis of a model in which a liquid is regarded as a mixture of vacancies and molecules of the liquid, he calculates that the liquid in an isolated capillary is in a different state from that in a capillary in contact with bulk liquid, and postulates that changes in the proportion of vacancies at the liquid–solid interface could lead to a liquid state which might be much more chemically reactive than ordinary water.

2. The material consists of a solution of some relatively involatile substance or substances in water. By evaporation, freezing, or solvent extraction of the water it can be concentrated and eventually obtained as a crystalline or glassy material. The lowering of the vapor pressure is a simple Raoult's law effect. Thermodynamic considerations indicate that more than one component must be present in addition to the water solvent, and that some at least are of relatively high molecular weight.

3. These components might consist of: (a) organic or inorganic contaminants; (b) material resulting from the chemical attack and dissolution of silica surfaces by water; (c) some new polymeric form of water (or indeed all of these).

4. From the practical point of view it does not seem to matter which of these solutes are responsible for the anomalous physical properties. Solutions containing inorganic salts, silica sols, or protein sols can be made to simulate the behavior of anomalous water.

5. For most of the samples which have been prepared so far the anomalous properties are due to contaminants or material leached out of the capillary walls. Certainly many preparations with the characteristic properties of anomalous water have been shown to contain substantial amounts of elements other than hydrogen and oxygen. The fact that the properties of the anomalous water differ somewhat from laboratory to laboratory could be due to the presence of different impurities. However, it remains just possible that one of the many solutes which can impart anomalous properties to water is a polymeric form of water.

6. Clearly it is important to find out whether or not a polymeric form of water exists, and, if it does exist, to determine its structure. This is a matter of great importance to the understanding of chemical bonding in general and in particular to concepts relating the structure and properties of ordinary water.

7. Further measurements of physical properties would be of little use since they are ambiguous and can be attributed either to contaminants or to silica sols or to water polymers. Spectroscopic measurements and nmr measurements are also ambiguous since here, too, the bands observed could be due to any or all of the three possible types of solute. The most useful technique for this problem would appear to be mass spectrometry. So far only normal water and a variety of contaminants, silicone polymers, silicate ions, lipids, etc., have been found to be present and no sign of a water polymer detected. The defenders of the water polymer idea have suggested that polywater could be a weakly bonded complex that disintegrates under electron impact and does not appear in the mass spectrum. This is possible.

Thus, there is clearly a need to prepare the material by methods more carefully designed to eliminate contamination and to examine material prepared in this way by modern analytical techniques. However, since the lack of definite results has caused interest in the subject to wane, it may be some time before these things are done. Nevertheless anomalous water remains a matter of great importance, particularly to scientists studying water structure and properties. If two forms of liquid water could exist together either in a stable or a metastable system, existing ideas of water structure and binding would have to be modified.

To those studying the behavior of water in capillaries and thin films the existence of anomalous water has an additional, important practical consequence. Whether the properties of anomalous water are due to contaminants, a silica hydrosol, or a polymeric form of water, it has certainly been established that water condensed onto silica or glass surfaces has unusual properties. Consequently conclusions about the temporary structure of water near surfaces which are based on experiments involving glass or silica surfaces must be regarded with considerable doubt, unless it can be shown that conditions for the formation of anomalous water did not exist.

The anomalous water investigation has led to useful and constructive speculation. There has been a great deal of thinking about the nature of bonding in water and the stabilities of many different kinds of model for water structure have been calculated. If, as seems possible, networks containing silica atoms are responsible for the properties of anomalous water, then a new type of silica gel structure may have been discovered and certainly the chemical attack by water vapor on surfaces not corroded by liquid water is a phenomenon of great practical and theoretical importance.*

7. LIQUID FILMS ON ICE

A thin film of aqueous liquid which is of particular interest and importance is that which is postulated to exist on ice below its melting point. In 1850 Faraday[234] explained many of the peculiar mechanical properties of ice by suggesting that at temperatures just below the melting point ice crystals are covered by a thin film of liquid water. This hypothesis

---

* A recent publication by the discoverer of anomalous water[162] has confirmed that the properties of anomalous water are in fact due to the chemical attack of water vapor on surfaces and the formation of some kind of gel containing elements other than oxygen and hydrogen.

has been questioned, but while it is undoubtedly true that many of the phenomena concerned can be accounted for in terms of pressure melting,[664] frictional heating,[74] or crystalline creep,[291] some cannot be so explained. In particular, observations of the aggregation of ice crystals to form snow seem to be best explained in terms of a liquid layer on ice down to $-25°$.[358] Consequently in recent years a number of experiments have been performed to determine if ice is covered by a liquid film. This work has been reviewed by Jellinek.[380]

These experiments have been concerned with two types of systems. One consists of pure ice which, if polycrystalline, might possess liquid layers both at the ice–vapor interface and at the interface between ice crystals. In the other type of system, a water–solid disperse system is frozen. In this case additional liquid films between the ice and the solid surface may also exist.

Most of the early evidence was for liquid films at the ice–vapor interface and was rather indirect. Nakaya and Matsumoto[515] and later Hosler et al.[359] measured the forces between spheres of ice suspended like pendulums and allowed to come into contact. The force required to separate two spheres was measured as a function of temperature and the results are shown in Fig. 20. These results were explained by assuming that the liquid layer at the point of contact is transformed into ice. The pressures involved

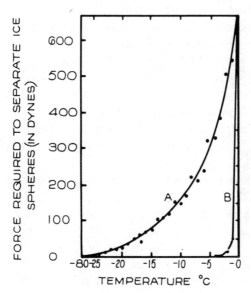

Fig. 20. Separation force as a function of temperature: (A) water vapor-saturated atmosphere, (B) dry atmosphere.

are several orders of magnitude too small to permit an explanation based on pressure melting and regelation (though it is not clear what effect taking into account surface roughness would have on this conclusion).

Kingery[396] measured the rate at which ice spheres grow together (sinter). Processes involved are viscous flow in a mobile surface layer, evaporation, condensation, and diffusion. The variation of the rate of sintering with sphere radius and temperature led Kingery to conclude that a mobile surface layer existed on ice. He suggested that it was one or two molecules thick, though in fact his experiments do not permit a decision between a very mobile thin layer and a more viscous thick layer. Hobbs and Mason,[352] on the other hand, found that contaminated surfaces sintered much more quickly than pure ones and clearly this observation must be considered to cast at least some doubt on other work on ice surfaces, as indicating that here, too, is a surface system in which small amounts of impurities can have a decisive effect on the results of experiments. Nevertheless recent careful work[381] still shows effects which are best explained in terms of a mobile layer involved in a plastic flow process.

The second kind of system, in which a mobile layer between ice and a solid is involved, has also been investigated in recent years. One technique[378,379,594] used was to freeze water on a solid surface, e.g., stainless steel, and measure the forces required to disrupt the system, by tensile stress or by shear stress. When tensile stress was used cohesive breaks within the ice occurred while shear stress led to adhesive rupture at yield values critically dependent on temperature, surface roughness, and rate of shear. From the results liquid layers of 10^{-6} cm thick with viscosities of hundreds of poises were calculated to exist between ice and stainless steel and between ice and quartz. Similar results have been obtained from measurements of the passage of metal wires through ice[662] and the motion of small particles in ice and water in frozen porous systems under a temperature gradient.[353] X-ray diffraction measurements on frozen clays[9] indicate that on freezing, normal ice is formed and that layers of unfrozen water from 15 Å thick at 0° to 3 Å thick at $-180°$ exist between the ice and the clay.

Nuclear magnetic resonance spectroscopy has been used to investigate this phenomenon in ice. Pure single crystals of ice give no proton resonance signal other than the broad line due to protons in solid ice. Spectra of polycrystalline samples, however, include a narrow line, suggesting that a small proportion of the protons are in a more mobile state.[121] It has been shown that impurities cannot be a direct cause of the narrow line[45] but it is possible that some special effect of impurities at ice crystal boundaries

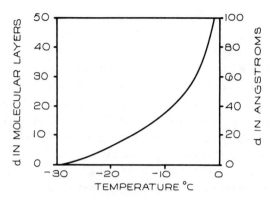

Fig. 21. The thickness d of the equilibrium liquid film on the surface of ice as a function of temperature.

may be responsible. Vapor-deposited ice and fresh snow[560] give narrow lines which are observed to be transient and to disappear on sintering. Here as in so many other studies of thin water films it cannot be said with certainty that the effect of impurities has been eliminated.

The study of thin water films on ice differs from other aspects of the subject in that an attempt has been made to treat surface structures of ice in a quantitative way. Fletcher[243] has calculated that the free energy of an ice crystal is reduced if the surface is covered with a thin liquid layer and has calculated the depth of the layer and how this should vary with temperature on a thermodynamic basis, by making assumptions about the surface structure of water. His results are shown in Fig. 21. He calculates that near 0° ice has a surface layer tens of molecules thick in which the outer monolayer is completed, oriented with the oxygen atoms outward, and the layers between this and the regular ice crystal are disordered.

So far no adequate experimental confirmation of this prediction has been achieved, but there is evidence for a disordered layer at the ice surface. However, the present lack of knowledge concerning the effect of impurities at an ice interface, and the surface properties of ice in general, make other explanations of observed phenomena possible.

8. WATER IN BIOLOGICAL CAPILLARIES

The state of water in living tissues is a matter of the most fundamental importance to biology and will clearly influence the rates and effectiveness of such processes as the transport of materials and energy in the cell and most biochemical reactions. Consequently, there have been many review articles on the state of water in biological systems.[203,657] There are two opposing points of view in this matter. The first[110] states that most of the water in the cell is in a free liquid state resembling the bulk water outside the cell, and considers such phenomena as differences between ionic concentrations inside and outside as being determined by membrane effects. The second is based on models in which the water, ions, and biopolymers inside the cell form a single, highly ordered system and in which the aqueous part must differ very much from normal liquid water.[439,656]* The evidence on which the arguments between the two views have been based is mostly indirect and inconclusive, and it is possible that neither is correct.[332] The complexity of biological systems leads to much uncertainty in the interpretation of results and the current lack of firm knowledge of the structure even of bulk water and of water near simple surfaces makes a precise understanding of water in biological systems impossible at present.

Much of the evidence for highly structured water in biological systems has been obtained by studying the temperature variation of the rates of a great number of processes, e.g., enzymatic reactions,[192] conduction of membranes,[137] hemolysis of erythrocytes,[294] liposome permeability,[383] nerve transmission,[443] and many others (see Ref. 203). Often abrupt changes in the variation of rates with temperature are observed at certain temperatures and these are ascribed to phase changes in surface water structures as described in Section 5, by analogy with similar results obtained in simpler systems. There is no doubt that such temperature anomalies occur and it is likely that they are associated with water, but there is no real reason to conclude that they are due to extensive water structuring throughout the system.

In recent years it has been realized that a meaningful study of water in such complex systems can only be carried out by using techniques which can be made to respond only to the state of the water and not to the other components in the system. Also it is essential to study the properties of water in relatively simple systems, e.g., solutions and gels of biological macromolecules and isolated membranes, as a basis for the understanding of true living systems.

Recent infrared absorption studies of water in membrane materials

---

* It should be pointed out that it is not necessary to assume long-range ordering in pure water layers as a basis for theories describing cell properties in terms of ordered water. The water in question contains substantial amounts of material in solution and in suspension, and in many situations gel-like networks exist. Quite short-range ordering in a complex water–protein–polysaccharide-ion system would be sufficient.

and membranes,[395] in DNA,[232] and in a variety of biological systems[395] indicate that the interaction between biological materials and water is essentially short range, and that most of the water in a biological system is indistinguishable from bulk water. However, it is possible that this technique might be insensitive to some possible structural changes because of the broad overlapping bands which occur in infrared spectra of aqueous systems.

Nuclear magnetic resonance spectroscopy has given some more definite information, although inadequacies of technique and interpretation, in particular the failure to realize the importance of the concept of exchange processes between water in different states, led to false conclusions being drawn from early work. Investigations of the proton relaxation times of water in solutions of biological macromolecules—proteins, polysaccharides, and nucleic acids—give results which can only be adequately explained in terms of small amounts of bound water on the polymers which exchange with molecules in bulk water with normal water properties.[699] Studies of biological gels[285] lead to similar conclusions and show that the high viscosities of these systems are caused only by the macromolecular framework and not by any unusual properties of the water. Similarly, self-diffusion coefficient measurements on gels indicate that most of the water is in a normal bulk state[698] and that the small differences between the self-diffusion coefficients of water in gels and that of bulk water can be accounted for in terms of a tortuosity effect.

This work has been extended to a number of living systems by Abetsedarskaya et al.[2] and Walter and Hope.[698] Here, too, the results appear to rule out extensive structuring of water in the cell and can be best explained in terms of an exchange averaged behavior of small amounts of bound water ($\sim$20%) and normal water. Usually exchange processes are rapid with short lifetimes for water in any one biological situation, but a few systems with slowly exchanging water have been observed.[102] Results on suspensions of erythrocyte membranes[126] indicated that a whole range of types of bound water exist within the membrane structure but that surface effects on water near the membranes were too small to be measured except possibly for situations where distances between membranes are very short.

Quite different conclusions have been drawn by Cope[133] from nmr studies of the behavior of D_2O and sodium in biological systems and his results seemed to indicate that considerable amounts of firmly bound sodium and water were present in the systems to be investigated. However, Civan and Shporer[115] have shown that an alternative explanation in terms of quadrupole effects is likely, and that only a small fraction of the sodium ions and water need be bound to account for Cope's results.

The systems referred to so far have contained far more water than nonaqueous material and it seems that in these most of the water possesses its bulk properties. In systems with much less water a very different situation exists. Following the work of Berendsen,[53] many investigators have used such techniques as nuclear magnetic resonance to show that in these systems, in which the water is adsorbed on protein or polysaccharide surfaces or is present in very small pores, the structure and properties of the water are quite different from those of the bulk liquid.

Thus, it seems that some indication of the state of biological water is beginning to emerge from recent work, though it must be said that at present it cannot be considered as more than a tentative hypothesis. Three types of water can exist: (i) water firmly bonded (mainly on active sites) to proteins, polysaccharides, lipids, etc.; (ii) water present in small pores of molecular dimensions, e.g., in membranes, between closely spaced membranes, and between polymer chains adsorbed on surfaces; and (iii) normal water present as the major component of a complex solution of small molecules and polymers, which may under some circumstances be a gel.

There can be no clear line drawn between the three types, particularly since water molecules must usually be able to exchange rapidly from situation to situation. Studies of the adsorption of water on a variety of biological macromolecules indicate that the first type of water is likely to be present to at most about half of the amount of the nonaqueous components in the system. The second type of water, in general less mobile (but less hydrogen-bonded) than bulk water, is perhaps the most interesting. It is suggested that it is this water present in small pores and "fuzzy" surfaces which gives rise to the temperature anomalies. Together the first two stages comprise perhaps one-fourth of the water in most living systems. The bulk of the water is in the third state. It should be pointed out that one of the main requirements for aqueous phases of living systems is that while molecular processes and transport should be as rapid as in a normal solution, the motion of large bodies should be prevented or slowed down. Thus, the cytoplasm of a cell must permit the motion of ions while limiting the motion of the organelles.

Similarly the ground substance of the dermis must hold organs and cells in position while permitting the free passage of nutrients, etc. It would seem that this is achieved by systems such as gels in which the rheological properties are determined by a macromolecular network (interactions with normal water no doubt play a vital role in the formation of junction zones) through which water and other small molecules can diffuse freely. Thus, in some circumstances the water would have properties very like those of "anomalous water." Indeed the presence of large amounts of biological contaminants in some samples of "anomalous" water indicates that these at least must have been very like water in biological systems.

9. GENERAL DISCUSSION

Little work has been done on the development of quantitative theoretical models for the behavior of water under the influence of a surface. In view of the present lack of understanding of the structure and behavior of bulk water this is reasonable. The theoretical attempts made to account for anomalous water indicated that the present knowledge of the basic interactions between water molecules in the normal liquid state is insufficient to provide an adequate basis for the calculations of the relative stabilities of different structures even in bulk water. Until this can be done it would seem impossible to predict the quantitative effect of various types of surface on the structure and properties of adjacent water. Also, the uncertainty of existing experimental evidence about water near surfaces makes exact and detailed predictions pointless at present. Consequently although there has been much speculation about surface effects on water, it has very little predictive value. However, a brief qualitative consideration of possible effects is of some interest.

In thin films and pores the dimensions of the film or pore may affect water directly by not allowing sufficient space for the development of bulk water structure. Clearly there are two types of situations as shown in Fig. 22. In A there is simply an interface between bulk water and another phase, solid, liquid, or vapor. In B there is a thin layer of water between two interfaces, possibly with partly overlapping effects, or spatial effects on the water film.

For situation A, Drost-Hansen has proposed[202] that a surface structure, determined by the interaction of water molecules with the other phase, merges with bulk water structure through an intermediate disordered zone. This is plausible, but at present both quantitative prediction and experimental confirmation of the effect of such a system on water properties are lacking.

The interaction of the water molecules with the surface will depend primarily on whether the surface contains active sites which can impose a structure on water in competition with bulk water structure. If such sites are absent, i.e., if the surface is hydrophobic, its effect on the water is

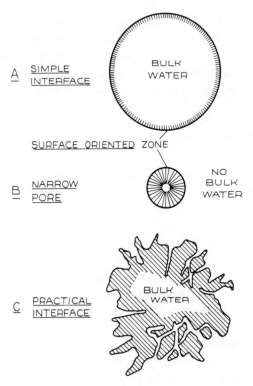

Fig. 22. Possible surface environments: (A) surface in contact with bulk water, (B) thin films and narrow pores, (C) practical surface.

likely to be relatively weak, although not necessarily of a short-range nature. (This is clearly shown by the way in which the adsorption of water on hydrophilic surfaces is dominated by its interaction with points rendered hydrophilic by impurities or dislocations.[724]) The liquid–vapor interface is the only one to be treated even semiquantitatively.[243] The water molecules were considered as being oriented with the oxygen outward at a water–air interface as a consequence of the lowering of energy due to the positioning in this way of the polarizable lone-pair electrons, an effect modified by dipole–dipole interactions, and a change in entropy due to ordering. It was originally suggested that bulk water structure was not reached until after ten layers of water molecules from the surface, but this is likely to be a gross overestimate.[700]

For solid hydrophobic surfaces it has been considered that the effect must be related to the ordering of water around nonpolar molecules in aqueous solutions. Whether this is due to dispersion effects on hydrogen bond energy levels,[527] the effect of steric hindrance in decreasing hydrogen bond flexibility leading to the formation of clathrate structures,[292] or the absence of forces on one side of water molecules adjacent to the nonpolar entity, is still a matter for conjecture. The effect is likely to lead to a less dense packing of water molecules (this has been observed[17]), is known to be weak compared to the effect of, for example, ions, and is short ranged. It is important to note that not only strictly nonpolar surfaces are hydrophobic. The hydrogen bond between water and the oxygen of a Si–O–Si group is weak and the interactions between silica and water are determined by the hydroxy groups on the surface. If these are removed by heating, silica becomes temporarily hydrophobic until it is rehydroxylated by chemisorption. It is the relative importance of cohesion of water molecules to each other to form bulk water, and adhesion to the surface to form a surface-oriented film, which is important.

The electrostatic orientation of water molecules around surface charged groups and bound counterions on surfaces is likely to be dominant where it occurs. For example, most of the energy involved in water adsorption and swelling in clays is considered to arise from the hydration energy of the cations.[296] There can be no doubt that the structure of the hydration layer of water immediately adjacent to ions is determined by electrostatic forces, and is quite different from that of bulk water, but here, too, it can be said, by analogy with what is known about the effect of ionic solutes on bulk water, that the range of the effect is likely to be quite small. Studies of aqueous solutions of molecules with an ionic group and a hydrocarbon chain indicate that the ionic group disturbs water structure sufficiently to modify its interaction with the chain for a distance of 6–10 Å.[124] Work on micellar systems has indicated that arrays of ions are more effective in ordering water molecules than isolated ions, but even so the influence on water mobility seems to be confined to three or four layers of water molecules at the most with rapid exchange between bound and bulk water.

If water molecules can form stronger hydrogen bonds to surface groups than among themselves, other possibilities exist for long-range effects. If hydrogen bonding in water is a cooperative phenomenon, as is generally supposed, then the formation of a very strong hydrogen bond—for example, to a surface Si–O–H group—may lead to increased overall hydrogen bonding near the surface. However, there is no reason to suppose that molecules containing strongly hydrogen-bonding groups have very long-range effects on water structure in their aqueous solutions.[657] The –OH groups on surfaces are stationary and are often ordered into patterns which might, for instance, fit into ice structure or some other possible form of water.

That the spatial arrangement of polar groups is important in determining, for example, the solubility of carbohydrates in water is well known,[657] but here, too, there is no reason to suppose that the effect on the water component is large or long range. In all circumstances covered by Fig. 22A the surface structure has to compete with the stable normal water structure. So far no very convincing mechanism has been suggested which would explain how it could do so over a range of more than a very few water molecules.

If, however, normal water structure is prevented from forming, or is rendered less stable by lack of space—the situations illustrated in Fig. 22B —then surface interactions have more chance of affecting the arrangement of water molecules. The question which should be asked first is, "How many H_2O molecules are required to give a liquid with the properties of bulk water?" Attempts have been made to calculate the size of clusters in liquid water.[49,263,526,668] Estimates of average cluster sizes vary from about six to 10,000 water molecules (i.e., distances of from 6 to 60 Å are involved), with more recent studies favoring the lower figures. It is possible that in spaces too small to hold such clusters the result bulk structure of liquid water possesses less stability, and calculations of the effect of pore size on water free energy seem possible.

The situations illustrated in Figs. 22A and 22B are greatly simplified models put forward to aid discussion. In practice, the interface between water and a surface is usually more like Fig. 22C. Pores of various sizes, surface imperfections, attached macromolecular chains, and impurities ensure that the surface area available for adsorption is many times the superficial surface area and that water molecules exist in spaces comparable with molecular dimensions. In effect, the surface is diffuse and extends over a distance, not only because of the water but because of the substrate. In fact much water at a type A surface is in a type B situation and will not be like bulk liquid water. It is tentatively suggested that much biological water is of this kind. When the "imperfections" are detached from the surface in sufficient quantities and held in solution or as a gel in the water, "anomalous" water is the result.

In many practical situations, i.e., in biological membranes, the pores etc., are essential properties of the surface. In other cases they can be formed by the adsorption of surface-active materials from the aqueous phase. However, water can also create its own pores and capillaries particularly in silicas and glasses. As Kitchener has pointed out,[403] in discussions of the unique physical properties of water and its physical interactions at aqueous interfaces the equally important and unique chemical properties of water and its chemical interactions with the substrate material are often forgotten. There is much evidence for extensive chemical hydration of many surfaces, including many used in thin film studies and it is likely that some long-range effects, particularly on glass surfaces, are due to the formation of a gel layer. The molecular properties of the water in the gel may or may not be those of normal water—this will depend on the size of the inter-stices—but the bulk properties of the gel will depend on the macromolecular network, and the rheological behavior in particular may be quite complex and altogether different from that of water.

10. CONCLUDING REMARKS

It will be very clear that at present there is little agreement about the state of water in thin films. The classical capillary theories—represented by the Kelvin equation—are adequate for many practical purposes and repre-sent an important effect influencing the behavior of water in pores and capillaries. For example, the hysteresis in the adsorption isotherms of water vapor on porous solids can be explained in these terms. Because of the high surface free energy of water this effect will be more important in aqueous than in most nonaqueous systems. Also, it can be regarded as certain that electrostatic double layer and van der Waals forces create a disjoining pressure which is vitally important in films less than 500 Å thick, and which affects many aspects of their behavior. Here, too, water will behave in a completely different way from other liquids because of its different bulk properties.

What is still in dispute is whether there is an important extra effect due to a change in the molecular properties of water in the vicinity of a surface. Half a century of research has not settled this matter. It has be-come quite certain that if a film of water to a thickness of two or three molecular layers exists on a surface or in a pore, then its behavior and structure will be determined by interactions with the surface and will be entirely different from those of bulk water. It is also clear that if water layers thicker than 1 μm are being considered, it can be assumed that their properties are those of the bulk liquid. It is likely that anomalous effects seen in layers thicker than 1 μm have been due to impurities. However, the properties of water in layers between 10 and 10,000 Å thickness is not established yet, though recent evidence indicates that it is likely that the influence of the surface on water structure does not extend to more than 100 Å at the most.

The reasons for the uncertainty are clear enough. The effects to be measured are very small. Special techniques must be used, with resulting uncertainties in the interpretation of the results. The effect of impurities, particularly of surface-active impurities, is enormous and makes necessary far more rigorous precautions to exclude contamination than is needed in work on bulk liquid properties. Quantitative theoretical prediction of the probabilities of existence and the properties of surface-ordered structures in water is not possible at present.

Recent work has indicated that the properties of water at highly disordered porous and diffuse interfaces are of the greatest importance. Such interfaces exist in biological systems. In fact, because of the tendency for water to create pores and impurities in many substrates by chemical attack and dissolution, such surfaces may occur in most practical situations. It is suggested that spatial limitation of the formation of normal water structure may be a factor in determining the behavior of water in this kind of situation.

While it cannot be said that the structure and properties of water in thin layers are understood at present, the associated problems are not insoluble and the application of modern experimental methods to the study of water in thin films and capillaries seems likely to lead to considerable progress in the near future.

CHAPTER 3

Hydration and the Stability of Foams and Emulsions

M. C. Phillips

Biosciences Division, Unilever Research Laboratory Colworth/Welwyn
The Frythe, Welwyn, Herts, England

1. INTRODUCTION

Our present understanding of the role of hydration and interfacial water structure in determining the properties of foams and emulsions is outlined in this chapter. A chapter on colloidal systems, if there is one at all, is usually the last in physical chemistry texts because the authors cannot think of much to say. Fortunately, today this difficulty does not apply to colloid science as a whole because since 1940 a quantitative theory to describe the van der Waals forces of attraction and electrical forces of repulsion between colloidal particles has been available (see Chapter 1) and it is still being refined. However, as far as the role of solvation and hydration in particular is concerned, there is justification for having this review near the end of a treatise on water. This aspect is poorly understood and it is only recently that research has been concentrated on improving the situation.

Traditionally, colloids have been differentiated as hydrophobic or hydrophilic. As depicted in Fig. 1, hydrophilic colloids have no tendency to flocculate at the isoelectric point because their stability is primarily due to their hydration interactions, with their electrical charge being of secondary importance.[326] An absolutely clear distinction between hydrophilic and hydrophobic systems does not exist because they can be interconverted (see Fig. 1). Thus, the role of hydration has been recognized for many years and hydration theories of colloid stability have been in existence for

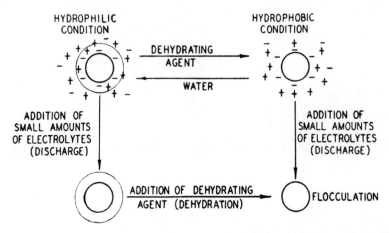

Fig. 1. Electric neutralization and dehydration of colloidal particles (from Hauser[326]).

a long time.[678] However, the shortcomings of such theories have also been pointed out[678] and in 1963, Kitchener[401] referred to the "still mysterious" forces of repulsion arising from solvation of any lyophilic chemical groups on the surfaces of colloidal particles. Overbeek[552] has also discussed the role of the solvent as a protective agent and concluded that "unfortunately, the whole situation is still somewhat ambiguous." It is generally agreed[130, 401,552] that more theoretical and experimental work on this subject is desirable.

Necessarily, this chapter cannot contain an extensive review of experimental results on the effects of hydration in foams and emulsions, because such data do not exist. However, theoretical descriptions of the role of solvation are becoming available. Therefore, after outlining possible effects of interfacial water and the potential roles of water in determining the properties of foams and emulsions, the review concentrates on theoretical aspects and quotes the few experimental results which relate directly to the theory.

2. WATER ENVIRONMENTS IN FOAMS AND EMULSIONS

Because of their technological importance, there is a vast literature covering both fundamental and applied aspects of foams and emulsions. General background material on foams can be obtained from the writings of Kitchener,[402] Bikerman,[65] Ross,[605] and Lemlich,[426] while emulsions have been covered by Becher[40] and Sherman.[634] In this section we are concerned solely with the different possible environments in aqueous foams and emulsions in which water molecules may find themselves, and how the presence of these different water domains is reflected in the properties of the dispersions.

There have been no direct investigations of the intermolecular order of water in foams and emulsions, but from a series of nmr investigations by Clifford and co-workers (Chapter 2) of water in aqueous disperse systems a coherent pattern of behavior of water in these circumstances emerges. This behavior has been summarized by Clifford and Pethica[125] and is illustrated in Table I. Where water exists essentially as single molecules or

TABLE I. The Behavior of Water in Disperse Systems[a]

Type of system	Size of water domain		
	I Essentially as single molecules	II Small groups of molecules	III Large enough for bulk water structure
(A) Water–solid particle	Water physisorbed on solid surfaces, less than two monolayers or water in micropores	Water in the smaller "intermediate pores" or in concentrated colloidal dispersions	Water in dilute dispersions of particles or in dilute colloidal dispersions
(B) Water–macromolecule	Water adsorbed in small amounts on macromolecules	Systems containing similar amounts of water and macromolecules	Dilute solutions of macromolecules
(C) Water–oil emulsions	Dilute solutions of water in oil	Water-in-oil emulsions; very small droplets	Water-in-oil emulsions (large droplets) and oil-in-water emulsions
(D) Water–soap systems	Hydrate water in solid soaps	Water in concentrated micellar systems and in reverse phase micelles	Dilute solution of single molecules or micelles in water
(E) Small molecules	Dilute solutions of water	Solutions of intermediate concentration	Dilute solutions in water

[a] From Clifford and Pethica.[125]

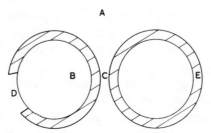

Fig. 2. Various possible environments for water molecules in foams or emulsions. (A) As aqueous continuous phase; (B) dissolved in disperse phase; (C) in thin film between two flocculated droplets; (D) at interface with immiscible fluid; (E) solvating adsorbed layer of surfactant.

monolayers (situation I) its general properties bear little resemblance to the properties and behavior of bulk water; its properties are largely determined by interactions between water molecules and the surface groups of the other components of the system. Before bulk water can form (situation III), there is an intermediate situation (II), where the water molecules do not form bulk water structure, possibly because there is a minimum size of water domain required for the formation of characteristic bulk water. Table I shows how all three situations can exist in water–oil emulsions.

   In systems containing droplets of one fluid dispersed in another and stabilized by a layer of surface-active material the water molecules can find themselves in each of the situations outlined above. These cases for water-continuous systems are shown in Fig. 2 and listed below in terms of the three situations given in Table I.

   (A) In bulk water or in dilute aqueous solutions as the continuous phase—situation III.
   (B) Dissolved in the disperse phase—situation II, III.
   (C) In thin aqueous films between droplets of disperse phase in close contact—situations II, III.
   (D) At clean interface with an immiscible fluid—situations I, II.
   (E) Solvating either charged or uncharged, surface-active, stabilizing agent—situations I, II.

Cases A and B form the subjects of Volumes 1–3 of this treatise, while C is covered by Clifford in the preceding chapter and E is treated in the various chapters of Volumes 4 and 5 which deal with the hydration of surfactants, lipids, synthetic polymers, and biopolymers. Because case D has not been considered elsewhere in this treatise, and because the air–water and oil–water interfaces are central to any consideration of aqueous foams and emulsions, the modification of water structure at these two interfaces is considered briefly below.

The problem of structural orientation near a water interface has been the subject of much controversy. Drost-Hansen[198,203] has discussed the relevant data and ideas at length and has highlighted the fact that many determinations of the variation of surface tension with temperature show the occurrence of kinks or "thermal anomalies." Drost-Hansen has interpreted this anomalous surface entropy data to indicate that below a certain critical temperature (28°) one particular orientation of water is stabilized adjacent to the air–water interface, whereas above a certain temperature (32°) another type of water structure predominates. However, two recent determinations of the temperature dependence of the surface tension of water have shown a linear dependence and failed to demonstrate any thermal anomalies.[114,283] On the basis of these later data Drost-Hansen[204] has modified his views and now accepts that sudden thermal transitions in interfacial water structure do not occur at the air–water interface. Randles[593] has reviewed the structure at the free surface of water and aqueous electrolyte solutions and also concludes that thermal anomalies do not exist. There has been disagreement about the range over which the presence of an interface affects water structure. Randles points out that it is fairly safe to assume that the liquid–vapor transition zone at $T = T_c/2$ (where T_c is the critical temperature) is two or three molecules thick. The surface of a liquid is not quiescent but is subject to thermal motion and frequent exchange of molecules with the vapor phase.[198] Surface irregularities of molecular dimensions rarely reach more than $\frac{3}{2}$ molecular diameters in height or depth,[593] so that the surface is smooth at the molecular level. The low surface excess entropy of the air–water interface[293,593] indicates that there is a considerable degree of preferred orientation of water molecules in the surface zone. At present there is no direct method of assessing the average orientation relative to the plane of the interface of the water molecules near the surface, and indirect evidence must be resorted to.

There is a dipolar contribution to the surface potential of aqueous solutions due to the average orientation of water molecules in the surface zone. It is impossible to directly measure the actual surface potential of water, but it is probably only a small fraction of the potential difference which would arise from a completely polarized single layer of water molecules; this indicates that there is only a small bias in the average orientation of water molecules in the surface zone.[593] Estimates of the surface potential of water based on its temperature dependence indicate that the potential is positive. Randles[593] has collected the various estimates and concludes that the surface potential lies in the range 0.03–0.13 V. This positive potential indicates a preferred orientation of water molecules with the oxygen atoms directed toward the air. Calculations of the potential field by classical electrostatic methods, assuming that the water molecule can be idealized as a point dipole plus a point quadrupole encased in a spherical exclusion envelope, leads to the same orientation.[649] At higher temperatures the molecular disorder in the interface is increased and the temperature coefficient of the surface potential is $(-)$ 0.39 $\pm$ 0.04 mV K^{-1}.[618]

Mingins and Pethica[507] have investigated the effect of temperature on the water dipole contributions to the surface potential of sodium octadecyl sulfate monolayers spread at the air–water interface. Over the temperature range 5–18° increasing temperature disorients water molecules less at the film-covered interface than at the clean air–water interface. The authors suggest that the presence of the insoluble monolayer causes the water molecules to be oriented with the oxygen atoms on average pointing more to the air phase than in the case of the clean interface. Comparison of the monolayer surface potentials at air–water and n-heptane–water interfaces indicates that at the latter interface the oxygens are more up, on average, and that their orientation is less easily disrupted by increases in temperature. Drost-Hansen[204] has postulated that at oil–water and solid–water interfaces there is long-range ordering of water, the "melting" of which can give rise to "thermal anomalies." However, as Clifford (Chapter 2) has pointed out, thermal anomalies are more frequently observed in systems containing complex interfaces which are irregular and porous. In view of Clifford's evidence and conclusion that "the influence of a surface on water structure does not extend to more than 100 Å at most" it seems likely that, as for the air–water interface, the orientation of water molecules at a clean oil–water interface does not extend over many molecular diameters. The situation may be quite different, when, as depicted in Fig. 22 of Clifford's article (Chapter 2), bulk water is prevented from forming through lack of space caused by, for example, adsorption of a macromolecular surfactant at the air– or oil–water interface.

We can conclude that at fluid interfaces which are smooth, even at the molecular level, there will be no long-range ordering of water molecules. The surface dissolution of polar material into water[162] and effects of surface roughness which can cause apparent long-range ordering of water at the interface with lyophilic solids[159,160] will not occur in foams and emulsions. In foams and emulsions where the fluid interfaces are covered with small molecule surfactants there will be no long-range, interfacial orientation of water molecules, but if loops of polymeric surfactant extend into the aqueous phase, then perturbation of water structure is possible at considerable distances from the fluid–fluid boundary.

The long-range interparticle repulsions which control flocculation in colloidal systems are primarily due to electrical double layer effects. Water molecules in this region are influenced by interaction with both the interface itself and the counterions. As discussed in Sections 3.2.1 and 3.3, these effects are all incorporated into the permittivity (dielectric constant) term in expressions for electrical double layer interaction. The flow properties of particulate dispersions are changed by the presence of a double layer. This so-called electroviscous effect arises because shear distorts the interaction between the counterions and the charged interface and this leads to an extra dissipation of energy and an increased viscosity.[634] In water-continuous foams and emulsions the stability of the thin film of water separating two flocculated gas bubbles or oil droplets dictates whether or not coalescence occurs. Because a major point of interest in relation to the stability of foams and emulsions is the protection of the dispersed droplets against coalescence, the properties of the thin aqueous films are of prime importance. The water structure in these films can affect their lifetime because, first, drainage is affected by the viscosity of the water and second, hydrated layers contribute to the short-range forces and the total free energy, which determines the equilibrium thickness of the film. The forces in thin films and solvation/steric effects in particular are considered in Sections 3 and 4, respectively.

3. STABILITY OF THIN LIQUID FILMS

3.1. Thinning and Rupture

When two droplets in a liquid environment approach one another (as the result of external forces such as gravity or hydrodynamic flow) the behavior of the system is governed by the interplay of hydrodynamic forces and surface forces. The hydrodynamic forces lead to viscous flow of the liquid medium from between the droplets and to distortion of the droplet shape because of the pressure developed between them. The mechanics of the approach of two drops in a fluid medium has been reviewed by Kitchener and Mussellwhite.[404]

With relatively large drops, when inertial forces are large compared to surface forces, the flow pattern known as "dimpling" is observed; liquid is temporarily trapped near the line of approach because of concave de-

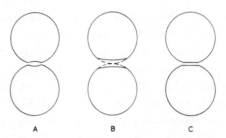

A   B   C

Fig. 3. Approach of a pair of drops. (A) Formation of a "dimple"; (B) distortion without dimpling; (C) formation of a flat lamella. (From Kitchener and Mussellwhite.[404])

pressions in the droplet surface (Fig. 3A). In such a system the thinnest regions of liquid separating the drops are in the ring forming the periphery of the dimple and it is here that coalescence tends to occur. With small droplets (or large droplets brought together very slowly) dimpling does not occur; the faces of the drops are compressed while the liquid is squeezed out from between them, but they remain convex (Fig. 3B). Thus the place of closest approach of the surfaces of the two drops is on the line of centers and it is here that coalescence tends to start. When surface-active agents are present, which is always the case in emulsions and foams, the time for coalescence is greatly extended, compared with that for "pure" liquids, and a thin liquid lamella is formed between the drops (Fig. 3C). The formation of such lamellae indicates the operation of surface forces opposing the thinning, for a parallel sheet of liquid could never arise simply from hydrostatic forces and an invariant surface tension.

In a foam or emulsion such lamellae are in contact with Plateau borders and their behavior is analogous to that of isolated, thin liquid films.[130,628] When the films are more than 1000 Å thick their drainage is controlled by the forces of gravity and capillary suction. For a vertical film the hydrostatic pressure or capillary suction π_H operating at the Plateau border is, according to the Young–Laplace relationship,

$$\pi_H = -\gamma\left(\frac{1}{a_1} + \frac{1}{a_2}\right) \tag{1}$$

where a_1 and a_2 are the principal radii of curvature at the border and γ is the bulk interfacial tension.

The rate of thinning of these fairly thick films is proportional to the capillary suction pressure and inversely proportional to the viscosity of the film fluid.[130,628] Agreement between theory and experiment is found when the viscosity is taken to be that of the bulk aqueous medium; this indicates that no unusual water structure effects are involved in films of thickness >1000 Å. Liquid films may be classified[512] according to their drainage behavior into two main categories, "rigid" and "mobile." Mobile films thin in a matter of minutes and exhibit turbulent motion along the plateau borders and motion of discrete patches of the film, whereas rigid films exhibit little motion and take hours or days to drain completely. The differences in drainage behavior are associated with the types of surface monolayers that are used to stabilize the films. Monolayers of high surface viscosity give rise to rigid films, whereas mobile films have monolayers of low surface viscosity. The high surface viscosity prevents gravitational convection from operating in the thinning by "marginal regeneration" and thereby slows the rate of drainage. Rigid films thin by the outflow of interlamellar fluid from between their immobile but flexible surface layers. The rate of this drainage can be decreased by increasing the bulk viscosity and in foam systems such conditions lead to the formation of "sphere foams."[605] These foams contain spherical gas bubbles which are widely separated from each other and they are "wet" in the sense that they contain large amounts of liquid. The bubbles change size chiefly because of gas diffusion from smaller to larger bubbles. On the other hand, relatively fast-draining, mobile films favor the formation of "polyhedron foams"; these foams are "dry" and consist of polyhedral bubbles separated by thin, plane films of liquid. Because the pressures inside contiguous cells are nearly equal, there is no significant spontaneous diffusion of gas between foam cells. The liquid films are subject to powerful capillary suction which tends to thin them.

When the film thickness is reduced to below about 1000 Å other forces begin to influence the drainage behavior. These "other" forces, which become more important as the film thickness decreases, are various types of intermolecular forces. These forces are discussed in more detail in Section 3.2, but a consequence of the existence of attractive forces between macroscopic bodies is that the thinning of a film under a constant external pressure can accelerate sharply when the thickness reaches the "range of action" of these forces. As a result there is a critical thickness H_{cr} for the onset of collapse. Because of the existence of H_{cr}, thin liquid films are transient and rupture unless the film is maintained in metastable equilibrium at thicknesses greater than H_{cr}, or unless it can form a stable structure of thickness less than H_{cr}. In the latter case H_{cr} is manifested by a sudden transition to these stable states, and the process is known as "black-spot formation" since at these thicknesses the film appears grey or black.[130,628] Such films are black because their thickness is less than that of the wavelength of visible light and therefore their reflectivity is very low.

The thinnest black films, which are bimolecular leaflets, rupture by spontaneous formation of a "nucleus" hole.[189] Thicker, metastable films require some external agent (e.g., thermal or mechanical shock) to provide the energy for their destruction. Scheludko[628] proposed that waves or corrugations exist in the surface of the film as a result of thermal motion, and when amplified by van der Waals forces, these can be responsible for the rupture. Vrij[687] further explored the stability of these surface corrugations and the kinetics of their attenuation and amplification. He found that a free, thin liquid film is unstable with respect to small, spontaneous thickness fluctuations with wavelengths larger than a critical wavelength

$$\Lambda_{cr} = \left(\frac{-2\pi^2 \gamma}{d^2 V / dH^2} \right)^{1/2} \qquad (2)$$

where γ is the interfacial tension and $V(H)$ is the free energy of interaction per unit area of film as a function of the film thickness H. A fluctuation with $\Lambda < \Lambda_{cr}$ varies around a metastable equilibrium, whereas a fluctuation with $\Lambda > \Lambda_{cr}$ grows in amplitude and makes the film unstable. By making assumptions about the contributions to $V(H)$ it is possible to derive expressions for H_{cr} (see Refs. 130, 628, 687). The critical thickness H_{cr} is independent of the liquid viscosity in the film. Ivanov et al.[370] have extended the theory by taking into account the influence of the surfactant on the hydrodynamics of the film. Measurements of H_{cr} of aniline films stabilized by dodecanol confirm the theory at low surfactant concentrations. However, at higher surfactant concentrations discrepancies occur which may arise either from an additional force opposing thinning or from increased viscosity in very thin films.

Additional stabilizing mechanisms due to the elasticity of the interfacial film are possible in foams and emulsions. Surface elasticity only occurs with solutions which are subject to adsorption (or interfaces with a spread film) and arises from nonuniformity of surface or interfacial tension. The surface dilatational modulus E can be considered as the response of the surface tension to a fractional increase of the area A of a small surface element:

$$E = d\gamma / d \ln A = A \, d\gamma / dA \qquad (3)$$

The immediate response of surface tension to any change in area is due to elastic properties of the interface, while delayed response is due to viscous components.[451] The modulus E reflects the ability of surfaces to support tangential stresses originating from the flow of liquid adjoining the surface. An initial deformation causes a surface tension gradient and this gradient resists any further deformation. Thus, any surface waves in a thin liquid film are damped out more rapidly when the film surfaces have elastic properties. The general equation relating frequency and wave number for vibrations in liquid films of arbitrary thickness has been derived.[452] Symmetric waves involve a locally varying film thickness; a sharp decrease in the growth rate of thickness fluctuations with increasing E occurs at very low values of E.[688] As a result of this damping, thin films are stabilized because lamellae of more or less uniform thickness are formed and steady drainage proceeds. In order for the thin film to achieve metastable equilibrium, the above effects of surface elasticity must be complemented by additional repulsion forces. Thus, the possibility of having a thin film stabilized by short-range interaction forces of electrical or entropic origin is a necessary condition for a stable foam or emulsion.

In summary, the role of water structure in the thinning and rupture of thin water films has not been treated explicitly. The involvement of water is expressed in terms of thermodynamic quantities such as surface tension and it is not possible, at present, to express this tension in terms of molecular interaction forces. However, for thin films in metastable equilibrium the role of water has to be considered more explicitly. This is done in the following treatment of short-range forces in thin films (Sections 3.2 and 4).

3.2. Forces in Thin Films

Derjaguin and Kusakov[168] suggested that the properties of thin films are different from those of the bulk phase and introduced the parameter disjoining pressure π_D as a measure of the corresponding change in the thermodynamic properties. π_D is a function of film thickness and acts at right angles to the plane of the liquid film; it is assumed to be positive when it resists film thinning. Thus the total specific surface energy γ_F of a thin, plane parallel lamella of liquid of thickness H is given by

$$\gamma_F = \gamma + \int_H^\infty \pi_D \, dH \qquad (4)$$

The work of bringing the two surfaces of a film together (second term) is very small compared with the work of forming the interfaces (first term). But because, by definition, γ is not affected by H, one is only concerned with π_D in considering equilibrium films. In an equilibrium film the disjoining pressure π_D is equal and opposite to the hydrostatic pressure π_H, or

$$\pi_D + \pi_H = 0 \qquad (5)$$

The various components of π_D have received considerable attention in recent years and three main types of intermolecular forces are considered as operating in thin films.[130,404,628] Along with the ever-present London–van der Waals attraction, which is a negative contribution (π_V) because it acts to thin the film, there is the relatively long-range electrostatic force π_E that arises from the repulsion of the similarly charged surfaces of the film as the diffuse parts of their respective double layers overlap during thinning.[170,680] Finally, there is another, less well-defined repulsion π_S associated with the properties of the adsorbed layer of surface-active material and its solvation layer. We are particularly concerned with π_S for aqueous systems in this review (Section 4). Incorporating the components of π_D into eqn. (5) gives the condition for equilibrium as

$$\pi_V + \pi_E + \pi_S + \pi_H = 0 \tag{6}$$

The effects of the contributions to π_D from cohesion in the liquid film and adhesion between the liquid and the surrounding medium have been described by Ross[606]; the latter term favors a more stable film.

Quantitative evaluation of π_V and π_E was first achieved by Derjaguin and Landau[170] and Verwey and Overbeek[680] in the classical "DLVO" theory of the stability of lyophobic colloids. The theory involves a calculation of the net interaction energy between dispersed particles by summing the London–van der Waals forces of attraction and the electrostatic repulsion between them. Kitchener and Mussellwhite[404] have pointed out that the theory cannot be expected to apply to the rate of coalescence of emulsion drops but that it may reasonably describe the electrostatic contribution (e.g., by ionic surfactants) to stabilization of emulsions against flocculation. The essentials of the theory for elucidation of π_V and π_E between hydrophobic spheres or semiinfinite parallel plates (cf. Fig. 3) are presented now, before the effects of adsorbed layers are discussed.

Quite apart from the original presentations,[170,680] the basic theory has been outlined many times before. Overbeek[550] has given a description for hydrophobic colloids, while the applications to emulsions,[404] foams,[402] and thin films[130,628] have been treated separately. Because the interaction between colloidal particles has to be compared to the kinetic energy of the particles, it is usual to calculate the potential energy of interaction as a function of the distance separating the particle surfaces (the following expressions for interaction energies per unit area can be converted to pressures by differentiating with respect to distance H).

3.2.1. Electrostatic Repulsion π_E

The potential energy of repulsion per unit area of surface (V_R) acting between two particles of like charge depends upon the size and shape of the particles, the ionic strength, the permittivity of the medium in which the particles are suspended, and the surface potential of the particles. It is assumed that the electrical double layers can be treated by the classical Gouy–Chapman theory. In terms of this review it is important to note that the solvent (water) is treated as a continuous medium influencing the double layer only through its permittivity. To a good approximation,[550] the repulsive energy between two flat plates arising from overlap of double layers is given by

$$V_R = \frac{64nkT}{\varkappa}\left(\frac{\exp(Ze\psi_0/2kT) - 1}{\exp(Ze\psi_0/2kT) + 1}\right)^2 \exp(-\varkappa H) \tag{7}$$

where n is the number of counterions per cubic centimeter in the bulk solution; $\varkappa$ is the Debye–Hückel reciprocal length ($1/\varkappa$ is the thickness of the diffuse part of the double layer) and $\varkappa = (4\pi e^2 \sum nZ^2/\varepsilon kT)^{1/2}$; e is the electronic charge; Z is the valence of symmetric electrolyte; ε is the permittivity of solvent; k is Boltzmann's constant; T is the absolute temperature; H is the distance between the two plates; and ψ_0 is the equilibrium surface potential. An important refinement of the original theory is to allow for the specific adsorption of counterions, taking the Stern layer into account. Stern adsorption has the effect of reducing the effective surface potential and then eqns. (7) and (8) are written in terms of the potential at the Gouy–Stern interface ψ_d. This quantity cannot be measured easily and for practical purposes is equated with the zeta potential. The corresponding expression for the interaction of two, identical, spherical double layers is

$$V_R = \tfrac{1}{2}\varepsilon a\psi_d^2 \ln[1 + \exp(-\varkappa H)] \tag{8}$$

where a is the radius of the particles (the center-to-center distance of the particles R can be taken as $2a + H$).

Equations (7) and (8) are only strictly valid for the case of weak interaction (i.e., $H \gg \varkappa^{-1}$), low potentials ($\psi_d < 50$ mV), and spheres whose radius is large compared to the thickness of the double layer (i.e., $\varkappa a \gg 1$). However, it is generally considered that the equations can be used as an approximation for most practical systems. Further refinements for the emulsion situation, where two oil droplets are approaching each other in water, have been discussed by Overbeek[550] and Kitchener and Mussellwhite.[404] The presence of an internal diffuse layer in the oil phase reduces the interfacial potential, thereby reducing the repulsive energy.

3.2.2. London–van der Waals Attraction π_V

The main source of attraction between colloidal particles arises as a consequence of van der Waals forces. Although these are short-range forces when interactions between atoms are considered, summation of the London[447] contributions over all atoms in a particulate system leads to long-range attractive forces. For the interaction between two plates of thickness t in a vacuum the mutual energy of attraction (V_A) per unit area of surface is[550,680]

$$V_A = -\frac{A}{12\pi}\left[\frac{1}{H^2} + \frac{1}{(H+2t)^2} - \frac{2}{(H+t)^2}\right] \tag{9}$$

A is the Hamaker constant (see below). When the distance between the plates is small compared with their thickness (i.e., $t \gg H$), the energy of attraction per unit area simplifies to

$$V_A = -A/12\pi H^2 \tag{10}$$

It should be noted that the attraction between two sides of a planar liquid lamella of thickness H in a vacuum is the same as the attraction between two semiinfinite parallel masses of liquid separated by a gap H.[402] Casimir and Polder[101] improved London's theory by making allowance for the electromagnetic retardation of the dispersion forces resulting from their finite velocity of propagation. Retardation causes the attractive energy to decay more rapidly with distance than shown in eqn. (10), and when $H \sim 10^4$ Å it was calculated that V_A would decay as H^{-3} instead of H^{-2}. When H is of the same order of magnitude as the wavelengths of the dispersion forces (10^2–10^3 Å) the London energy is reduced by a factor F given by[404,550]

$$F = 1.01 - 0.14p \qquad \text{for} \quad 0 < p < 3$$

$$F = \frac{2.45}{p} - \frac{2.04}{p^2} \qquad \text{for} \quad 3 < p < \infty$$

where $p = 2\pi\nu l/c$ (l is the distance between atoms, ν is the London frequency, and c is the velocity of light).

The attractive interaction between two spherical particles of radius a in a vacuum with a distance R between their centers was first treated by Hamaker,[320] who obtained the result

$$V_A = -\frac{A}{6}\left[\frac{2a^2}{R^2-4a^2} + \frac{2a^2}{R^2} + \ln\frac{R^2-4a^2}{R^2}\right] \tag{11}$$

The Hamaker constant $A = \pi^2 q^2 \lambda$, where q is the number of atoms per cm^3

of material, $\lambda = \frac{3}{4}h\nu\alpha^2$, with α the polarizability of the material and $h\nu$ a characteristic energy corresponding to the chief dispersion frequency. For small distances between the spheres ($H \ll a$), eqn. (11) becomes

$$V_A = -Aa/12H \qquad (12)$$

This energy is proportional to the size of the particles and it does not decay very sharply with the distance between them, whereas, of course, for large distances between the spheres V_A decays as R^{-6}. Equations (11) and (12) are only valid if retardation is neglected; the retardation correction for this case has been discussed by Overbeek.[550]

Equations (9)–(12) describe interaction between particles in a vacuum. When liquid is present between the particles, allowance must be made for this in calculating the resultant value of A. Thus, for material 1 dispersed in material 2 the resultant value of A is given by

$$A = A_{11} + A_{22} - 2A_{12} \qquad (13)$$

where A_{11} refers to the interaction of material 1 with itself, A_{22} to the interaction of material 2 with itself, and A_{12} to the interaction of material 1 with material 2. It is generally assumed that A_{12} can be taken as the geometric mean of A_{11} and A_{22} and so

$$A = (A_{11}^{1/2} - A_{22}^{1/2})^2 \qquad (14)$$

It follows that A must always be positive and therefore colloidal particles are always subject to London–van der Waals attraction. Unfortunately, A is given by the difference between two similar quantities and estimates of its magnitude are subject to large error. Recently, Visser[683] has given a compilation of A values; estimates of A for water range from 3 to 30×10^{-13} erg, although most are closer to the lower value. Kitchener and Mussellwhite[404] have shown by the use of the above formulas that for oil droplets ($a = 0.5~\mu$m) in water, at a separation $H = 0.1~\mu$m, V_A is about equal to the thermal energy kT.

Finally, before concluding this section it should be noted that the classical treatment of van der Waals attraction given above contains several assumptions which limit its validity. As emphasized by Lifshitz and co-workers,[210] calculations of van der Waals forces of attraction between condensed media on the basis of assumptions which are valid for dilute gases is unsound. Lifshitz developed a theory which overcomes these difficulties by (1) including all many-body forces through a continuum picture, (2) retaining contributions from all interaction frequencies, and (3) dealing correctly with the effects of intermediate substances (e.g., hydrocarbon between aqueous regions). Ninham and Parsegian[530] have presented a general method of calculating van der Waals forces based on Lifshitz' theory. The values of A derived from this theory for materials separated by water are similar to those from the classical treatment[303] but the effects of retardation are quite different. Retardation effects are such as to damp out high-frequency fluctuation contributions to the interaction and for lipid–water systems the van der Waals energy increases with temperature and has a nonretarded form even at large distances.[561,562] For example, the ratio of A(retarded)$/A$(nonretarded) is 0.75 when $H = 500$ Å, whereas Casimir and Polder's treatment gives this ratio as 0.28. As a result, eqn. (10) applies over a wider range of H. Calculation of van der Waals forces involves knowledge of the dielectric permeabilities of the interacting media as a function of frequency. Nir et al.[533] have commented on the relative contributions from the ultraviolet, visible, and infrared frequencies in the interactions between two water phases separated by a hydrocarbon film. Recently, Bargeman and van Voorst Vader[32] have attempted to reconcile the differences between the London–Hamaker and Lifshitz theories by correcting for the nonadditivity of atom–atom interactions.

3.2.3. Combined Interaction Energy Curves ($\pi_E + \pi_V$)

The energy of interaction V between hydrophobic colloidal particles is found by addition of the repulsion V_R and the attraction V_A curves. For the interaction of two plates, combination of eqns. (7) and (9) gives curves of the form shown in Fig. 4. This figure actually shows the total potential energy as a function of film thickness for a planar thin film and includes the contributions from hydrostatic pressure and steric hindrance. In the case of hydrophobic colloids the vertical rise in potential energy at very short distances originates from Born repulsion due to overlap of the electronic clouds of the atoms in the particles. Beyond this, the attraction predominates at short distances and at very large distances. At intermediate distances the repulsion may predominate but whether this occurs or not depends on the magnitudes of V_A and V_R. Provided the double layer repulsion is strong enough to overcome the London–van der Waals attraction at some distances of separation, there will be a secondary minimum at greater distances. As the ionic strength of the system is increased, the secondary minimum moves to smaller distances. As long as the potential energy barrier is considerably larger than the mean translational energy of colloidal particles ($=kT$), the particles will be stable against practically irreversible flocculation (i.e., coagulation) into the primary minimum. If

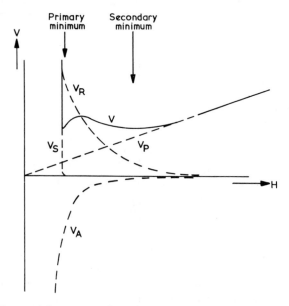

Fig. 4. Total potential energy V of a thin aqueous film as a function of thickness H. V_A is the potential energy of van der Waals attraction; V_R is the potential energy of repulsion between two electrical double layers; V_S is the potential energy of repulsion originating from steric hindrance and hydration; V_P is the potential energy due to hydrostatic pressure. (From Duyvis and Overbeek.[208])

the depth of the secondary minimum is several times kT, particles can be stabilized at this separation; such flocculation is usually reversible (e.g., by shaking).

Combination of eqns. (8) and (11) enables potential energy curves to be calculated for the interaction between two spherical particles of equal size. The form of the curve depends on particle size and, more particularly, on $\varkappa a$. The shapes of the curves at small $\varkappa a$ and large $\varkappa a$ values are shown schematically in Fig. 5. It is clear that the secondary minimum becomes more significant with larger particles ($a \sim 10~\mu$m). Ottewill[543] has summarized the main categories of conditions under which the potential energy maximum is reduced and flocculation ensues. First, an increase in ionic strength and/or specific ion adsorption can decrease the repulsive potential energy by decreasing $\varkappa^{-1}$ and/or ψ_d. Second, specific flocculation can be induced by cross-linking agents such as polyelectrolytes which carry an opposite charge to the colloidal particle.

In conclusion, it is apparent that the DLVO theory does not take explicit account of microscopic ordering of water. When the theory is

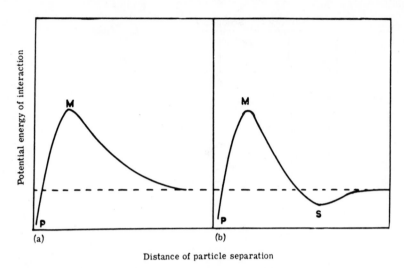

Fig. 5. Schematic diagram of potential energy of interaction against distance for inter-action between two spherical particles. (a) Small $\varkappa a$; (b) large $\varkappa a$. P, primary minimum; M, stability maximum; S, secondary minimum (from Ottewill[543]).

applied to aqueous systems, the effects of water on the electrical repulsion and van der Waals attraction are accounted for in the permittivity and Hamaker constant, respectively. Foams and emulsions are always stabilized by an adsorbed layer of surface-active material and this has to be taken into account in any application of the DLVO theory to these systems. The influence of the adsorbed layer is discussed in the next section.

3.3. Effects of Adsorbed Layers on π_E and π_V

Adsorption of charged surfactant molecules to the air–water or oil–water interface has a dramatic effect on the interfacial potential. This causes a change in ψ_d and in the resultant electrostatic repulsion [eqns. (7) and (8)]. It is possible that when two colloidal particles bearing similar double layers approach one another the surface potential will remain constant as the particles interact; this implies that adsorption equilibrium for both the surfactant and counterions is established very rapidly. Alternatively, if desorption is slow, the surface charge will remain constant. In reality, both effects may occur but, fortunately, assumptions of constant surface potential or constant surface charge do not lead to very different estimates of the interaction energy. At a highly charged aqueous interface the field strength in the double layer may be so high as to cause dielectric saturation of the water. However, the effect of this field strength on experimentally measurable properties is negligible and the permittivity ε in eqns. (7) and (8) may safely be taken as a constant.[550] In the DLVO theory[550] it is assumed that the dielectric properties of water in the vicinity of and between particles are the same as in bulk (i.e., $\varepsilon = 78.55$ at $25°$). The value of ε for the region close to an interface is a matter for conjecture, but if one follows Bockris et al.,[68] then the primary layer of water molecules at the interface effectively has $\varepsilon = 6$, the secondary layer has $\varepsilon = 32$, and the remainder has the permittivity of normal bulk water. Hunter[366] has discussed the influence of the electrical field around a charged particle on the viscosity and permittivity of the dispersion medium. He concludes that both can be taken as constants and that electrokinetic potentials can be interpreted satisfactorily on the basis that the viscosity and permittivity near the surface do not depart significantly from their bulk values. In the presence of adsorbed neutral polymer the double layer on charged particles can expand due to excluded volume effects[83]; this increases the surface and zeta potentials, thereby increasing π_E [eqns. (7) and (8)].

In 1961, Vold[684] published an analysis of the effect of solvation sheaths, and of adsorbed layers of varying Hamaker constant, on the van der Waals attraction between spherical particles. She showed that the attractive energy V_A for spherical particles surrounded by any number of concentric shells of adsorbed material has the form

$$-12V_A = \sum_{i,j} f(A_i)h_j \qquad (15)$$

where $f(A_i)$ is a function of the various Hamaker constants involved, and h_j is a function of the geometry of the system. Because the treatment was concerned with particles at minimum separation, the full, but unretarded Hamaker functions h were used; the same assumption was made in recent analyses[542,681] of the effects of adsorbed layers on V_A. In the general case of two spherical particles of radii a_1 and a_2 and Hamaker constants A_{p1} and A_{p2}, having homogeneous adsorbed layers of thickness δ_1 and δ_2 and Hamaker constants A_{s1} and A_{s2} in a medium of Hamaker constant A_m (see Fig. 6), the van der Waals attraction energy is given by

$$
\begin{aligned}
-12V_A = {} & h_{s1s2}(A_{s1}^{1/2} - A_m^{1/2})(A_{s2}^{1/2} - A_m^{1/2}) \\
& + h_{p1p2}(A_{p1}^{1/2} - A_{s1}^{1/2})(A_{p2}^{1/2} - A_{s2}^{1/2}) \\
& + h_{p1s2}(A_{p1}^{1/2} - A_{s1}^{1/2})(A_{s2}^{1/2} - A_m^{1/2}) \\
& + h_{p2s1}(A_{p2}^{1/2} - A_{s2}^{1/2})(A_{s1}^{1/2} - A_m^{1/2})
\end{aligned}
\qquad (16)
$$

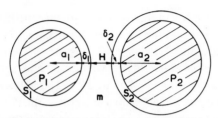

Fig. 6. Definition of symbols used in the derivation of Eqn. (16) for the attractive energy between two spherical particles coated with homogeneous adsorbed layers.

where the unretarded h functions[320] are given by

$$h(x, y) = \frac{x}{x^2 + xy + x} + \frac{y}{x^2 + xy + x + y}$$
$$+ 2 \ln \frac{x^2 + xy + x}{x^2 + xy + x + y} \tag{17}$$

The separation of the interacting surfaces is Δ, $x = \Delta/2r_1$, and $y = r_2/r_1$, where the following hold:

for h_{s1s2}: $\Delta = H$ $r_1 = a_1 + \delta_1$ $r_2 = a_2 + \delta_2$

for h_{p1p2}: $\Delta = H + \delta_1 + \delta_2$ $r_1 = a_1$ $r_2 = a_2$

for h_{p1s2}: $\Delta = H + \delta_1$ $r_1 = a_1$ $r_2 = a_2 + \delta_2$

for h_{p2s1}: $\Delta = H + \delta_2$ $r_1 = a_1 + \delta_1$ $r_2 = a_2$

Equation (16) simplifies for the specific case of two identical, spherical particles[542,684] to

$$-12V_A = (A_m^{1/2} - A_s^{1/2})^2 h_s + (A_s^{1/2} - A_p^{1/2})^2 h_p$$
$$+ 2(A_m^{1/2} - A_s^{1/2})(A_s^{1/2} - A_p^{1/2}) h_{ps} \tag{18}$$

Whereas V_A for identical particles is always negative,[684] this does not hold for the more general case described by eqn. (16). Here V_A can become positive (i.e., there is net repulsion) if there is a suitable sequence for the various Hamaker constants involved. Since the term in h_{s1s2} is the dominant one in eqn. (16), one condition[542,681] for V_A to be positive is that A_{s1} and A_{s2} lie to either side of A_m.

Vold[684] has considered a special case which is of particular interest, namely the situation where the adsorbed layer is simply bound medium (i.e., it is a solvation shell). Equation (18) then reduces to

$$-12V_A = (A_m^{1/2} - A_p^{1/2})^2 h\left(\frac{\Delta + 2\delta}{2a}, 1\right) \tag{19}$$

where δ is the thickness of the layer of bound medium, a is the particle radius, and h is given by eqn. (17). The interaction energy of solvated particles may be reduced by factors of the order of 5–50 compared to that of unsolvated particles. However, flocculation can be inhibited by solvation alone only for small particles ($a \lesssim 500$ Å) and quite thick layers ($\delta \gtrsim 20$ Å).

Calculations have also been carried out for the case of particles surrounded by double sheaths[542,684]; this is particularly relevant to the case of adsorbed amphipathic molecules. V_A has been evaluated for the two situations where the outer sheath is or is not equivalent in Hamaker constant to the medium. In the former case it is unlikely that an overall increase in attraction can be produced, and indeed, the maximum reduction in V_A occurs when the outer layer is chosen to have a Hamaker constant more or less identical to the medium (solvent).

In many applications (e.g., emulsions) where the particle radius is large in comparison with the thickness of the adsorbed layer and the interparticle distance it is more fruitful to consider the interaction of parallel flat plates. The appropriate equations for the case of thick plates covered with a homogeneous adsorbed layer have been derived by Sonntag et al.,[641] Vincent,[681] and Becher.[41] The equation is similar in form to (18) and shows that the effect of the adsorbed layer effectively disappears at distances of the order of ten times its thickness.[41] Duyvis[130,207] has considered the case of a heterogeneous film in air where the film consists of two identical hydrocarbon layers, each of constant thickness, enclosing an aqueous core of variable thickness.

Vincent[681] has considered the effects of employing retarded geometric (h) functions in equations such as (16) and compared his results with those of earlier studies. As expected, retardation decreases V_A more as the particle separation increases and its effect is greater with larger particles.

In all of the above considerations of the effects of adsorbed layers the layers were assumed to be homogeneous and discrete. In reality, an adsorbed layer will normally contain solvent and in the case of adsorbed macromolecules will be heterogeneous because of a distribution of segment densities normal to the interface. The presence of solvent can be allowed for by employing a composite Hamaker constant for the layer.[681] To a good approximation, the Hamaker constant for a layer containing materials i and j (A_{i+j}) can be expressed as

$$A_{i+j} = [D_i(A_i^{\circ})^{1/2} + (1 - D_i)(A_j^{\circ})^{1/2}]^2 \tag{20}$$

where D_i is the volume fraction of component i and the superscript $^{\circ}$ denotes pure materials. Vincent[681] has derived expressions for V_A for a system of two semiinfinite flat plates with homopolymer adsorbed with either uniform or nonuniform segment densities. In the former case V_A is given by

$$-12V_A = h_f(A_f^{1/2} - A_m^{1/2})^2 + h_i(A_x^{1/2} - A_y^{1/2})^2$$
$$+ 2h_{if}(A_{fx} + A_{my} - A_{fy} - A_{mx}) \qquad (21)$$

where the A and h terms have their usual significance:

$$A_x = A_a\phi^2, \qquad A_y = A_m(P\phi)^2 \qquad (22)$$

where A_a and A_m are the Hamaker constants of the pure polymer adsorbate and solvent, respectively, ϕ is the mean volume fraction of the segments in the adsorbed layer, and P is the ratio of the partial molar volumes of the segments and the solvent molecules.

So far in this account of forces in thin films steric effects and solvation have not been discussed. Because these topics are central to the subject of this review, they are described at length in the next section.

4. THE ROLE OF STERIC EFFECTS AND SOLVATION

The term π_S in eqn. (6) describing the equilibrium condition in thin films is the least well understood. It is usually invoked in a qualitative fashion to explain discrepancies between theory and experiment.[130] It is generally considered to be short range and to arise from steric effects associated with the close approach of adsorbed layers. The hydration of the adsorbed layers becomes critical in such a phenomenon.

Figure 7 shows schematically the overlap of adsorbed layers on the approach of two spherical particles with adsorbed layers of thickness δ. Initially, the particles are an infinite distance apart and are then brought together such that the distance between the surfaces of the two particles is H and there is an interaction between the adsorbed layers. Ottewill[543] has discussed the change in Gibbs free energy ΔG_R which arises as the particles come together:

$$\Delta G_R = G_H - G_\infty \qquad (23)$$

ΔG_R can be composed of several components which vary according to the nature of the adsorbed layer. Ottewill considered the most likely contributions to be the change in surface free energy on close approach, the excess chemical potential caused by overlapping of layers, and an elastic

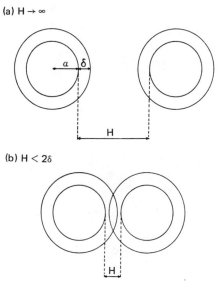

Fig. 7. Overlap of adsorbed layers on the approach of two spherical particles coated with adsorbed layers of thickness δ (from Ottewill[543]).

energy arising from compression of the adsorbed layers. If ΔG_R is suitably positive, then a net repulsion between the particles can arise. Napper and Hunter[522] have summarized the three different ways in which a positive value of ΔG_R can arise from its constituent enthalpy change ΔH_R and entropy change ΔS_R on close approach.

(i) *Enthalpic stabilization.* If both ΔH_R and ΔS_R are positive and $\Delta H_R > T \Delta S_R$, then a positive value of ΔG_R results. Thus, the net enthalpy change on close approach, which opposes flocculation, outweighs the net effect of the increase in entropy, which promotes flocculation. Enthalpic stabilization is characterized by flocculation on heating.

(ii) *Entropic stabilization.* Entropic stabilization arises when both ΔH_R and ΔS_R are negative but $T | \Delta S_R | > | \Delta H_R |$. In principle, entropically stabilized dispersions are characterized by flocculation on cooling.

(iii) *Combined enthalpic–entropic stabilization.* Both the enthalpy and entropy changes oppose flocculation if ΔH_R is positive and ΔS_R is negative. The term steric stabilization has been coined to describe these effects. Of course, the above thermodynamic considerations do not depend on the molecular origins of ΔG_R and so do not give information on how ΔG_R is generated or allow its magnitude to be predicted.

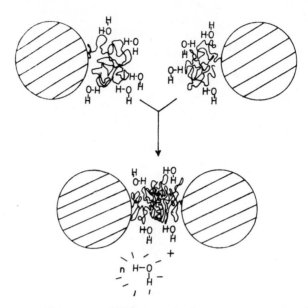

Fig. 8. Schematic representation of steric stabilization in an aqueous medium showing interpenetration and compression of polymer chains and displacement of water of hydration of the adsorbed layers (from Napper[519]).

To overcome the above shortcomings, models of the molecular events occurring during the situation depicted in Fig. 8 have been proposed and treated by statistical thermodynamic methods. Because of the well-known "protective action" of natural macromolecules in stabilizing colloids against flocculation and the stabilizing abilities of adsorbed nonionic macromolecules for dispersions in aqueous and nonaqueous media, the usual model consists of a particle with irreversibly adsorbed polymer chains. There are two mechanisms by which the adsorbed polymer can counteract van der Waals attraction between the core particles. Hesselink *et al.*[347] have described these as the "volume restriction effect" due to loss of configurational entropy on the approach of a second particle, and a local "osmotic pressure effect" arising from interpenetration (mixing) of the adsorbed layers; the higher polymer segment concentration between the particles causes a change in the free energy of mixing of polymer segments and solvent. The theoretical treatments assume that the stabilizing chain configuration is independent of surface coverage and that the polymer layers are irreversibly adsorbed. If the adsorption were reversible (i.e., small molecule surfactants) the volume restriction effect would cause the adsorbed molecules to desorb and gain entropy by moving into the bulk of the solution; this would increase the interfacial free energy according to Gibbs' law. These considerations are analogous to the question of constant surface charge or constant surface potential in treating the overlap of electrical double layers. In the case of fluid interfaces another contribution to ΔG_R may arise from deformation of the droplets with a concomitant change in interfacial area. Jäckel[374] has given a formula, based on the theory of elastic collisions, for the energy change involved in the deformation of a compressible adsorbed layer. The energy is expressed in terms of the elastic modulus of the layer, but so far no estimates of this parameter are available.

Evans and Napper[224] have summarized the various treatments of steric stabilization and compared the theories with experiment. Mackor,[466, 467] Clayfield and Lumb,[118,119] Ash and Findenegg,[17] and Bagchi and Vold[21] assumed that the adsorbed molecules alone contribute to ΔG_R and calculated the loss of their configurational entropy ΔS_{conf} on close approach of two colloidal particles. Then ΔG_R was calculated from $\Delta G_R = -T \Delta S_{conf}$. However, these theories cannot be complete because they ignore the solvent and it has been shown that the solvent power of the dispersion medium can be critical in determining the stability of sterically stabilized dispersions.[522] Fischer[240] was the first to point out the critical importance of the quality of the solvent power that the dispersion medium displays toward the stabilizing chains. He postulated that when two dissolved polymer sheaths interpenetrate (Figs. 7 and 8) the chemical potential of the solvent in the interaction zone decreases, thereby establishing a gradient in chemical potential between solvent molecules in the interaction zone and those in the external dispersion medium. An excess osmotic pressure is then generated by the diffusion of external solvent molecules into the interaction zone; this ingress of solvent molecules forces the colloidal particles apart and stabilizes them. Fischer related the repulsive potential energy to the second virial coefficient B of the polymer in free solution using the Flory–Huggins theory:

$$\Delta G_R \sim 2BRT\langle Cg\rangle(\Delta V) \tag{24}$$

where $\langle Cg\rangle$ is the mean segment concentration and (ΔV) is the overlap volume. As Evans and Napper[224] have pointed out, this theory is approximate because use of a mean segment concentration ignores the fact that the segment concentration is a function of distance from the particle. Also, compression of the polymer chains when the minimum distance between the particles is less than the contour length of the polymer chains is not considered.

Ottewill and Walker[546] and Napper[518] have evaluated ΔG_R for the overlap and mixing of two adsorbed layers and allowed for variation in segment density. Thus when $L \leq H < 2L$, where L is the contour length of the longest stabilizing moiety,

$$\Delta G_R = 2kT(\psi_1 - \chi_1)(V_s^2/V_1) \int \varrho_i \varrho_j \, dV \tag{25}$$

where $\chi_1 \; (= \frac{1}{2} + K_1 - \psi_1)$ is the polymer–solvent interaction parameter, K_1 is the enthalpy of dilution parameter, ψ_1 is the entropy of dilution parameter, V_s is the volume of a stabilizing segment, V_1 is the volume of a solvent molecule, and ϱ_i and ϱ_j are the segment density contributions, at a separation H, in the volume element dV of the stabilizing moieties attached to the two particles i and j, respectively. The integral is taken over the entire volume of stabilizer overlap. Doroszkowski and Lambourne[193] have pointed out that the above models do not take into account the case where $H < \delta$ (Fig. 7). In this situation the volume of overlap of the adsorbed layers is decreased as part of the overlap volume is occupied by solid core. These workers allowed for this effect and found that ΔG_R increases more rapidly with overlap than is the case if eqn. (25) is utilized. Napper[518] also discussed the additional volume compression terms required in eqn. (25) for the case where $H < L$.

Meier[496] evaluated the segment density distribution at planar interfaces for linear chains terminally attached at one end and combined the entropy approach of Mackor[466,467] and Clayfield and Lumb[118,119] with Fischer's[240] solvent theory to produce a generalized hybrid theory of steric stabilization. Hesselink et al.[347] have corrected the segment density function used by Meier[496] and extended the hybrid theory to stabilization by loops. Evans and Napper[224] have discussed the predictions and shortcomings of the latter theory. In particular, the theory fails to predict that the θ point $(\theta = K_1 T/\psi_1)$ corresponds closely to the limit of stability of sterically stabilized dispersions. The theory also fails to predict that the critical flocculation point for a dispersion will be independent of the molecular weight of the polymer. The theory of Hesselink et al.[347] contains an erroneous volume restriction term which does not account for the enthalpy of compression and which incorrectly fails to become small at the θ point. Evans and Napper[223] have extended Fischer's solvent power theory to include this enthalpy term and their approach is described below.

The basic dissolution equation for polymer chains attached to particles is derived from the Flory–Krigbaum theory[249] for the mixing of solvent with randomly oriented polymer chains whose centers of gravity are fixed in space:

$$\delta(\Delta G) = kT(V_s^2/V_1)(\tfrac{1}{2} - \chi_1)\varrho_2^2 \, \delta V - kT(V_s/V_1)(\tfrac{1}{2} - \chi_1)\varrho_2 \, \delta V$$
$$- \tfrac{1}{2}kT(V_s/V_1)\varrho_2 \, \delta V \tag{26}$$

where ϱ_2 is the number of polymer segments per unit volume and the other parameters are the same as those given in eqn. (25). The third term describes the dissolution of randomly oriented polymer segments into a θ solvent in a volume element δV, while the second term describes the effect of increasing the quality of the solvent from a θ solvent to some value of $\chi_1 < \tfrac{1}{2}$, while retaining the random coil θ-solvent conformation. The first term describes the free energy change associated with the conformational change of the polymer that occurs in a solvent better than a θ solvent. If there are ν mono-disperse chains (i segments) attached irreversibly at complete surface coverage to unit area of the flat plates in Fig. 9, then for step X envisaged in this figure, ΔG_∞ for one plate is given by integration of eqn. (26) as

$$\Delta G_\infty = kT(V_s^2/V_1)(\tfrac{1}{2} - \chi_1) \int_{x=0}^{\infty} \varrho_\infty^2 \, dV - kT(V_s/V_1)(\tfrac{1}{2} - \chi_1) \int_{x=0}^{\infty} \varrho_\infty \, dV$$

$$- \tfrac{1}{2}kT(V_s/V_1) \int_{x=0}^{\infty} \varrho_\infty \, dV \tag{27}$$

where ϱ_∞ is the segment density distribution function at infinite separation

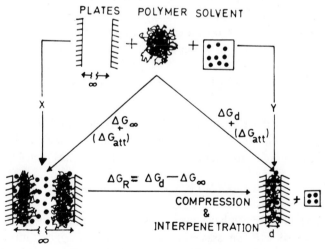

Fig. 9. Schematic representation of the close approach of two sterically stabilized plates (from Evans and Napper[223]).

of the plates. The free energy of attachment ΔG_{att} of the polymer to the plates is not included because it cancels out if it is assumed that ΔG_{att} is independent of the distance of separation of the plates. Since ϱ_∞ is only a function of the distance normal to the plate x, the total number of segments in the volume normal to unit surface area is

$$vi = \int_{x=0}^{\infty} \hat{\varrho}_\infty \, dV \tag{28}$$

where $\hat{\varrho}_\infty$ is the normalized segment density distribution function. An analogous procedure for step Y in Fig. 9 gives ΔG_d for the dissolution of polymer when the plate separation is d. The Gibbs free energy of compression per unit area ΔG_c for two plates is then given by

$$\Delta G_c = \Delta G_d - \Delta G_\infty$$
$$= 2kT(V_s^2/V_1)(\tfrac{1}{2} - \chi_1)v^2 i^2 \left(\int_0^x \hat{\varrho}_d^2 - \int_0^\infty \hat{\varrho}_\infty^2 \, dx \right) \tag{29}$$

Equation (29) makes no allowance for interpenetration of the polymer chains, but the free energy of interpenetration per unit area ΔG_I is given by an equation like (25):

$$\Delta G_I = 2kT(V_s^2/V_1)(\tfrac{1}{2} - \chi_1)v^2 i^2 \int_0^d \hat{\varrho}_d \hat{\varrho}_d' \, dx \tag{30}$$

where $\hat{\varrho}_d'$ is the mirror image of $\hat{\varrho}_d$. The total free energy change per unit surface area on close approach ΔG_R is given by adding (29) and (30). Hesselink et al.[347] have tabulated the sum of the integral terms for loops and tails, which results as a dimensionless function W:

$$W = \langle r^2 \rangle^{1/2} \left(\int_0^d \hat{\varrho}_d^2 \, dx + \int_0^d \hat{\varrho}_d \hat{\varrho}_d' \, dx - \int_0^d \hat{\varrho}_\infty^2 \, dx \right) \tag{31}$$

where $\langle r^2 \rangle^{1/2}$ is the root mean square end-to-end length of the chains in free solution. The fundamental relationship for evaluating the repulsive Gibbs free energy in the steric stabilization of flat plates then becomes

$$\Delta G_R = (2kTV_s^2/\langle r^2 \rangle^{1/2}V_1)(\tfrac{1}{2} - \chi_1)v^2 i^2 W \tag{32}$$

Evans and Napper[223] have also considered ΔG_R for the interaction of spheres coated with an adsorbed layer of nonionic macromolecule. The total interaction free energy V_R between two plates is obtained by summing eqns. (10) and (32). Vincent[681] has shown how the residual attraction V_A

between two plates can be calculated after the overlap of adsorbed layers. Potential energy curves for sterically stabilized particles do not usually exhibit the maximum that is so characteristic of electrostatically stabilized dispersions.[223] As a result, flocculated, sterically stabilized dispersions redisperse spontaneously under suitable solvent conditions.[518] The theory correctly predicts the presence of a sharp transition from stability to instability very close to the θ point and an independence of these effects on the particle radius and the polymer molecular weight.[223]

Evans and Napper[223] have suggested that the London–van der Waals attraction between the core particles can be relatively unimportant in determining the flocculation of sterically stabilized dispersions. Flocculation of latices is induced as a result of the attraction between the stabilizing sheaths, which in turn derives from the solvent power of the dispersion medium being made marginally worse than a θ solvent. Doroszkowski and Lambourne[194] have investigated the repulsive forces generated by steric barriers on small particles under varying conditions of solvent power, including the θ condition. They conclude that when the stabilizing polymer is in a good solvent and present at high concentrations the osmotic pressure term only need be considered. The volume restriction effect increases linearly with the polymer concentration, while the osmotic pressure term (which comes into operation before the configurational term[347]) increases with the square of concentration.[496] The above findings highlight the importance of solvation effects in steric stabilization and the role of solvation (and hydration in particular) is discussed below.

Evans and Napper[223] have shown that eqn. (32) can be expressed in a more tractable form as

$$\Delta G_R \sim [2(2\pi)^{3/2}/27]\langle r^2\rangle \nu^2(\alpha^2 - 1)WkT \tag{33}$$

where ν is the molecules per unit area of interface and α is the intramolecular expansion factor introduced by Flory and Fox.[248] If $\langle r^2\rangle_0$ is the mean square of the end-to-end distance of the polymer chain in the absence of perturbation of the distribution by long-range effects and α is the factor by which a linear dimension of the average configuration is altered as a result of long-range effects (i.e., excluded volume effect), then

$$\langle r^2\rangle = \alpha^2\langle r^2\rangle_0 \tag{34}$$

where $\langle r^2\rangle$ is the actual mean square of the end-to-end distance.[247] The volume effect depends not only on the actual volume of the chain unit but also on its interaction with the solvent. Thus, the perturbation of the configuration is determined by the effective covolume for a pair of segments immersed in the solvent and not by the actual volume of the segment. The covolume for a segment can be enhanced by use of a good solvent for the polymer and diminished in a poor solvent which is barely capable of dissolving the polymer. With a particular combination of solvent and temperature the finite volume of the segment can be compensated exactly by the mutual attractions between segments when immersed in a poor solvent. The covolume is then zero and the excluded volume effect vanishes; $\alpha = 1$ and $\langle r^2 \rangle = \langle r^2 \rangle_0$. This point is the θ point and the solvent is the θ solvent for the polymer. At the θ point the osmotic pressure of the solution obeys the van't Hoff law up to concentrations of several percent and the second coefficient B [see eqn. (24)] vanishes in the virial expansion of the osmotic pressure. The factor α is clearly a measure of the solvent power of the dispersion medium (α is larger for good solvents where segment–solvent interactions are preferred to segment–segment interactions) and it is apparent that through eqn. (33) the role of the solvent in steric stabilization is described. The factor α is related to the radius of gyration of the molecule[659] and can also be determined experimentally from viscosity η measurements. The mean hydrodynamic expansion factor $\langle \alpha_\eta \rangle$ is given by[246]

$$\langle \alpha_\eta \rangle = ([\eta]/[\eta]_\theta)^{1/3} \tag{35}$$

where θ denotes the value measured in a θ solvent. Viscosity measurements can also be used to determine $\langle r^2 \rangle$.[247,659] In the Flory–Krigbaum theory[249] the interaction between polymer segments and solvent molecules is characterized by enthalpy K_1 and entropy ψ_1 of dilution parameters such that $\theta = K_1 T/\psi_1$ and

$$\alpha^3(\alpha^2 - 1) = 2C_M \psi_1 (1 - \theta/T)M^{1/2} \tag{36}$$

In eqn. (36), M is the molecular weight of the polymer and C_M is a constant relating the partial specific volume of the polymer, the molar volume of the solvent, and the unperturbed mean square end-to-end length of the polymer.[246,651,659] This relationship was substituted for χ_1 ($= \frac{1}{2} + K_1 - \psi_1$) in eqn. (32) by Evans and Napper[223] to derive eqn. (33). It should be noted that K_1 and ψ_1 are assumed to be unaltered by adsorption of the polymer.

Bagchi[20] has also derived an expression for the enthalpic repulsion between two identical spherical particles coated with a polymeric adsorption layer. The calculation requires the use of only experimentally accessible parameters. The increase in enthalpy ΔH_R arising from an increase in polymer concentration C between the particles and the interaction plane on close approach is given as

$$\Delta H_R = 2n' \int_{C_\infty}^{C_H} [\partial(\Delta \bar{H})/\partial C]_{\tilde{n}} \, dC \tag{37}$$

$[\partial(\Delta \bar{H})/\partial C]_{\tilde{n}}$ is the differential heat of dilution of the polymer in the solvent at constant number of moles $\tilde{n}$ of polymer; n' is the number of moles of polymer adsorbed on the spherical section of particle surface involved in overlap divided by $\tilde{n}$; and C_∞ and C_H are the average concentrations of polymer in the adsorption layer at particle separations of infinity and H, respectively. $[\partial(\Delta \bar{H})/\partial C]_{\tilde{n}}$ can be obtained by differentiating the experimental integral heat of dilution curve of the polymer as a function of concentration.

The effect of changes in α (i.e., solvent power) on the total interaction energy between spherical latex particles is depicted in Fig. 10. The curves show that stable ($\Delta G_R > nkT$, where n is a small integer) dispersions are

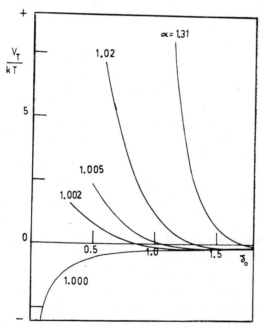

Fig. 10. The total potential energy curves at 30° for two equal, spherical, poly(vinyl acetate) particles stabilized by poly(ethylene oxide) loops of molecular weight 9.6×10^4 for various values of the intramolecular expansion factor α. Here $\delta_0 = d_0/\langle r^2 \rangle^{1/2}$, where d_0 is the minimum distance of surface separation of the spheres and $\langle r^2 \rangle^{1/2}$ is the root mean square end-to-end length of the chains in free solution. The radii of the particles is 10^3 Å and the effective Hamaker constant is 3×10^{-14} erg (from Evans and Napper[223]).

expected if $\alpha = 1.002$ but that instability should be evident if $\alpha = 1.000$.[223] As has been confirmed experimentally,[520] a sharp transition from stability to instability is therefore expected in this region, which corresponds to the θ point. A temperature change of only a few degrees or addition of small amounts of nonsolvent can induce sudden destabilization in this region.[520] From eqns. (32) and (33) it can be seen that ΔG_R is zero for a θ solvent dispersion medium $(T = \theta)$, is positive when $T > \theta$, and is negative when $T < \theta$. The ideal behavior of polymer chains in dilute θ solutions means that they cannot "see" one another and repulsion is precluded. Of course, in many practical situations the polymer forms a concentrated adsorbed layer and then the theory breaks down. The requirement of both aqueous and nonaqueous sterically stabilized dispersions for the solvent power of the dispersion medium for the stabilizing moieties to be better than a θ solvent indicates that the segmental excluded volume must be positive.[520] This positive excluded volume can originate in either entropic effects or enthalpic interactions, or a combination of both. When entropic effects are stabilizing and enthalpic effects are destabilizing, stability arises when $\psi_1 > K_1$ and $T > \theta$. Enthalpic stabilization arises when there is a positive enthalpy of interpenetration of adsorbed layers which outweighs the attractive forces. Napper[520] has summarized various possible combinations of K_1 and ψ_1 and described the effects on the stabilities of the dispersions (see Table II). Ottewill[544] has tentatively concluded that enthalpic contributions are the most important in aqueous dispersions, whereas entropic contributions are most important in nonaqueous dispersions.

It is clear from the above that a molecular understanding of entropic repulsion is reasonably straightforward. The net entropy change that opposes flocculation is derived from a decrease in the configurational entropy of the chain on interpenetration and/or compression, coupled with a decrease in the entropy of mixing of polymer segments with solvent. A molecular understanding of enthalpic stabilization is much less clear.[522] As depicted in Fig. 8, water molecules can be displaced from the surface of the polymer by overlap of the adsorbed layers. Removal of this water of hydration requires energy and this corresponds to an enthalpy opposing overlap. Release of the strongly bound water molecules is accompanied by an increase in entropy, which may or may not outweigh the decrease in entropy due to both loss of configurational entropy and entropy of mixing. The first possibility results in enthalpic stabilization, the second in combined enthalpic–entropic stabilization. A more quantitative approach requires knowledge of the hydration numbers and water-binding energies of the stabilizing species in the adsorbed state.

TABLE II. Possible Values of the Entropy (ψ_1) and Enthalpy (K_1) of Dilution Parameters and Their Effect on the Stability of Sterically Stabilized Dispersions[a]

Case	Sign				Stability	Type
	ψ_1	K_1	$\mid K_1 \mid / \mid \psi_1 \mid$	θ		
A	+	+	<1	+	Yes	Entropic
B	+	+	≥1	+	No	—
C	−	−	>1	+	Yes	Enthalpic
D	−	−	≤1	+	No	—
E	+	−	≷1	−	Yes	Enthalpic–entropic
F	−	+	≷1	−	No	—
G	+	0	0	0	Yes	Entropic
H	−	0	0	0	No	—
I	0	+			No	—
J	0	−			Yes	Enthalpic
K	0	0			No	—

[a] From Napper.[520]

The influence of ions on this type of hydration interaction is of interest and Napper[521] has investigated the effects of ions on the flocculation of poly(vinyl acetate) sols stabilized by poly(ethylene oxide) chains. For these enthalpically stabilized dispersions, the flocculation potency of cations is in direct conflict with the postulate that more highly hydrated ions are more effective flocculants. The order for cations is

$$Rb^+ = K^+ = Na^+ = Cs^+ > NH_4^+ = Sr^{2+} > Li^+ = Ca^{2+} = Ba^{2+} = Mg^{2+}$$

The orthodox explanation of the Hofmeister series couched in terms of the ions competing with the hydrophilic layer for the water seems inapplicable. The results have been discussed in terms of water "structure" around the ions[521] and in terms of the formation of pseudopolyelectrolytes in aqueous salt solutions of nonionic polymers.[220] Glazman[286] has shown that inorganic sols stabilized by nonionic surfactant are flocculated by high concentrations of electrolyte because of disturbance of the solvation layers around the colloidal particles. The electrolytes have parallel effects in salting out solutions of the nonionic surfactant and flocculating the sol.

It is apparent from the above theory that a basis for understanding hydrophilic colloids is emerging. Hydration theories of stability have been popular in colloid science for a long time. Van Olphen[678] has discussed these ideas and pointed out the fallacies. Repulsion between particles has been attributed to the formation of a solvation shell containing oriented water molecules; when these diffuse hydration shells overlap, the mutual repulsion between oppositely oriented water dipoles stabilizes the system against flocculation. However, the ordering effect of the particle can be expected to be significant only up to a few water molecule diameters away from the surface (Section 2 and the preceding chapter by Clifford). For square, finite plates of a cm side the total dipole repulsion energy is

$$E(x) = \frac{8\mu^2 a}{\varepsilon} \left[0.773 - \ln a - \frac{s}{a} - 1 - \frac{x}{a} - 0.5 \right.$$
$$\left. + \frac{x^2}{2s^2} \ln \frac{(x+2s)x}{(x+s)^2} - \frac{2x}{s} \ln \frac{x+2s}{x+s} + \ln \frac{(x+2s)^2}{x+s} \right] \qquad (38)$$

where μ is the dipole moment per unit area, x is the separation of the plane of the closest poles of the dipole, s is the pole distance of the dipole, and ε is the permittivity.[281] Van Olphen[678] and Overbeek[553] have pointed out that such a repulsive energy is normally small and the range is not larger than the lateral distance between the dipoles in one layer. In general, such a repulsion would not be expected to be strong enough to cancel out the van der Waals attraction between particles, unless the particles are very small. However, hydration repulsion should not be entirely discarded because the required work of desorption of any strongly bound water manifests itself as a short-range repulsion (cf. Fig. 8). The magnitude of this repulsion depends upon the hydration energy of the surface. In the cases of lecithins (see Chapter 4, Volume 4) and certain clays (Chapter 4, this volume) the hydration energy is sufficient to cause swelling and separation of the unit layers. As a dramatic example of the contribution that hydration energy can make to π_S, Van Olphen[678] has calculated that the pressure required to remove one monomolecular layer of water from clay surfaces can be as high as 4000 atm. Surface-initiated polarization of water molecules does not stop at the first layer, because Harkins[322] has shown that the excess free energy $(= \Delta G_{\text{desorption}} - \Delta G_{\text{vaporization}})$ for removal of water from titanium dioxide is 6.5 kcal mol^{-1} for the first layer, 1.4 kcal mol^{-1} for the second, and 0.2 kcal mol^{-1} for the third. Clearly, an important aspect of future work on the role of hydration in colloid stability must be concerned with determining hydration energies. As mentioned above, for stabilizing polymers these must come from measurements of heats of solution, mixing or wetting, and for other surfaces from water adsorption isotherms.[678]

To summarize the present position with regard to the role of hydration in stabilization of disperse systems, the physical basis of entropic repulsion is on a sound footing but that of enthalpic stabilization is less clear. Traditionally, polymer–solvent interactions and the χ parameter have been associated with the heat of mixing of polymer and solvent liquid and with the difference in contact energies. It is now becoming clear that χ is usually mainly entropic and associated with the difference in free volume between polymer and solvent[214,565]; these effects will have to be incorporated in future treatments of steric stabilization. The theoretical treatments only consider random homopolymers, so that the specific structures of biopolymers are outside the scope of the above discussion. Heteropolymers containing both polar residues which can hydrogen bond with water and apolar residues which cause water to form a clathrate structure around them will have to be considered in terms of good and bad solvent power. Such macromolecules adsorb to form loops with the polar segments preferentially in the aqueous phase and the appropriate intramolecular expansion factor is then that for these parts of the polymer molecule.

5. EXPERIMENTAL EVIDENCE FOR THE INFLUENCE OF HYDRATION IN FOAMS AND EMULSIONS

The air–water and oil–water interfaces intrinsic to the systems being considered here are smooth, even at the molecular level, and of relatively low energy. Therefore, only data from colloidal systems which contain clean interfaces that satisfy these criteria are directly pertinent to our discussion. This excludes consideration of solid surfaces which are rough and which may have multimolecular layers of structured water adsorbed on them (see Section 2 and Chapter 2).

Van den Tempel[672] has studied the flocculation and coalescence of oil-in-water emulsions. The emulsion droplets were paraffin oil stabilized with sodium dodecyl sulfate and the aqueous phase contained sodium chloride. In the flocculated state the oil globules in an aggregate are separated by a water film which is of the order of 100 Å thick. Coalescence of oil globules occurs by rupture of this thin water film and is a first-order process. The rate of coalescence is identical with the probability of rupture of the water film. The thickness of this film is determined by the type of emulsifier and the electrolyte content, and not by the amount of water

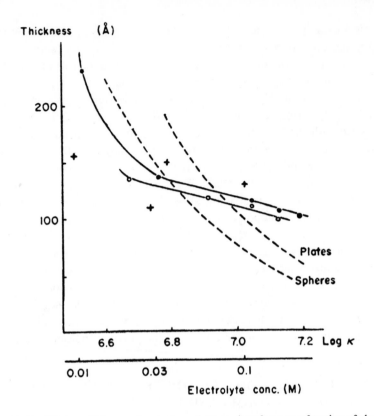

Fig. 11. Equilibrium thickness of water layer between interfaces as a function of electro-lyte concentration. (○) Two drops; (●) single drop at plane interface; (+) results of Derjaguin and Titjevskaja[178]; (- - -) theory (from Van den Tempel[673]).

available. In a further study Van den Tempel[673] measured by interferometry the equilibrium thickness of these films as a function of electrolyte concentration. The results are shown in Fig. 11 together with Derjaguin and Titjevskaja's[178] equivalent data for air bubbles. The theoretical curves obtained by equating π_E [eqns. (7) and (8)] and π_V [eqns. (10) and (12)] are also shown in Fig. 11. It is apparent that the equilibrium distance decreases less rapidly with increasing electrolyte concentration than the theory predicts. The experimental results suggest the presence of additional repulsive forces which become operative at distances between the interfaces of less than about 125 Å. Van den Tempel[673] suggested that these might be of a steric nature (i.e., $\pi_S \neq 0$).

Direct measurement of π_S in a foam or emulsion system has not been achieved, but has been attempted in another system where the interfaces are smooth. Barclay and co-workers[30,31] have measured the forces between platelike particles of sodium montmorillonite in a dispersion as a function of the distance of separation of the surfaces. The force was measured down to distances of the order of 10 Å. The contributions of π_V and π_E to the disjoining pressure [eqn. (6)] were calculated from DLVO theory and the curve of the resultant total pressure $\pi_V + \pi_E$ against distance between surfaces H is shown in Fig. 12. At a distance of about 15 Å the theoretical curve shows a maximum and hence at distances less than 15 Å the van der Waals attractive forces should be predominant. No evidence for this was found and at the shortest separation distances achieved experimentally the

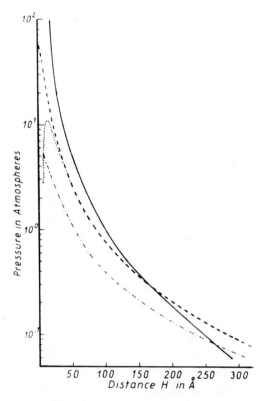

Fig. 12. Pressure against distance between the surfaces for two plates in 10^{-4} M sodium chloride solution. (- - -) Theoretical curves for interaction between two diffuse double layers at a constant $\psi_0 = 250$ mV; ($\cdots$) theoretical curve for interaction between two diffuse double layers at constant $\psi_0 = 250$ mV including the attractive interaction (plate thickness 8.0 Å); (-----) theoretical curve for interaction between two Stern layers with constant $\psi_0 = 250$ mV; (——) experimental curve for sodium montmorillonite in 10^{-4} M sodium chloride solution. (From Barclay et al.[30].)

pressure remains highly repulsive. It can be seen from Fig. 12 that the pressure increase at short distances observed experimentally exceeds that expected in terms of DLVO theory. Barclay and co-workers conclude that the high gradient of the curve at short distances is a consequence of strong repulsive forces arising from the interaction of hydrated layers (i.e., $\pi_S > 0$).

Davies[143,144] has pointed out that before oil droplets can coalesce, any water of hydration of the surfactant must be displaced. The energy required for this displacement also forms a contribution to π_S. The total energy barrier E_h required for this displacement depends on the total number and type of the hydrated groups on each molecule of the surfactant and on β, the fraction of the interface covered. As a result, the energy barrier Q to coalescence due to hydration can be expressed as

$$Q = \beta \sum E_h \tag{39}$$

and Davies[143] has discussed some possible magnitudes for $\sum E_h$. This energy barrier should affect the rate of coalescence but there has been no experimental verification of this yet. Sonntag and his collaborators[639–641] have studied the thinning of oil/water/oil and water/oil/water lamellae using a microreflectivity technique for the former and capacity measurements for the latter and have derived disjoining pressures. Both positive and negative disjoining pressures were measured, with positive disjoining forces only evident in very thin black films. The effects of adsorbed layers were investigated and it was found that by increasing the concentration of surfactant, the attraction between the oil droplets is increased and consequently the stability reduced.[641]

There has been much more work[130,628] on the drainage and stability of thin water films in air because it has been recognized that such films provide an excellent experimental system for studying the nature and range of action of surface forces. In particular, investigation of very thin (black) films offers a direct method for studying the interaction of hydration layers. Lyklema et al.[459] have studied the thickness of the interlamellar film core in the region immediately adjacent to the lower Plateau border as a function of the rate of vertical withdrawal of large films from bulk solutions of sodium dodecyl sulfate, dodecanol, and lithium chloride. With these rigid films a layer of 16 ± 8 Å thickness on each side of the film does not participate in viscous flow. Since the surfactant monolayer thickness was estimated as 16 Å, this shows that any rigidified solvation layer associated with the layer is likely to be less than about 10 Å thick.

The equilibrium thickness of thin liquid films is governed by eqn. (6). If the expressions for π_H, π_E, and π_V [eqns. (1), (7), and (10)] are substituted into (6), a theoretical equilibrium thickness can be calculated if it is assumed that $\pi_S = 0$. The validity of this assumption can be checked by comparison of experimental and theoretical equilibrium thicknesses. This approach has been used by many authors and their results have been reviewed elsewhere.[130,402,628] Since there are two minima in the potential energy vs. thickness plot for thin films (Fig. 4) there are two possible equilibrium thicknesses.[208] The "first black" or common black film corresponds to the secondary minimum, while the "second black" or Newton black film corresponds to the primary minimum. Jones et al.[387] have studied the second black films formed from aqueous solutions of sodium n-alkyl sulfates with added sodium chloride. The final thicknesses of the thinnest films of sodium dodecyl sulfate were 42–48 Å for sodium chloride concentrations up to 1 M. For the C_{10}, C_{12}, and C_{14} homologs, thicknesses of approximately 40, 44, and 48 Å were obtained. The thickness of the water core in all cases proved to be about 25 Å. This is presumably equivalent to the thickness of two hydration layers. For different mono- valent metal ions the tendency to favor second black films is inversely correlated with the size of the hydrated ion; this suggests that the structure of the second black film can only accommodate ions below a certain size and does it preferentially as they become smaller. Overbeek[551] has summa- rized the earlier findings on the interaction of double layers and thicknesses of aqueous cores. There are discrepancies among these data but a low-angle X-ray scattering study[128] has shown that the aqueous core of thin black decyltrimethylammonium decyl sulfate films has a limiting thickness of about 20 Å. In such films there are only three to four layers of water mole- cules associated with each surface monolayer. Further thinning would require dehydration of the polar groups of the surfactant molecules and the free energy required for this dehydration gives rise to the repulsive potential π_S which stabilizes the film.[131,511]

As yet, the effect of temperature on disjoining pressure or its conse- quences has not been elucidated for systems related to foams and emulsions. In view of the likely sensitivity of the structure of hydration layers to changes in temperature, this is an aspect which warrants investigation. There is some evidence[637] which suggests that temperature-induced changes in the hydra- tion of hydrophobic sols stabilized by nonionic polymer are reflected in the stability of the dispersion. The distribution of ions at an aqueous interface can also have dramatic effects on the structure of any hydration layers. Apart from double layer effects, in the absence of specific adsorption, ionic solutes can be repelled from the water region adjacent to a neutral inter- face.[357] It has been shown that changes in the free energy of ion solvation play an important role in cation adsorption at oxide–water interfaces.[708] Thus a detailed analysis of cation adsorption requires consideration of both solvation energy effects and discrete ion effects (see Chapter 1). Hydrophobic effects can be of importance when considering the effects of ions on disperse systems. For example, hydrophobically hydrated quaternary tetraalkylammonium ions do not flocculate dispersions of negatively charged lipid particles, whereas electrostrictively hydrated inorganic cations cause flocculation; the zeta potential is the same in both cases.[327] Thus, a hydrophobic colloid becomes hydrophilic on addition of the quaternary tetraalkylammonium ions.

6. SUMMARY AND CONCLUSIONS

This review outlines our current understanding of the role of hydration and interfacial water structure in determining the properties of foams and emulsions. The various possible environments for water molecules in foams and emulsions are described. Hydration and steric effects can contribute to the disjoining pressures which stabilize the thin water films which prevent coalescence in water-continuous foams and emulsions. In systems where the fluid interfaces are covered with small molecule surfactants there is no long-range interfacial restructuring of water and hydration effects only influence the interaction between two droplets when the distance of separation is approximately 20 Å. However, in the case of stabilization by polymeric surfactants the steric and hydration effects associated with the overlap of two adsorbed layers can be of longer range. In these cases the van der Waals attraction between droplets which approach closely can be counteracted by a repulsion arising from loss of configurational entropy of the polymer due to volume restriction, and an osmotic repulsion arising from a change in the free energy of mixing of polymer segments and water on interpenetration of the adsorbed layers. For stabilized dispersions in aqueous media, the most important part of this free energy change seems to be that associated with alterations in enthalpy due to displacement of water molecules solvating the adsorbed layer. The hydration contribution to the free energy of repulsion can be described theoretically for polymeric molecules in terms of an intramolecular expansion factor. The limited experimental evidence which relates directly to the above effects has been presented.

CHAPTER 4

Clay–Water Systems

Erik Forslind

Division of Physical Chemistry
Royal Institute of Technology
Stockholm, Sweden

and

Arvid Jacobsson

Division of Geology
University of Stockholm
Stockholm, Sweden

1. INTRODUCTION

Three recent excellent and extensive reviews[296,450,476] regarding adsorbed water on clay are available. Graham,[296] being the latest of the three, takes advantage of his position to review the reviewers in a very stimulating and clear survey of the research situation some ten years ago—or today.

There is certainly no need to retrace the ground covered by these authors. Of course, we disagree with much of what is written in those three reviews, but it should be obvious how much we are indebted to our precursors in this attempt to sum up a few of the research problems that, according to our beliefs, require particular attention and extended effort.

Clay–water interactions determine to a great extent the clay mineral genesis[506] and subsequent transformation processes, and a thorough understanding of the relevant molecular mechanisms will find direct application in the solution of practical problems arising in the handling of natural clay deposits[93] as in geotechnology, soil and rock mechanics, foundation engineering, and oil drilling. Thus the strength and deformation properties of clay beds[93] are largely governed by the amount and type of clay minerals present, the clay crystallite and crystal aggregate sizes, their ionic composition, and the amount of intraaggregate and interaggregate water. These various components and parameters interact as a unit, the full complexity of which must be admitted and understood for efficient and reliable control of the material. To exemplify, the enhancement of the natural thixotropy of a clay is utilized to stabilize the walls of deep excavations[679] and oil drillings[423,676] to prevent percolation of water into the excavation and out of the drilling mud. The method is in practical use but could very likely be improved through a better understanding of the intraaggregate water contribution to thixotropy.

Although we are not directly concerned with the clay mineral organic complexes which are of fundamental importance in the field of agriculture, it should be pointed out that the transport and exchange processes that are vital for the uptake of nutrients[282] through the root fibrils of plants depend critically on the clay–water interaction and are very little understood.

The properties of the clay minerals used in ceramics[308] and in the brick and cement industries are essentially dependent on the clay–water interactions for the achievement of optimal plasticity and coherence needed in the handling of the ceramic material prior to firing, to assure shape retention during the precursory drying processes and homogeneity with respect to density and heat transport during firing. Most of the knowledge fundamental to the present manufacturing processes in these industries is empirical and to a great extent derived by trial and error techniques. The technical and economical problems that nevertheless remain to be solved represent a vast field of very badly needed fundamental clay–water research.

Of the clays used in the paint and graphic industries only a few are applied in water-based systems and processes. In the majority of cases the clays are made oleophilic but in both types of application their uses depend on the gel-forming and thixotropic properties of the lyophilic compounds. We shall see that the study of the hydrogel-forming properties of the hydrophilic clay mineral like montmorillonite and the ambiphilic clay minerals like kaolinite is still in its infancy and requires a concerted effort if the complex rheological effects of the clay–ion–water interactions are to be satisfactorily elucidated.

Similar considerations apply to the uses of clay minerals in foundries to impart coherence and plasticity to casting mould sands and many of the problems associated with the precursory drying of ceramic clays reappear in this context. Thus, it is known[308] that halloysite hydrated to about two or three water layers is very efficient for obtaining high surface smoothness and quality of the cast. However, the relation of this property to the particular water transport processes that must obtain in this mineral, due to the coiledup structure, does not seem to have been clarified, which is perhaps a technical and economic loss to foundries, since halloysite is scarce but might be replaced.

Clays, mainly kaolins, are used on a large scale by the paper industry as filler materials and coatings.[462] As in the ceramics, the guiding knowledge is mainly empirically acquired and leaves plenty of room for fundamental research, in particular on questions related to the water balance between the cellulose network and the coating and in which the clay represents the least known partner. The interplay of hydrophilic and hydrophobic kaolinite properties similarly deserves more attention on the part of basic research before the joint problems of the paper and graphic industries can be regarded as satisfactorily elucidated.

2. MINERALOGICAL PROPERTIES OF LAYER SILICATE CLAYS OF INTEREST IN THE PRESENT CONTEXT[82,86,377]

For the convenience of the reader we shall briefly review the essential structural properties and mineralogical compositions of the kaolin, montmorillonite, and vermiculite groups together with pyrophyllite and talc for comparison with the above clay minerals. According to the general ideas set forth by Pauling in 1930,[567] the lattice formation of the layer

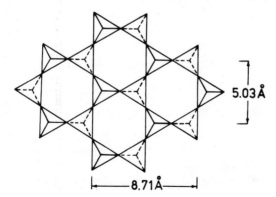

Fig. 1. Tetrahedral layer from β-cristobalite or β-tridymite. A silicon atom is located at the center of each tetrahedron and an oxygen at each corner.[567]

silicates can be described as a condensation process. If one side of the un-shared hydroxyl groups from a tetrahedral β-tridymite or β-cristobalite layer of silica is condensed on one layer of aluminum octahedra in gibbsite $Al(OH)_3$ (Figs. 1 and 2) with elimination of water, a 1 : 1 layer of crystal kaolinite $Si_4Al_4O_{10}(OH)_8$ will result (Fig. 3).

Similarly, the condensation of two layers on each side of a gibbsite or brucite sheet will give rise to the 2 : 1 layer pyrophyllite $Al_2 \mid (OH)_2Si_4O_{10} \mid$ and the 2 : 1 layer talc $Mg_3 \mid (OH)_2Si_4O_{10} \mid$ structures, respectively.

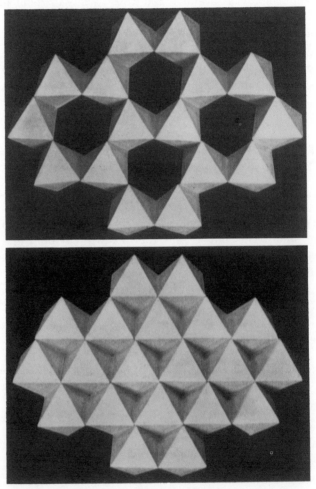

Fig. 2. (a) Layer of aluminum octahedra in hydrargillite (gibbsite) $Al(OH)_3$.[567]
(b) Layer of magnesium octahedra in brucite $Mg(OH)_2$.[567]

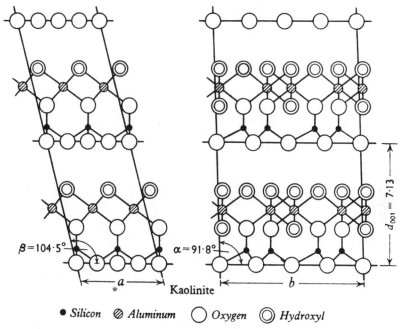

• *Silicon*   ⊘ *Aluminum*   ◯ *Oxygen*   ◎ *Hydroxyl*

Fig. 3. Two unit layers of kaolinite.[82]

Montmorillonite

$$(Si_8)^{IV}(Al_{3.33}Mg_{0.67})^{VI}O_{20}(OH)_4$$
$$\downarrow$$
$$Na^+_{0.67}$$

and hectorite

$$(Si_8)^{IV}(Mg_{5.33}Li_{0.67})^{VI}O_{20}(OH)_4$$
$$\downarrow$$
$$X_{0.67}$$

are usually supposed to have similar 2 : 1 structures with replacement of some silicon ions in tetrahedral coordination by other, mainly trivalent ions and/or some Al^{3+} or Mg^{2+} ions in octahedral positions replaced by Fe^{3+}, Fe^{2+}, Mg^{2+}, or Li^+ giving rise to lattice charge defects compensated by exchangeable cations. The structure described with the silicon atoms of each layer contained in one plane was first proposed by Hofmann *et al.*[354] (Fig. 4).

Vermiculite,[78,310,477] based on the talc structure, contains a magnesium–water complex in intercrystalline coordination between the silicate layers

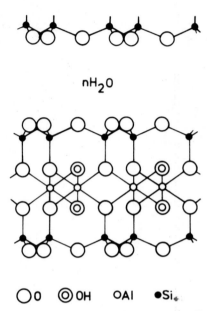

$$nH_2O$$

○ O   ◎ OH   O Al   ● Si,

Fig. 4. Hofmann–Endell–Wilms conception of the montmorillonite structure.[354]

(Fig. 5). The octahedral layer contains Mg, Fe, and Al and the tetrahedral layer Si and Al. The relations between water and these clay minerals will be discussed in the next section.

2.1. Hydrophobic Clays

In pyrophyllite and talc all tetrahedral apices are directed toward the central layer, the common atoms of the tetrahedral and octahedral layers being oxygens. The silica layers are said to be in cis coordination. The bonding force between the lattice units described is of the van der Waals type. It is weak and permits easy cleavage, which explains the soft and soapy touch of the minerals. Pyrophyllite and talc are substantially without net charge and display insignificant exchange properties. Water molecules and other polar molecules do not give rise to swelling.

The same is true of vermiculite saturated with magnesium as intercrystalline exchangeable ion. The reason for this behavior will be discussed in Section 4.4.2(a). The magnesium ions, which compensate for the usually high charge deficiency of the mineral, may, however, be exchanged for Li or butylammonium ions, which are able to disturb the mechanism that inhibits osmotic swelling.

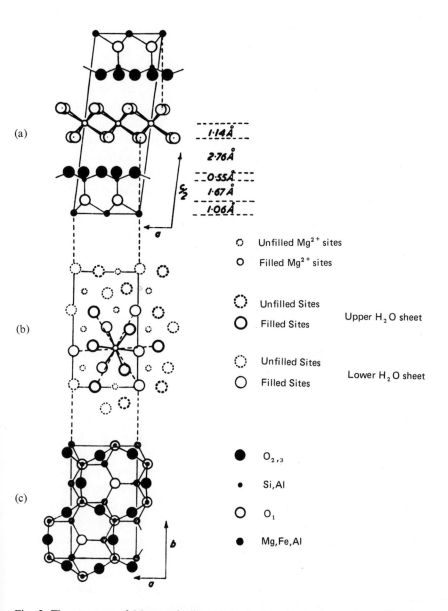

Fig. 5. The structure of Mg vermiculite. (a) Projection normal to *ac* plane; (b) projection normal to *ab* plane; interlayer region; (c) projection normal to *ab* plane; one-half a silicate layer ($z = 0$ to $c/8$).[692]

2.2. Hydrophilic Clays

2.2.1. Weakly Hydrophilic Clays

In the kaolin group the 1 : 1 structure of kaolinite gives rise to both hydrophobic and hydrophilic surface properties. The weak hydrophilic, almost hydrophobic character is caused by the cis coordination of the silica tetrahedra, the external surface oxygens of which carry the exchangeable cations[702] which may impart an apparent hydrophilic character to the aggregate surface. The hydrophilic nature of the crystallite is caused by the surface OH groups of the octahedral layer. Halloysite | $(OH)_8Si_4Al_4O_{10}$ ·4H₂O |, on the other hand, shows strong hydrophilic properties, which are apparently a consequence of a trans-coordinated silica layer of the type occurring in β-tridymite or β-cristobalite shown in Fig. 1. In such a structural framework both sides of the kaolin layer become hydrophilic but the misfit between the lattices causes the asymmetric crystallites to roll up to cylindrical shapes (cf. Ref. 703). The layers of the kaolin clay minerals are kept together by weak hydrogen bonding and van der Waals forces (Fig. 6).

2.2.2. Hydrophilic Clays

In a much debated paper published in 1940, Edelman and Favejee[213] pointed out that the generally accepted structures of both montmorillonite and halloysite are inadequate to explain the hydrophilic properties of these minerals. Discussing the similarity between the structures of pyrophyllite and montmorillonite postulated by Hofman et al.,[354] Edelman and Favejee[213] pointed out that the former mineral is almost entirely hydrophobic while the latter is extremely hydrophilic and conclude that the difference between the two minerals in their behavior toward water must be due to structural differences. They therefore proposed the structure model shown schematically in Fig. 7 with silica tetrahedra in trans coordination, the protruding apices carrying acidic hydroxyl group. One objection immediately raised against the model was the high ensuing exchange capacity, six times greater than the mean value observed by Ross and Hendricks.[607] The attack was countered by Edelman,[212] who modified his theory by assuming that the number of protruding hydroxyl groups could be smaller than postulated in the original model. We shall see that this additional assumption is probably correct but unnecessary for explaining the presumed excess exchange capacity.

Indirect evidence in support of the assumption of acidic hydroxyl groups and trans-coordinated silica layers was presented by Berger,[55] who proved

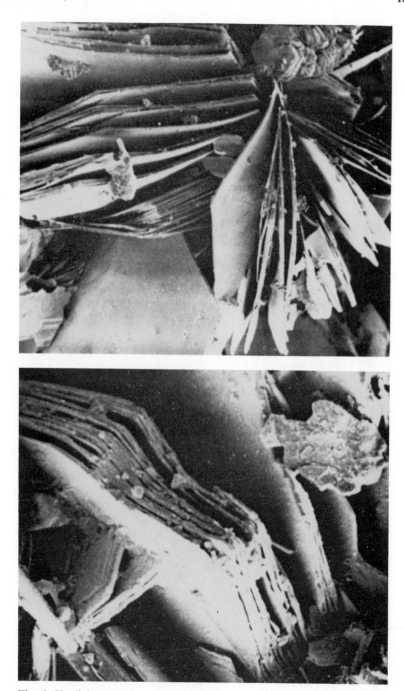

Fig. 6. Kaolinite crystal aggregates seen in a scanning electron microscope.

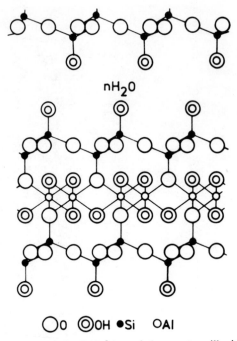

Fig. 7. Edelman and Favejee's conception of the montmorillonite structure.[213]

the existence of surface hydroxyls by methylation of montmorillonite, leading to a hydrophobic clay with a basal spacing of 14.1 Å, in good agreement with the expected value.

On the other hand, Fourier syntheses based on X-ray data obtained
by Brown[84] and Pezerat and Méring[578] were interpreted to be incompatible with the Edelman–Favejee model as originally presented. Similar conclusions were drawn from a recent X-ray investigation of the *hk* bands of
montmorillonite as influenced by the state of hydration and the nature
of the exchangeable cations, carried out by Longuet-Escard *et al.*[448]
However, their main argument with respect to the structure model hinges on
the previous study of Pezerat and Méring.[578]

2.3. Oleophilic Clays

Some naturally occurring clays have turned oleophilic due to adsorption of organic matter. Their mineralogical characteristics and general
behavior[307] are mainly outside the scope of the present review and only
a few points of interest should be mentioned.

Water replacement by ambiphilic organic liquids of low volatility may be used for diagnostic purposes, since their interlamellar adsorption gives rise to characteristic basal spacings.[77,463] Thus montmorillonites treated with glycerol show a basal spacing of enhanced intensity at about 17.7 Å and a series of higher orders usually well separated from other reflections of clay minerals. A H_2O halloysite similarly treated will give rise to a basal spacing of about 11 Å with the first, third, and fifth orders visible. Barshad[35] has made the interesting observation that a Mg, Ca, Ba, H, Li, and Na vermiculite will take up a single layer of ethylene glycol, producing a 14.3 Å spacing when immersed in the hot anhydrous liquid. However, an NH_4, K, Rb, or Cs vermiculite will take up no glycol and show no expansion, while a montmorillonite similarly treated will absorb the usual double layer associated with the 17.7 Å spacing. It will be shown in the following pages how this rather unexpected behavior can be related to the structural properties of water.

Walker[691] observes that the basal spacing of a glycol vermiculite of 14.28 Å is very close to that of the Mg-saturated vermiculite of 14.36 Å, reducing the value of glycol as a means of vermiculite identification. He also points out that the glycol molecule orients itself with a height of 3.9 Å in the montmorillonite complex, while in the vermiculite complex the corresponding dimension is 5 Å. Walker concludes that the glycol molecules in vermiculite are either less strongly bonded than in montmorillonite or differently oriented with respect to the silicate surfaces. There are reasons to believe that both conditions obtain.

3. THE GENERAL PHYSICOCHEMICAL PROPERTIES OF CLAY SUSPENSIONS IN WATER

By definition the clay fraction of a soil is composed of particle sizes smaller than 2 μm. This conventional size limit to the crystallite aggregates of the polydisperse clay is, as a rule, considered to be sufficiently low to secure an essentially colloidal behavior of the aqueous suspension. Thus, the hydrous micas that display both hydrophobic and hydrophilic properties, as do the kaolin minerals, are readily dispersed in water. Their interaction with water may, in essence, be understood from the study of two representative clay minerals, the strongly hydrophilic montmorillonite and the hydrophobic vermiculite. In our attempt to elucidate the fundamental principles of the clay–water interaction we shall therefore be mainly concerned with the properties of the water and these two minerals.

3.1. The Protolytic and Electrophoretic Properties of the Hydrophilic Clays

It is well known that a pure hydrophilic clay always takes on a negative charge in aqueous solution.[677] This is a consequence of the fact that the clay really goes into solution, i.e., interacts with the water by the formation of hydrogen bonds. In nonideal systems there is always a certain amount of proton transfer in the hydrogen bond[251,254] and in the case of the hydrophilic clay minerals this leads to a certain amount of proton expulsion leaving the clay with a charge deficiency, making the crystal negative independent of whether charge-creating isomorphous substitutions occur or not.

It has been shown[510] that the electrophoretic mobility of hydrogen montmorillonite aggregates increases with increasing concentration, attaining a maximum before macroscopic flocculation or gelation occurs. The phenomenon has been interpreted as due to the formation of larger aggregates prior to flocculation. This interpretation is corroborated by extinction measurements, which show a minimum at maximum change in electrophoretic mobility. At maximum electrophoretic mobility, on the other hand, a minimum in total conductivity occurs, as expected from the above interpretation.

3.1.1. Crystallite Dispersion and Aggregation

The separation of the crystal aggregates that occur in a natural clay can, as a rule, only be realized to a limited extent. In earlier laboratory practice a number of dispersion agents were used before it was realized that the clays so treated were irremediably contaminated by the agent. Similarly, use of ethylene glycol or glycerol to facilitate aggregate expansion and cation exchange is not to be recommended, as shown by nmr checks. It is particularly difficult to disperse a hydrogen montmorillonite for reasons that will become clear later. However, a very stable sol may be achieved by the use of artificially enhanced cavitation in an ultrasonic field, which will also become understandable as our discussion advances.

Reaggregation occurs after dispersion if the sol is left standing undisturbed. Below a certain concentration and initial aggregate size, however, the phase separation is extremely slow. We shall later discuss the aggregation mechanism from the molecular point of view taking due account of the effects produced by the exchangeable ions present in the clay–water system. The general macroscopic behavior will be described in the next section.

3.1.2. Critical "Micelle" Formation and the pH Dependence on the Clay Concentration

Starting from a dilute solution of an approximately monodisperse fraction of a hydrogen montmorillonite Mukherjee and Mitra[510] measured the hydrogen ion concentration as function of increasing clay concentration. Their observations at low clay concentrations are reproduced in Fig. 8. At first the hydrogen ion concentration c_H increases linearly with the clay concentration c, until the latter attains a value of about 0.2%. From this point on the slope of the curve changes, indicating an inhibition of proton release from the clay, which Mukherjee and Mitra attributed to the formation of larger "micelles," that is, larger aggregates of associated clay crystallites. We shall later consider this aggregation process, associating it with the organization and stabilization of intercrystalline water as an alternative to the, in our opinion, rather primitive concepts of classical colloid chemistry which must be considered unsuitable for a description of the important *structural* features of the clay–water interface.

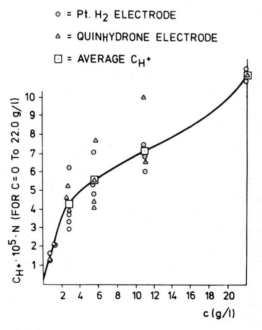

Fig. 8. Hydrogen ion concentration in a hydrogen bentonite fraction (Kashmir) sol as a function of the clay concentration (g/liter) (after Mukherjee and Mitra[510]).

3.1.3. General Effects of the Exchangeable Cations on the Dispersion Stability

From a macroscopic point of view the effects of electrolytes on the clay–water system would appear to follow a straightforward application of colloid chemistry.[524] However, an increase in clay concentration increases, as we have seen, the ion expulsion in contrast to the behavior predicted by the interionic attraction theory of strong electrolytes. There is, moreover, an energy barrier to flocculation and the addition of sodium chloride to a clay suspension gives rise to an increase in viscosity only after a certain critical value is reached, which depends on the type of exchangeable cation present. Below this point addition of sodium chloride may decrease the viscosity. In addition it is found[524] that the viscosity of hydrogen montmorillonites is extremely sensitive to electrolytes as long as the clay remains significantly acidic; small additions of, e.g., sodium chloride change the relative viscosity by more than an order of magnitude, indicating extensive gel formation.

Abundant work on electrolyte effects on the dispersion stability of kaolinite suspension have been carried out in the research laboratories of the ceramic industry. As a rule these investigations aim at immediate practical results needed for the handling of ceramic raw material and are adapted to specific natural clay deposits and processing techniques. Not infrequently the results are guarded as manufacturing "know how," although a wealth of pure scientific information might be derived from some of these studies. A few recent systematic investigations[389,411,475,645] regarding the interaction between the clay–water system and exchangeable ions have been carried out. The investigations of Martin[475,645] take on a special interest because the experimental results have been compared to a very interesting, but unfortunately unique, theoretical approach to the kaolinite–water–ion interaction problem. The attempt by Sposito and Babcock[645] to employ quantum mechanics and statistical mechanics to investigate the cation hydration and water hydrogen bonding to the clay represents, to our knowledge, the first step in a much needed direction. Even if the success of their approach is to some estent fortuitous—they seem, for instance, to be unaware of the fact that Weiss and Russow[702] have shown that the exchangeable cation of kaolinite can only be attached to the tetrahedral surface oxygen layer—it is unquestionable that the theoretical information obtained on the divalent ion–water complexes is extremely interesting, as is the theoretical verification of the weak monovalent ion–water interaction. Their ad hoc assumption of efficient hydrogen bonding to the oxygens of the silica layer in cis conformation is questionable and will be discussed in Section 4.4.3(a).

3.2. Some Optical and Rheological Properties of Clay Suspensions

We shall not be concerned with the purely mineralogical application of optical methods to the study of clays. However, some investigations have been performed in which birefringence and light scattering effects are associated with the clay–water interaction.

3.2.1. Flow and Seiche Birefringence

The effect in based[706] on the parallel orientation of rod-shaped or disk-shaped particles dispersed in a viscous medium. The parallelism is usually assumed to be brought about by hydrodynamic forces due to velocity gradients appearing as a consequence of flow or seiche effects.[422] Although seldom recognized, the phenomenon is often closely associated with the ordering of interaggregate as well as intraaggregate water in hydrophilic clay suspensions. The intraaggregate water layers contribute to the optical anisotropy that is enhanced by the interaggregate orientation. Bentonite suspensions are in fact used to demonstrate laminar flow and the appearance of turbulence in hydrodynamic experiments. Grim[307] points out that this orientational or form birefringence effect is likely to have contributed in a number of optical studies of clays without being recognized as an effect differing from the one produced by the anisotropy of the clay mineral lattice per se.

3.2.2. Dielectric Birefringence

Clay crystallite aggregates may be oriented in an electric field gradient made to alternate at low frequency to avoid electrophoresis. It is assumed that natural charge anisotropy due to isomorphous substitutions and a more or less inhomogeneous distribution of exchangeable counterions are responsible for the electrical moments of the crystal aggregate. Marshall[473] observed that the dielectric birefringence was influenced by the type of exchangeable cation present in the beidellite samples investigated, while no such effects were observed for kaolinite. This was taken as evidence for the exterior position of the exchangeable cations in kaolinite and for their intralattice positions in montmorillonite. To our knowledge no contribution from the intraaggregate water layers has ever been considered, although under certain conditions a much higher dielectric constant might be expected for the adsorbed phase as compared to the surrounding bulk water. From this point of view the situation is entirely different in the kaolinite case.

3.2.3. *Light Scattering*

M'Ewen and Pratt[502] used light scattering techniques to study the gelation of dilute, approximately monodisperse sols of montmorillonite in an attempt to clarify the geometry of the gel structure.

In fair agreement with earlier observations of a critical micelle concentration occurring at about 0.2% in a polydisperse hydrogen montmorillonite sol, they conclude that mechanical interference appears at a concentration of 0.17%, but fail to detect "any continuous resistance to shear at concentrations below 1%," using a coaxial cone viscometer designed by M'Ewen and Mould.[501] From this observation the conclusion is drawn that the random-mesh theory of gel structure appears to be inadequate for a description of the gelation of the montmorillonite sol. We shall return to this question in Section 3.2.4.

Making use of the theory of light scattering as presented by Doty and Edsall[195] and Zimm,[730] M'Ewen *et al.* claim to show that the basic structure of the montmorillonite gel is a linear aggregate, "one lattice layer" thick, in which the particles are oriented edge to edge in the form of flat ribbons. Although it is difficult to reconcile the idealized ribbon model with the actual crystallite habitus shown in the accompanying electron optical picture, the gist of the idea seems to be as follows. Take a pack (or deck if you prefer) of cards, place the long edge of the card pack on a table, cards vertical, cut the pack in two equal parts, lift up the second part and place it on top of the first, long edge to long edge, cards vertical. Rotate the the top half pack about 6° around a vertical axis through the center of gravity of the two half packs, replace the cards by clay monocrystallites, and expand the whole by introducing water layers to separate the crystals in each half pack. This is the linear chain of particles proposed by M'Even *et al.*, although they neglected to introduce organized water between the crystallites, disclaiming the existence of micelles, basal plane to basal plane, in the observed system, and refraining from a description of how such linear chains coexist in the sol. Seen from the present point of view, however, the model appears quite reasonable and can easily be reconciled with the general ideas of the clay–water interaction discussed in this review.

3.2.4. *Phase Separation and Planar Tactoids*

In a remarkable paper on plant virus preparations Bernal and Fankuchen[59] described the structural behavior of some plant virus hydrogels, making use of Zocher's[731] observation that colloidal particles sometimes aggregate in bundles, which he named tactoids. These aggregations have

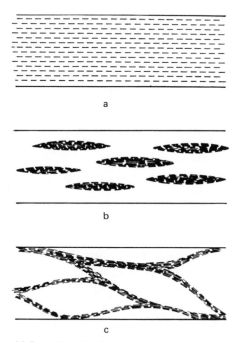

a

b

c

Fig. 9. Positive tactoid formation (b) from a nematic state (a) associating to a three-dimensional gel structure (c) (after Bernal and Fankuchen[59]).

characteristic, spindle-shaped bodies as shown schematically in Fig. 9. The tactoids are of two kinds, positive and negative, the latter representing a zone of the sol that is either partially or totally depleted of the disperse phase. The phenomenon can obviously be described as a phase separation, first giving rise to positive tactoids, producing a significant decrease in viscosity, followed by a drastic viscosity rise on interference between bundles to form a three-dimensional network of positive and negative tactoids. This is essentially the gelation process proposed by Bernal and Fankuchen,[59] who considered the mechanism to be quite general, although they did not, at the time (1941), associate it with an ordering of the interparticle water layers. They had, however, some doubts about the adequacy of the prevailing theory of particle interactions in ionic media, remarking that the model for the theory may be too simple in that *it does not allow for the finite molecular structure of the water medium* (our italics).

Comparing the ideas of Bernal and Fankuchen regarding the mechanism of gelation with the model proposed by M'Ewen et al.,[501,502] it seems clear that the concept of planar tactoids is entirely reasonable. It will be substantiated in what follows.

4. MOLECULAR PHENOMENA RELATING TO THE
MACROSCOPIC BEHAVIOR OF THE CLAY–WATER SYSTEM

4.1. Some Liquid Water Properties of Particular Relevance to the Understanding of the Clay–Water Interaction

The following description of liquid water is based on three principal ideas:

1. Liquid water shows strong affinities with crystalline matter.[60]
2. The crystalline lattice of ice is molecular in spite of intermolecular proton transfer.[566]
3. The "structure" of liquid water may be derived from that of ice Ih.[139,251,614]

The unit cell dimensions of the ideal ice Ih lattice as measured by Megaw[495] and corrected according to Owston and Lonsdale[554] are $a = 4.6226$ Å and $c = 7.3670$ Å at 0°C. The hexagonal unit cell of the oxygen lattice contains four H_2O molecules, in only approximately tetrahedral coordination, since the axis ratio $c/a = 1.62893$ deviates from the theoretical value $c/a = 8/3^{1/2} = 1.63299$ of the ideal tetrahedral structure (see Volume 1, Chapter 4). Since the deviation is larger than the reported error of observation, it is interesting to note that Bjerrum,[66] in a discussion of the structure of ice, suggested that the distance between atoms in neighboring basal layers, possessing mirror symmetry, should be slightly shorter than the bonds between atoms in a central symmetric arrangement. This would imply that the hydrogen bond along the hexagonal c axis is weaker, which is in good agreement with experimental data regarding the plastic deformation of ice single crystals.

The ideal ice Ih oxygen lattice, shown in Fig. 10, should then be modified by the introduction of slightly bent and consequently weaker bonds in the direction of the c axis. In such a modified model of the ice lattice, changes in hybridization of the orbitals may be expected to occur, implying the appearance of nonlinear restoring forces and coupling between valence and bending vibrations. With rising temperature the increasing aharmonicity of the intermolecular vibrations will lead to fluctuations in the distribution of the thermal amplitudes, producing localized amplitude peaks that eventually lead to disrupture of one or several hydrogen bonds. If the thermal amplitude is sufficient, the molecule that is made free will pass through the face of the surrounding coordination tetrahedron to take up an interstitial position in the lattice. The arising lattice defect, a so-called Frenkel defect,

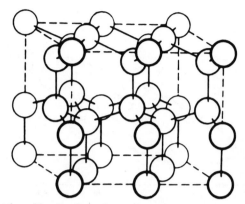

Fig. 10. Ideal lattice of ice Ih. Only the oxygen atoms are shown. The hydrogen bonds are indicated by lines joining the oxygens.

consists of one vacant lattice point and one interstitial, nonassociated water molecule.

Just below and at the melting point a certain number of Frenkel defects will appear in the ice lattice. In this discussion of principles, the arguments refer to the single crystal lattice, because in the polycrystalline case other types of lattice defects will predominate. If we designate the negative parts of the Frenkel defects, the vacant lattice points, as "holes," we find that the holes and the interstitial molecules may be dissociated and each part of the defect will be free to diffuse through the lattice. For this diffusion, however, a certain activation energy is required and, when provided, a number of holes will eventually reach the boundaries of the system, where they will be annihilated and leave behind lone interstitial molecules in the lattice. The annihilation of holes will produce a volume decrease and a corresponding increase in density due to the lone interstitial molecules. As long as both components of the Frenkel defects are equal in number, the volume change produced by the defects can be neglected in comparison to the volume changes due to annihilation of the vacant lattice sites. The annihilation of holes at boundaries of the system corresponds to the melting process and the latent heat of fusion of ice is essentially determined by the energy of formation of the lone interstitial molecules, which we henceforth shall call Schottky defects.

Making use of these ideas, it is possible to assess the equilibrium conditions of the molecules in lattice positions, the holes, and the interstitial molecules in pure water as functions of the temperature.[251,254] The total fraction m of interstitial molecules and the energy of activation for viscous flow thus computed are shown in Fig. 11. In order to take account of the

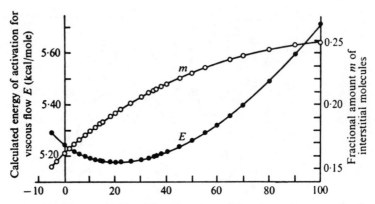

Fig. 11. Total fraction m of lattice defects in water and the computed energy of activation for viscous flow as functions of the temperature.[254]

contribution from the Schottky defects, we can use the Raman observations of Walrafen,[695] who concludes that the intensity decrease of the intermolecular Raman bands with rising temperature should depend on the change in concentration of the four-coordinated species in liquid water as shown in Fig. 12, in which the index on the fraction f indicates the coordination number. In these calculations[254] we have neglected to introduce the one-coordinated molecules as a separate phase. It is considered to appear spuriously within the interstitial phase. No two-coordinated water molecules appear in the bulk phase of water below about 25°C, which temperature is close to the minimum of activation energy for viscous flow.

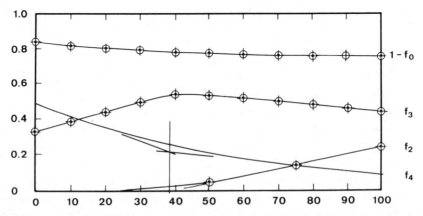

Fig. 12. Temperature dependence of the various coordination species in water computed on the basis of Walrafen's[695] Raman intensity data.[254] End point tangents to the f_4 curve cross at 38°C.

4.2. General Aspects of Water Uptake and Swelling of Clays

4.2.1. Two Conflicting Structural Models for Hydrophilic Clays

Of the various structural models suggested for the montmorillonite lattice, only two have been widely considered and discussed. The earlier and most generally accepted hypothesis was advanced by Hofmann et al.[354] in their pioneer work on montmorillonite. In conformity with the general ideas set forth by Pauling,[567] they conceived the lattice to be formed by condensation of a silica layer on each side of a gibbsite sheet (Fig. 2). The silica tetrahedra were supposed to be arranged in cis coordination, implying that the Si atoms of a silica layer were all contained in one plane parallel to the central aluminum layer of the gibbsite sheet. In its ideal form this structure is electrostatically balanced. However, in order to allow for the presence of magnesium in the natural montmorillonites, Hofmann and co-workers assumed that magnesium could replace aluminum in the gibbsite layer but then, on the other hand, they failed to provide a satisfactory explanation for the presence of other exchangeable (or not exchangeable) cations, as revealed by chemical analyses.

The formulation of a consistent working hypothesis to account for the base exchange capacity of the Hofmann structure, enabling a reconciliation of the contradictory conclusions drawn from chemical and X-ray data, is due to Marshall[474] and Hendricks,[336] whose ideas have been of the greatest importance for the further development of clay research. In 1945 Ross and Hendricks[607] presented a general survey of the minerals of the montmorillonite group, based on an imposing set of analytical data, and providing strong evidence in support of the structural hypotheses advanced by Hofmann and his collaborators.

In principle, the hypothesis of Marshall and Hendricks implies that the variable, water-filled intercrystalline space shown by Hofmann et al. to occur in the hydrated aggregates of single crystals can accommodate the exchangeable cations needed to compensate the charge defects, which are considered to arise from intracrystalline isomorphic substitutions.

In the structural model of montmorillonite proposed by Edelman and Favejee[213] the number of hydroxyl groups is, however, increased from four to 12, which introduces new aspects of the ionic exchange problem, if the protons of the hydroxyl groups can be regarded as exchangeable.

4.2.2. Two Origins of the Ion Exchange Capacity

Since the structural model proposed by Hofmann and his co-workers is electrostatically neutral, the origin of the observed ion exchange capacity was attributed to the presence of isomorphous substitutions in the gibbsite and silica sheets, giving rise to charge deficiencies that had to be compensated by external exchangeable cations, essentially supposed to be situated in the intercrystalline spaces of the aggregates. This hypothesis could, as a rule, be substantiated by a fair agreement between the observed exchange capacity and the one calculated from chemical analyses of the specimen. However, due to the statistical nature of the problem, the detailed balance between the isomorphous substitution and the exchangeable ions could, of course, never be verified by a small number of experiments, leaving open the possibility of a different exchange mechanism, as suggested by the Edelman–Favejee structure.

One of the early objections raised against the theory of Edelman and Favejee concerned the cation exchange capacity. Assuming that all the protruding hydroxyls contribute to the exchange, implying that four unit charges per unit cell may be compensated by foreign cations, the exchange capacity appears with a value six times larger than the average value observed, which corresponds to two-thirds of a charge unit per cell. In consequence of this argument, Edelman[212] modified his theory by supposing that the number of protruding hydroxyl groups could be smaller than that postulated in his first hypothesis. Although, as we shall see, recent experimental findings lend support to Edelman's assumption, the problem of the exchange capacity of the Edelman structure can be resolved without recourse to the additional hypothesis.

Since an electrically neutral montmorillonite crystal immersed in water always assumes a negative charge, one of the first questions that arises in an attempt to understand the exchange processes is concerned with the mechanism of charge transfer between the clay and the surrounding water lattice. Assuming specific water adsorption, taken in the sense suggested by Forslind[251] and implying the formation of a stabilized water lattice of reduced density coupled to the clay crystal surface, the transfer may obviously be expected to take place by hydrogen bonds, formed between the surface hydroxyl groups and adjacent water molecules. However, to make clear the conditions of proton transfer in the hydrogen bonds it is convenient to recall some of the essential properties of the water molecule in relation to the characteristics of the hydrogen bond.

Pauling,[567] in connection with a study of the residual entropy of ice at absolute zero, suggested a structural model for the ice lattice in which the proton of each hydrogen bond was assumed to oscillate between two equilibrium positions situated on the line joining the nuclei of adjacent oxygens. The shift of positions was assumed to take place in a manner to preserve the individuality of the water molecule in the lattice, each oxygen being always associated with two hydrogens. Obviously, this assumption requires the simultaneous shift of positions of a number of protons within a closed group of molecules in the ideal ice lattice. The situation is scarcely altered when, as in the case of liquid water, lattice defects are introduced in the form of vacant lattice points and interstitial molecules.[251,254] While the proton shifts in an ideal ice lattice will have to occur in a cyclic chain of molecules, the appearance of defects in the lattice will, on the other hand, make possible charge transfer by way of proton shifts in acyclic groups of molecules, beginning and ending in lattice defects at vacant lattice points.

The consequences of the preceding considerations are rather important for the problem at hand. We realize that the transport of charge in a water lattice takes place prefentially via the intermolecular hydrogen bonds and that the charge transfer between an electrostatically neutral hydrogen clay and the adsorbed water "lattice" will essentially depend on the hydrogen bonding between the two phases. It is also obvious that the state of order at the interface and in the water lattice will have a considerable influence on the probability of proton transfer.

Having discussed the mechanism of charge transfer, we may revert to the coupling between the water lattice and a transcoordinated silica layer, assuming this to occur in the manner shown in Figs. 13 and 14. The former represents a projection of the protruding hydroxyl groups on the (001) plane, together with the boundaries of the clay unit cell in the same plane, while in Fig. 14 the basal layer of the ice lattice has been superimposed on the projection of the hydroxyls.

It is seen that the fit between the two lattices is fairly good, although not perfect. Actually, the basal dimensions of the montmorillonites appear to vary with the chemical composition in both the dioctahedral and the triocta-hedral series. The lengths of the basal axes represented in Figs. 13 and 14 have been derived from densitometer records of powder diagrams of the <0.6 μm fraction of hydrogen Wyoming bentonite and are in excellent agreement with the values quoted by MacEwan.[464] They may serve, in the present discussion, to represent the dioctahedral montmorillonites. The cell dimensions of ice as measured by Megaw[495] and Owston and Lonsdale[554] lead to the basal dimensions of 9.04×15.32 Å^2 for the water lattice cell shown in Fig. 14 covering three clay unit cells with the total area of 8.94×15.48 Å^2.

The difference in basal cell dimensions thus amounts to about 0.9% in both axis directions, which is comparable to the probable error in the clay cell dimensions. On the other hand, the appearance of defects in the

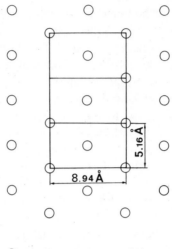

O Hydroxyl group of the
silica layer

Fig. 13. The protruding hydroxyl groups of a trans-coordinated silica layer in montmorillonite projected on the (001) plane. The basal clay unit cell dimensions are indicated.[251]

adsorbed water lattice will tend to smooth out the difference, making the already very small angular deviations of the bonding hybrids quite negligible with respect to the energy of the hydrogen bonds.

As may be inferred from Figs. 13 and 14, only four hydrogen bonds can be formed between three clay unit cells and the surrounding water lattice. The a priori probability of finding the proton of a bond in either position of equilibrium will be 50%, and half of the protons may accordingly be associated with the water lattice, while the other half may be expected to remain associated with the clay crystal. It follows that two units of charge will be transferred to the water lattice from three clay unit cells, implying that the charge defect of the clay crystal will attain a mean value of two-thirds of a unit per unit clay cell, which is the average observed exchange capacity of montmorillonite.

As mentioned above, it should be noted that the actual transfer probability of a proton will depend on the state of order in the adsorbed water lattice. No charge transfer will be possible in the case of the perfectly well-ordered system, since the proton shifts will have to be cyclic, prohibiting the participation of a proton not originally present in the chain. Consequently, an increase in the degree of order in the intercrystalline water layers may be expected to be accompanied by a reduction in the proton transfer from clay to water.

4.2.3. *Two Types of Intercrystalline Water Uptake and Swelling Mechanism*

Méring,[498] in a study of the hydration stages of montmorillonite, observed that the adsorption taking place in contact with liquid water gave rise to a gradual blurring and final disappearance of the (00*l*) reflections, while the apparent spacing remained fixed at about 20 Å. Similar effects were observed[57] in an nmr study of interlayer water in hydrate layer silicates. This led to suggestions that the swelling behavior of layer minerals might be divided into two regions, depending on the water content. At low water contents region I corresponds to a limited number of individual, intercrystalline water sheets, resulting in a number of distinct phases or crystalline hydrates. At the higher water contents of region II the quantity of interlayer water becomes continuously variable. The interlayer changeable cations are believed to form a diffuse double layer, subjected to pure os-

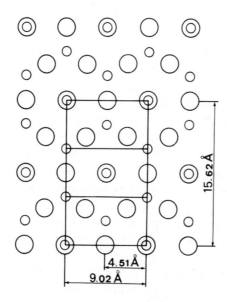

O Hydroxyl group of the silica layer.
◯ Water molecule belonging to a basal
ice lattice layer.

Fig. 14. The basal molecules of an ice lattice superimposed on the hydroxyl groups shown in Fig. 13.[251]

motic swelling. A transition region is assumed to exist in which both kinds of water uptake and swelling take place simultaneously. The term crystalline swelling is used for region I water uptake without any specification of the detailed molecular mechanism responsible for a crystallization phenomenon. Obviously, there are at least two possibilities: *either* the clay/water interface exerts an ordering influence on the water lattice, reducing the thermal amplitudes of the intermolecular vibrations and consequently the density of the water layers in the immediate neighborhood of the clay surface, *or* the same attenuation of the molecular motions is produced by the strong polarizing fields of the intercrystalline exchangeable cations, which, under certain conditions, can be expected to form stable ion–water complexes.

We shall, as a matter of fact, find it necessary to make a sharp distinction between hydrophilic and hydrophobic clays.

4.2.4. *The Mica–Montmorillonoid Transformation*

Accepting the terminology proposed by Brown,[85] all forms of clay mica, with or without additional interstratified layers, should be referred to as hydrous micas. According to this nomenclature, the hydrous micas represent all the intermediary stages in the transformation of micas to montmorillonoids, the transformation being due to a reduction of crystallite sizes below 2 μm, associated with diminishing alkali content and increasing water content. According to this description, no essential structural changes are assumed to occur in the crystallites during the transformation. The dioctahedral micas, represented by muscovite, are then transformed to montmorillonite having the pyrophyllite structure, while the trioctahedral micas, represented by biotite, transform either to vermiculite or to hectorite, both belonging to the talc type of structure. In the case of vermiculite, however, the intermediary hydrous mica has in many cases escaped the extensive size degradation characterizing the precursors of montmorillonite and hectorite. It is accordingly possible to find aggregates of single crystals of vermiculite sufficiently large to permit a complete structure determination by X-ray techniques.[310,477] Results indicate a total absence of surface hydroxyls in the vermiculite crystallites, which accordingly possess external surfaces similar to those of the hydrophobic, nonswelling talc structure. In conformity with this behavior, true vermiculite is, indeed, a nonswelling material that shows all the characteristic features of a hydrophobic mineral in spite of low alkali and high water content in the intercrystalline spaces of the clay crystal aggregates. This apparent paradox is due to the particular way in which magnesium counterions are incorporated in the system to compensate for the original alkali metal ion content and lattice charge defects.

Since, however, the surface oxygen structure of vermiculite just described has been taken as an argument against the Edelman–Favejee structure model of montmorillonite, we shall consider the vermiculite case in some detail.

The Vermiculite Hydration. According to Grudemo[310] and Mathieson and Walker,[477] the intercrystalline magnesium counterions form ion–water complexes of octahedral water coordination. As may be inferred from the observed internuclear distances, the direct interaction between the overall dipolar moments of the water molecules and the ionic field is sufficiently strong to orient the water molecules and to prevent their mutual interaction by hydrogen bonding. In fact, the orientation of the water molecules appears to be unfavorable even for the interaction with intercomplex water molecules.

Release of the complexing water may, however, be achieved by rapid heating of the mineral to about 300°, when exfoliation takes place due to the vapor pressure of the suddenly released but occluded water. Heating to temperatures slightly above 100° does not lead to exfoliation, engaging only intercomplex water, which is easily lost.

It is nevertheless possible[297] to produce an abundant water uptake in the mineral by breaking down the natural magnesium–water complexes. This can be achieved by the introduction of lithium ions to redistribute the charge of the counterions, previously concentrated at the magnesium–water complexes. The small and mobile lithium ions[57] give rise to a depolarization of the complexing water molecules, increasing the probability of mutual hydrogen bonding that is stabilized as the number of lithium ions increases. Extensive replacement of the Mg^{2+} by Li^+ thus leads to a system capable of swelling due to osmotic effects but which does not depend on water adsorption on the substrate. We shall return to these questions.

4.3. The Clay Monocrystallite Regarded as Substrate for Intercrystalline Water Adsorption

4.3.1. *Aspects of the Surface Structure Controversy*

(a) *The Clay Crystal Lattice Stability at Elevated Temperatures.* Differential thermogravimetric analysis of montmorillonite[333] shows a distinct loss peak in the temperature region of 30–100°, which is usually attributed to desorption of intercrystalline water (Fig. 15). In the case of the Edelman–Favejee structure it could also involve surface hydroxyls. Differential thermal analysis, on the other hand, indicates a progressing structural reorganization extending to a temperature region of 150–200°. The extent of the lattice reorganization depends on the type of exchangeable cation present, and leads

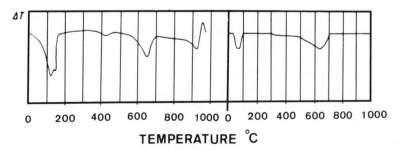

TEMPERATURE °C

Fig. 15. Differential thermal (left) and differential thermogravimetric (right) curves for NH$_4$-saturated normal montmorillonite (original material, central layer not dehydroxylated).[333]

eventually, to irreversible dehydration, when the temperature exceeds 200°.[464] Potassium-saturated montmorillonite may, however, show irreversibility after repeated wetting and drying at temperatures considerably below 200°. Bradley[77] observed that montmorillonite treated in this way may still be able to form complexes with polyhydric alcohols, such as ethylene glycol. Since the dehydroxylation of the central lattice occurs at temperatures around 800°,[384,385,465] and subsequent rehydroxylation is optimal around 400° but does not restore the original clay properties[333] with respect to the water uptake, we shall not be concerned with the clay properties above 400°.

The lattice reorganization indicated by differential thermal analysis is also revealed by an nmr technique,[376] observing proton wide line resonances as function of the temperature. Since the bandwidth may be taken as a measure of the degree of interaction of the resonating protons, mutually and with the clay lattice, the bandwidth plotted as a function of temperature will reveal structure changes appearing as changes in the proton environment. The purpose of the investigation was to look for temperature-dependent structural differences between hydrophilic and hydrophobic, swelling and nonswelling clay minerals belonging to the dioctahedral and trioctahedral series. The results are presented graphically in Fig. 16. The bandwidth is expressed in gauss, indicating the change in magnetic field needed to sweep over the resonance half-width at half maximum amplitude. Low bandwidth implies low interproton coupling or weak coupling between protons and the lattice or vice versa in both cases.

It is to be noted that talc and hectorite, belonging to the trioctahedral series, give rise to narrow bands in agreement with the orientation of the central lattice hydroxyls in a direction perpendicular to the basal planes of the clay crystallites. This orientation, which corresponds to the case of no hydrogen bonding between the hydroxyl groups, is characteristic of a weak interproton coupling, both mutually and to the lattice, of the trioctahedral series. Talc, which is hydrophobic and below 100° contains no water, shows a constant bandwidth over the whole observed temperature range. The structure is perfectly stable and undergoes no transformations influencing the proton resonances of the central layer hydroxyls. The same type of behavior is expected and also observed for pyrophyllite, which belongs to the dioctahedral series. In this case, however, the central layer hydroxyls are mutually coupled by hydrogen bonding, making them interdependent and increasing the probability of several proton relaxation mechanisms. (See, e.g., Abragam.[3]) The bandwidth is accordingly observed to be about doubled as compared to that of talc. The hydrophobic pyrophyllite undergoes no transformation affecting the central layer hydroxyls in the observed temperature range.

In contradistinction to talc, sodium hectorite absorbs water below to 100°, giving rise to a superposition of the adsorbed water proton bands on the central layer proton bands between 0 and 100°. We shall return to a detailed analysis of the band shape of the water proton resonances over an

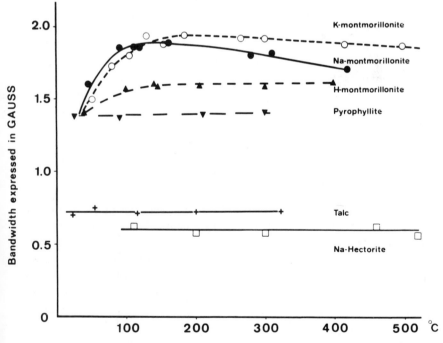

Fig. 16. Proton magnetic resonance bandwidths of dioctahedral and trioctahedral clay minerals as function of the temperature.[376]

extended temperature range. Here we shall only note that the adsorbed water molecules, fixed in secondary or higher adsorption sites further removed
from the primary sites of the highest bonding energy, will be subject to less
efficient relaxation mechanisms than those in the primary sites and will
show corresponding decreases of the bandwidths.

As the temperature of sodium hectorite is raised, an increasing number
of the adsorbed water molecules that contribute to the resonance signal will
come closer to the primary adsorption sites and give rise to a correspondingly broader band. At around 100°, when practically all the adsorbed water
has been removed, the bandwidth is determined by the intracrystalline
hydroxyl protons of the central layer. It then remains constant over the observed temperature range as for talc. Since the hectorite lattice contains a
numer of isomorphic substitutions of Li^+ for Mg^{2+}, the bandwidth is expected to be slightly reduced, as observed.

If we now turn to sodium montmorillonite, we find a different behavior
of the central layer proton band with rising temperature. We recall that the
thermogravimetric measurements show the intercrystalline water to be
completely removed at about 100°, eventually including surface hydroxyls,
while the differential thermal analysis indicates a lattice reorganization extending into the temperature range of 150–200°. The observed change in
bandwidth for the sodium montmorillonite is in good agreement with these
findings. At around 100° all the montmorillonite samples containing different exchangeable cations show bandwidths larger than that observed for
pyrophyllite. Above 150° the general development of the changes in bandwidths clearly depends on the type of exchangeable ion present.

(b) *The Hypothesis of Silica Layer Conversion.* The dehydration of montmorillonite by heating below 200° is, as a rule, reversible. A thermal disor-
ganization of the central lattice does not seem to occur below 300° (cf.
Heller *et al.*[333]), indicating that the observed endothermal lattice reorganization around 100° must be associated with surface phenomena. In fact,
rehydroxylation after dehydroxylation of the central lattice does not restore,
even in part, the surface properties of the clay. In line with these observations the Edelman–Favejee structural model provides an interesting explanation of the bandwidth changes of montmorillonite. If we ascribe the
surface layer reorganization around 100° to a reequilibration of the SiO_2
lattice, disturbed by surface dehydroxylation, we may then understand the
observed difference between acid and basic montmorillonites.

The trans–cis conversion of the silica layers will affect the central layer
hydroxyls such as to shorten the proton relaxation times as a consequence

of lattice disturbance and ionic field effects. In this process the exchangeable ions will cooperate with the converted silicons, leading to similar results for sodium and potassium in the initial stages. We note, however, that the rise of the potassium curve is less steep, in agreement with the fact that potassium is less mobile in the cavities of the silica lattice. The bandwidth increase of the hydrogen montmorillonite would, on the other hand, only correspond to the effect of the silicon atom conversion if we were dealing with a pure H clay. However, the preparation of the acidic form of montmorillonite always entails the introduction of small amounts of intercrystalline aluminum ions that vary, depending on the method of preparation employed, and which contribute to the bandwidth increase.

The ordinate of the bandwidth curve of the sodium clay exhibits a decay with rising temperature. We ascribe this behavior to sodium diffusion into the octahedral layer.[376]

The potassium montmorillonite represents a rather interesting case due to the close fit of the ion into the holes of the silica layers. As mentioned earlier, the dehydration of potassium montmorillonite tends to become irreversible on repeated cycles of drying and wetting, even at temperatures considerably below 200°. It seems probable that the ions sink deeper into the silica layer as the lattice expands at rising temperature and tends to set up tensions and lattice distortions that prevent the surface rehydroxylation. In agreement with this ion fixation, the change in bandwidth for the potassium clay exceeds the one observed for the sodium montmorillonite.

We can now turn our attention to the question of the partial surface dehydroxylation mechanism and its consequences for the stability of the trans conformation of the SiO_2 lattice.

In Section 2 we saw that the fit between the *trans*-SiO_2 lattice and gibbsite lattice is very good indeed. The well-known difficulty, however, of producing montmorillonite crystallites of naturally occurring dimensions by way of hydrothermal synthesis points to a difference in the thermal expansion coefficients of the two component lattices of the mineral. The difference is in fact expected, considering the different types of interatomic bonding which occur in the lattices. Since the thermal expansion depends on the anharmonicity of the interatomic vibrations, it is to be expected that the dilatation of the component layers will depend on the types of exchangeable cations present and on the types of lattice defects and their distribution over octahedral and tetrahedral sheets. One way of compensating for an excess of central layer expansion would be to introduce a partial conversion of the SiO_2 lattice from trans to cis form. We actually suggest that this process occurs during the surface dehydroxylation stage between 30 and 100°, implying that some of the mass losses are due to surface hydroxyls, which, as a rule, may be partly or completely restored on rehydroxylation. The rate and degree of the rehydroxylation would depend on the environmental conditions and time.

Surface dehydroxylation would, moreover, be expected to depend on the size of the crystallites, which more or less determines the internal thermal stresses, and on initially occurring lattice defects. The conversion would also be essentially random, making a detailed structure analysis by diffraction techniques extremely difficult. A very interesting attempt in this direction was, however, made by Pezerat and Méring.[577,578] Their first investigation concerned a Wyoming montmorillonite dehydrated over P_2O_5 in vacuum at room temperature, while the second was performed on the same material sodium-saturated and rehydrated at 28% relative humidity, correspondng to a monomolecular intercrystalline layer of water at a basal spacing of 14.2 Å (according to current beliefs). Méring and Pezerat, however, show that the layer is split in two parts of slightly overlapping electron densities, indicating the tetrahedral coordination, characteristic of water. Figure 17, taken from their second paper,[578] shows the one-dimensional Fourier synthesis in a direction normal to the planes of the crystallites. The electron densities have been calculated from 18 (00l) reflections for the hydrated and from 14 reflections for the dehydrated sample. The coordinates, shown along the normal to the basal plane, have been estimated from an enlargement of the original diagram and added to Fig. 17, in which the peaks have been designated by roman numerals and estimated electron distributions in order to facilitate the discussion. The calculated number of electrons belonging to the various atomic layers according to the Edelman–Favejee structure of the crystallites is shown in Table I and in the fourth row beneath the diagram.

It should be noted that the electron sums of the peaks I–V in the two bottom rows of Fig. 17 are equal. The formation of the crystal lattice only implies a redistribution of the atomic charges.

Pezerat and Méring base their interpretation of the diagram on the Hofmann structure, which, at a first glance, seems to be confirmed by the observed electron distribution. They ascribe peak V to the exchangeable sodium ions and observe that peak VI contains 65 ± 2 units, which is in excess of the expected number of 40–50 electrons. To explain the last observation, the authors tentatively suggest that the electron excess is due to the presence of $Al(OH_3)(H_2O)_3$ complexes in the intercrystalline spaces. We shall return to this hypothesis in Section 4.4.3(b) in connection with a discussion of the dielectric properties of hydrogen montmorillonite.

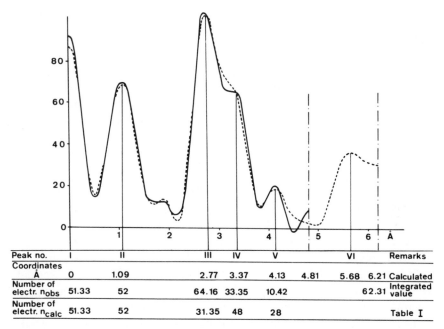

Peak no.	I	II		III	IV	V		VI		Remarks
Coordinates Å	0	1.09		2.77	3.37	4.13	4.81	5.68	6.21	Calculated
Number of electr. n_{obs}	51.33	52		64.16	33.35	10.42		62.31		Integrated value
Number of electr. n_{calc}	51.33	52		31.35	48	28				Table I

Fig. 17. Linear Fourier synthesis of a sodium montmorillonite according to Pezerat and Méring.[578]

The data of Pezerat and Méring[578] raise a number of interesting questions. The first concerns the magnitude of peak V, which corresponds to one sodium ion. The unit cell would then contain two exchangeable sodium ions, which is far in excess of the mean exchange capacity of about two-thirds. In order to satisfy the hypothesis of Méring and Pezerat, the ions

TABLE I. Calculated Electron Density with Symmetrically Distributed Exchangeable Cations in the Third Peak

Peak number	Ions involved	Calculated n
I	3.33 Al + 0.67 Mg	51.33
II	4 OH + 2 O	52
III	2 Si + 0.335 Na	31.35
IV	6 O	48
V	2 Si	28
VI	4 OH	36

would have to be bonded to the lattice in the position of peak V in a manner to render two-thirds of their number inaccessible for exchange, which seems hard to accept.

The second problem concerns the electron distribution between the three peaks III, IV, and V. In Table I the corresponding distribution has been calculated for the Edelman–Favejee structure disturbed by the introduction of lattice defects in the central layer. The resulting charge deficiency is supposed to be compensated by exchangeable sodium ions concentrated in or near the layer corresponding to peak III. The reason for this assumption is as follows.

Accepting for a moment the Hofmann model, which places all the Si atoms in peak III and leaves peak V empty to receive the 3.35 electrons deriving from the sodium ions corresponding to the normal exchange capacity, there is, however, still a charge deficiency of about eight electrons in peak III which might be filled by charge delocalization from the oxygens corresponding to peak IV. The remainder of the excess electrons to be removed from this atomic layer, 6.64 units, can be added to peak V, thus raising its charge content to the ten units observed. This charge redistribution more or less restores the formal balance. It implies that the four Si atoms per half unit cell should each receive only two additional electrons in the tetrahedral covalent bonds to the external oxygen layer and that the supposedly exchangeable sodium ion, belonging to about three unit cells, should receive 3×6.64 or about 20 electrons, which is obviously inadmissible. The situation does not change very much in the dehydrated state, so that the charge delocalization cannot be ascribed to adsorbed water. The electron distribution given by the Fourier synthesis of Pezerat and Méring clearly invites an attempt at an electron redistribution on the basis of the Edelman–Favejee model with partially dehydroxylated surface layers.

To simplify the discussion, let us assume that the charge delocalization between the central and the silica layer atoms can be neglected to a first approximation. The assumption does not affect the main argument and a correction can easily be applied afterward.

Placing the exchangeable cations at the "bottom" of the holes in the silica layers, as suggested in Table I, corresponds to a closest approach between the counterions and the assumed charge deficiencies.

In order to approximate the observed charge distribution, let us assume that a fraction of the silicon atoms in the external layer are transferred to the internal layer during the conversion processes associated with the dehydroxylation and rehydroxylation. The corresponding charge redistribution between the atomic layers belonging to peaks III, IV, and V in Fig. 17 must then satisfy the equation

$$31.35 + x(28 + 14.65) = 64.16$$

for layers III and IV, assuming that the charge delocalization of the oxygens in bonding silicons is proportional to the number of silicons transferred. We find $x = 0.769$ and $1 - x = 0.231$, which implies that 77% of the protruding hydroxyls have been removed from the surface, leaving 23% to couple with water molecules and the protruding hydroxyls of adjacent layers. Since the total amount of electrons displaced from the oxygen layer IV is 14.65 units, 11.28 go to the inner Si layer and 3.37 to the outer layer.

In the first case, referring to charge transfer to the inner Si layer, each silicon forming three bonds to the oxygen layer receives about 1.06 unit charges per bond, which is satisfactory.

In the second case, i.e., charge transfer to the outer Si layer, there are about 1.39 Si–O bonds per half unit cell to the oxygen layer. Assuming for simplicity the same charge displacement of 1.06 units per bond, the difference, $3.37 - 1.39 \times 1.06 = 1.9$ electrons, can be ascribed to some very weak hydrogen bonding between water molecules intercalated in the hydration layer and oxygens adjoining the protruding hydroxyl groups, as will be presently discussed.

As suggested by Méring and Pezerat, the splitting of the hydration layer water peak into two distinct components points to the characteristic tetrahedral coordination of associated water molecules. In Section 4.3.2 we shall discuss the polymerization of the adsorbed water and its relation to the intercrystalline basal spacings. Here it shall only be noted that the observed basal spacing of 12.4 Å corresponds to the spacing of directly interacting protruding hydroxyls, slightly expanded in the c-axis direction due to intercalated interstitial molecules, needed, as we shall see, for layer stabilization.

Since the hexagonal spacing of the surface hydroxyls is 5.16 Å, the height of the central molecule over the bases (in a regular coordination tetrahedron this is equal to the edge divided by $\sqrt{24}$) will equal 1.05 Å, in satisfactory agreement with the splitting of the central peak VI obtained from the Fourier synthesis.

As may be inferred from Fig. 18, at least two nonbonded, interstitial molecules may be incorporated in the unit cell of the 12.4 Å spacing. We have seen that 23% of the protruding hydroxyls remain intact and that 77% have to be replaced by water molecules to fill the matrix set by the protruding hydroxyls. Together they contribute 39 electrons to the intercrystalline space.

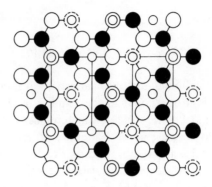

Fig. 18. A double layer of water molecules adsorbed on montmorillonite. Projection on the basal plane of the clay. Small circles, surface hydroxyls; large, filled circles, second water layer; large, open circles, first water layer; broken circles, second water layer, showing hydroxyl positions.

Adding the contribution from the interstitial molecules, i.e., 20 electrons, we arrive at a total of 59 electrons in the intercrystalline layer, which is again in satisfactory agreement with observations.

Thus the Edelman–Favejee structure seems to provide a satisfactory basis for the interpretation of the montmorillonite Fourier synthesis of Pezerat and Méring, which also lends support to the hypothesis of the silica layer conversion due to surface dehydroxylation.

4.3.2. Water "Polymerization" at the Hydrophilic Clay Interface

(a) *Density Effects.* Making use of the Edelman–Favejee structural model as a convenient means of characterizing a partially or wholly hydrophilic clay mineral surface, we have encountered a surprising phenomenon. While in normal cases of specific adsorption, where the attachment to the heavy mass of the substrate leads to an attenuation of the thermal lattice vibration amplitudes of the water molecules and in consequence to a density reduction, we now observe, in the case investigated by Méring and Pezerat, a pronounced density increase of the intercrystalline hydration layer. A fair number of nonassociated, interstitial water molecules seem to be entrapped in the structure of associated molecules. It is, however, not hard to find an explanation for this anomalous behavior.

While the edge length of the normal water coordination tetrahedron in the adsorbed state approaches that of ice Ih at the transition point, or about 4.5 Å. we have seen that the basis of the hydroxyl coordination tetrahedron in the montmorillonite surface attains the value 5.16 Å. There is a corresponding expansion of the intergroup hydrogen bond length from 2.76 to 3.16 Å (or slightly less due to ionic contraction effects to be discussed in Section 4.4.2) which also implies a much weaker bond. Now, the rigidity of the surface hydroxyl groups does not easily promote stress relaxation in the interaction between the crystallites and only the interpolation of the flexible, adsorbed water lattice, stabilized against the imposed local distortions around the protruding hydroxyls, may serve as an efficient medium for stress redistribution and relaxation. In view of the normal function of interstitial water molecules which is to balance the mechanism of defect formation and assure the lattice stability, it is understandable that the heavily distorted lattice needs a maximum of interstitial molecules to maintain the organized state of equilibrium observed.

The situation is radically changed as the water uptake becomes sufficient to build a fairly complete layer of a normal water lattice between the hydrophilic clay crystal surfaces. The structure is easily derived from the ideal, defect-free lattice model shown in Fig. 10. It is schematically represented in Table II. The mean number of water molecules per cell is 5.33 (16 molecules per three cells) or 0.119 g H_2O/g clay, corresponding to a theoretical intercrystalline spacing of 17.81 Å. A projection on the basal plane is shown schematically in Fig. 18 to illustrate the low density of the adsorbed water phase and the intermittent coupling by hydrogen bonding to only two hydroxyls per three clay unit cells.

It is obvious that the present arrangement of water molecules and its attachment to the substrate provides a strong, but very flexible medium joining two clay crystallites. The low sensitivity of the hydrogen bond energy to fairly large angular deviations from the interoxygen direction makes the closely associated central layer of water molecules easily adapted to the protruding hydroxyls at practically full hydrogen bond strength. The disturbing effect of the counterions is, as we have seen, reduced to a minimum in the sodium montmorillonite case considered by Méring and Pezerat, but we shall return to the question of ionic effects in Section 4.3.2(d). For the moment we may disregard the ionic effects and assume complete trans conformation of the silica layers.

The next lower state of hydration is then the one schematically represented in Table II, corresponding to 2.66 H_2O per clay unit cell or 0.059 g H_2O/g clay and a theoretical spacing of 15.05 Å. It is obviously less stable than the above-mentioned 17.81 Å conformation as a consequence of the weaker bonding to the hydroxyls. Conditioning of the clay in humid air far above the "equilibrium" vapor pressure of the 17.81 Å spacing is therefore expected to leave this latter adsorption level stationary until the less

TABLE II

Schematic interlayer structure	H_2O molecules per unit cell	Basal spacing, Å	g H_2O/g clay	mM H_2O/g clay	Remarks
0	0	12.30	0; 0,084[a]	0; 4.667[a]	Unstable; no hydration; four OH groups per unit cell
1	2.66	15.05	0.059	3.278	Unstable
2	5.33	17.81	0.119	6.661	Stable monolayer
3	8.0	18.73	0.179	9.944	Unstable
4	10.67	21.49	0.238	13.222	Stable; two layers
5	13.32	22.41	0.297	16.5	Unstable
6	16.0	25.17	0.357	19.833	Stable; three layers

[a] At complete dehydroxylation.

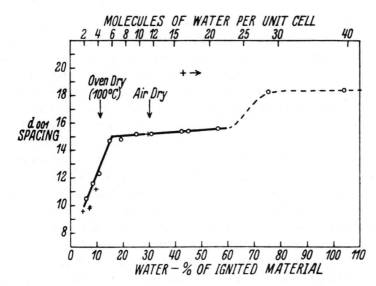

Fig. 19. Montmorillonite d_{001} spacing as a function of observed water content (see text).[514]

stable level has been significantly reduced in number within the specimen. This effect is beautifully illustrated by the measurements of Nagelschmidt[514] reproduced in Fig. 19. When the total water content of the sample reaches a certain level the observed spacing sharply reaches an almost constant value, in spite of the rising water content, which obviously cannot be simply related to the basal spacing. The transition from steep to very slow change of the spacing with the water uptake takes place at the 15 Å level. After this point the stable 17.8 Å spacing is not exceeded, only the reduction of the remaining 15 Å spacings progresses linearly until the populations of the two levels attain a ratio of 3 : 1 at the end of the full trace.

Further increase in the sample water content leads to a new plateau at the 18.73 Å level, which also represents an unstable state.

Méring[498] made similar observations, showing that the apparent spacing remains fixed at 19.5–20 Å while the (00l) reflections gradually vanish, when the specimen is in contact with liquid water. To conclude this discussion of density effects, which will be briefly reopened in Section 4.3.3(c), it is desirable again to emphasize the importance of the substrate influence on the defect formation in controlling the density of the adsorbed water layers. The next section will therefore be devoted to the defect formation mechanism discussed in relation to two different measurements of the density of the intercrystalline water sheets.

(b) *Frenkel and Schottky Defect Densities.* The two components of a Frenkel defect in water, the vacant lattice site or "hole" and the interstitial, nonbonded molecule, may separate by a diffusion process. The hole may eventually reach a phase boundary to be annihilated and produce a corresponding volume reduction. The interstitial molecule is then left behind as a lone nonbonded molecule and is usually described as a Schottky defect. While the intact Frenkel defect does not, to a first approximation, change the average density of the system, the separation and annihilation processes significantly increase the density. The thermodynamic balance set up among the lattice molecules, the Frenkel defects, and the Schottky defects may obviously be shifted in either direction to increase or diminish the average density of the system under the influence of changes in the internal and external boundary conditions of the liquid. In the clay–water system the external boundary conditions, at water contents of interest to this discussion, are essentially determined by the intraaggregate crystal–water interfaces. The internal boundaries are mostly determined by dissolved gases, electrolytes—simple or complex—and by ambiphilic, neutral or charged, organic molecules. We shall leave the discussion of the internal boundary conditions to a later section and concentrate on the external boundaries.

As may be inferred from Fig. 12, the total number of interstitial molecules in pure bulk water at 20° amounts to some 20%. About 45% of these are Schottky defects and the remaining 55% are Frenkel defects. From the same Fig. 12 it is seen that the temperature range of 0–40° is characterized by a steady increase of three-coordinated water molecules. Since the generation of a Frenkel defect depends on the presence of three-coordinated molecules, which appear as a precursor stage in the thermal lattice degradation, [254,695] it is to be expected that the three-coordinated state, henceforth referred to as the fraction f_3 in Fig. 12, will play a prominent role also in the clay–water system. To see this clearly we shall consider the density properties of bulk water in some detail.

When the fraction f_3 is low the a priori probability of a water molecule leaving the lattice and entering an interstitial position is about 0.5, so that the number of f_3 states neeced to balance the interstitial molecules is about twice the number of the latter. Under these conditions, which actually prevail at 0°, the change in lattice energy per degree due to the formation of a Frenkel or a Schottky defect determines the choice of event. Up to the point of maximum density at 4° the lattice energy change favors the Schottky defect formation. However, as the temperature continues to rise the number of f_3 states rapidly increases due to anharmonicity effects, which counteract the vibrational attenuation produced by the interstitial molecules.

TABLE III. Fraction of Frenkel m_1 and Schottky m_2 Defects in Water Calculated at 0, 4, and 20°C from Refs. 254 and 695

	0°	4°	20°
m_1	0.0701	0.0769	0.1033
m_2	0.0915	0.0917	0.0897
$m_1 + m_2$	0.1616	0.1686	0.1930

It is now easy to see how these conditions, ruling the behavior of water in bulk, are modified by the presence of the clay–water interface.

Part of the vibrational attenuation exerted by the interstitial molecules is obviously taken over by the surrounding crystal masses, to which the water lattice is coupled via hydrogen bonding at about two-thirds of the surface hydroxyls. A corresponding number of interstitial molecules may then return to lattice positions via Frenkel defects, shifting the equilibrium distribution in a direction to reduce the fraction of Schottky defects, as is even observed for water in bulk (cf. Table III) in the temperature region of augmenting crystallinity between $+4°$ and $0°$.

The defect reduction amounts to a density decrease that has been measured[12] in a sodium montmorillonite (Wyoming bentonite) ad 25°. Two

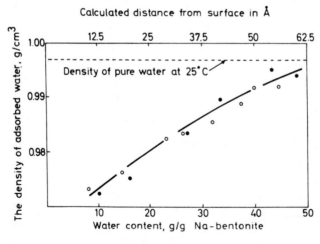

Fig. 20. Density of the adsorbed water as a function of the water content for a sodium bentonite at 25° (after Anderson and Low[12]). Filled circles, first series of measurements. Open circles, second series of measurements.

series of observations are reproduced in Fig. 20. The observed density
reduction occurs over a distance of about 70 Å from the clay surface. At
about 10 Å from the surface it amounts to 2.5% of the bulk water density
at 25°. It should be noted that this reduction is only a fraction of the ~8%
reduction that occurs on freezing water in bulk. It is desirable to emphasize
this difference, since the concept of water turned to ice at the clay surface
has now and then appeared in the literature. The freezing of water at a
hydrophilic clay interface is an entirely different process.[11] Figure 21 shows
that the freezing of a clay–water system leads to a rapid expulsion of complete, adsorbed layers of intercrystalline water, associated with a distinct,
uniaxial shrinkage of the clay crystal aggregates. A residual nonfreezing
layer of water has been observed,[10,11] the amount of which actually corresponds to two monolayers (cf. Section 4.4.3a).

It is well known that the plastic deformation of ice takes place in glide
planes parallel to the crystalline basal planes of the ice lattice.[66] It was
pointed out by Bjerrum[66] that the hydrogen bonding parallel to the c
axis of ice Ih should be expected to be shorter than in a similar lattice

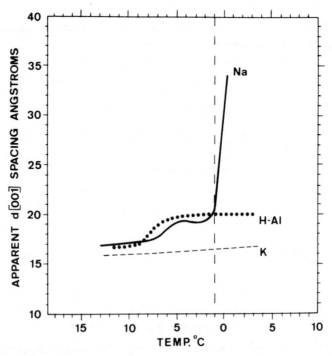

Fig. 21. Change in d_{001} spacing for Na-, K-, and H(Al)-saturated montmorillonite
during the heating period of a freeze–thaw cycle (after Anderson and Hoekstra[11]).

possessing a center of symmetry. The bonds normal to the basal planes should accordingly be slightly bent and weaker. The conditions for dislocation propagation along the clay–water interface are therefore extremely favorable, even at low temperatures, as demonstrated experimentally by Anderson and Hoekstra.[10] This is also in agreement with the slight misfit between the clay and ice surface cell dimensions.

Some doubts as to the correctness of the density data[12] have been reported[360] in an investigation of interlayer water in vermiculite. The argument is based on pyknometric density determinations[149] in which the apparent density was compared with the calculated crystallographic density of the mineral. The observed small difference was considered to indicate "an estimated 2.5% greater packing density of the water molecules in the interlayer space, *assuming the normal density of water beyond two adsorbed water layers in bulk*" (our italics). The assumption in question was supported by a separate determination of the apparent density of the mineral containing two preadsorbed layers of water and using *n*-decane as immersion liquid. The experiment gave an apparent density identical with the one observed in water as medium of immersion. Excluding the occurrence of abnormal densities in the *n*-decane surrounding the mineral sample, it was concluded that "the water density would indeed be normal beyond about 5 Å from the surface." The results are at variance with those of Anderson and Low,[12] if the results of the latter authors "were correct, and would also apply to the vermiculite + water system," since in that case the apparent density of sodium vermiculite in water should have been about 20% lower than that in the system prehydrated vermiculite + *n*-decane, exceeding by far the probable experimental error in the pykonometric data.

We believe that both groups of experimentalists have made correct observations but that the montmorillonite and vermiculite cases are not comparable. We shall give the reasons for our contention in Section 4.3.3, but first we shall have to consider some additional questions regarding water polymerization at hydrophilic interfaces.

(c) *Foreign Ion Effects.* The classical approach to the structure problem of aqueous electrolyte solutions as formulated by Bernal and Fowler[60] has only been slightly modified by recent development of experimental techniques, the fundamental concepts of the pioneering work being essentially verified by later investigations.* The development of the theory of water, as

---

* Such studies of aqueous electrolyte solutions form the subject matter of Volume 3 of this treatise.

reported by Frank in this treatise (Volume 1, Chapter 14), has to some extent helped to throw new light on the dissolution process; X-ray techniques[250,433] specifically developed for the study of the liquid state and laser Raman techniques[693-697] have contributed to the knowledge of the structure of water and ionic solutions. For the present purposes we shall essentially make use of information obtained from nuclear magnetic resonance studies of "structure-making" and "structure-breaking" effects on the water lattice.[57,187,217,218,242,340-344, 346,351] A survey of this field of research is given by Deverell.[187] The structure modifications in water induced by the dissolved electrolytes may be observed by way of changes in the average energy of the intermolecular hydrogen bonding as revealed by the shift of the resonance signal. Increasing strength of the hydrogen bond shifts the proton resonance, observed at constant radiofrequency, toward lower polarizing magnetic fields. The shift may be regarded as proportional to the change in energy of the hydrogen bond. Similarly, decreasing bond strengths result in corresponding upfield shifts of the resonance.

It should be noted that the structure-breaking effects induced by dissolved ions do not produce states of water disorder that are identical with those deriving from thermal excitation leading to the same change in average bond energy. The electrolyte-induced disorder can in other words not be described by a characteristic "structural temperature" as was suggested by Bernal and Fowler.[60]

While the balance between the various types of molecular coordination in pure water as a function of the temperature can be reasonably well assessed (cf. Section 4) due to recent Raman investigations,[694] the corresponding relationships for ionic solutions still await their elucidation (see Volume 3, Chapter 6).

We have repeatedly underlined the importance of the structural properties of the intercrystalline water in clays. However, most of the attempts so far made to describe theoretically the interaction between dissolved electrolytes and water have been based, more or less, on the classical concept of an "ideal" liquid, in which the molecules are free to rotate and keep the distance due to intermolecular collisions obeying some sort of statistical correlation function. In particular, most workers have been careful to emphasize the lack of long-range order in the water, more or less as an excuse for some quite inadmissible assumptions, and eager to neglect the fact that a molecular system may possess a very high degree of order without obeying the stringent third Laue condition defining the crystalline order *sensu stricto*. While perfectly willing to admit the reasonableness of a simplified approach in view of the formidable difficulties that meet the theoretician embarking upon this problem, we nevertheless believe that, in the attempts to find a manageable treatment, too little attention has been paid to the real situation.

Some inferences from the behavior of pure water may obviously be drawn, keeping in mind that short-range molecular interactions strongly affect the lattice defect balance of the entire system. Figure 22 exemplifies the observed molal proton resonance shifts due to alkali halides in aqueous solution as functions of the ionic crystallographic radius. Negative shifts correspond to structure-making effects, positive shifts to structure-breaking effects. In this context it should be made clear that the terms "structure-making" and "structure-breaking" are to be understood in the sense that the strength of the average intermolecular hydrogen bond is increased or decreased, respectively. This definition is compatible with the ones proposed by Hertz[346] and Narten and Lindenbaum.[523]

Steric lattice disturbances leading to increased lattice anharmonicity and vibrational amplitude fluctuations are in particular associated with the large anions which expand and distort the cavities of the water lattice. The fluoride ion, however, with a diameter close to that of the water molecule and with a similar electron configuration, represents an exception. We shall see (Section 4.3.3b) that the ion may take the place of a water molecule in tetrahedral

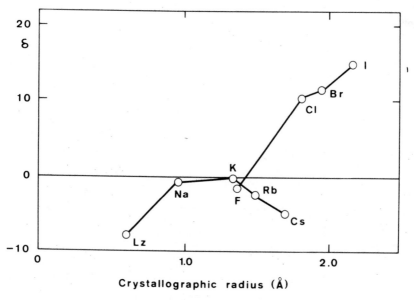

Fig. 22. Molal proton magnetic resonance shifts produced by the ions in aqueous alkali halide solutions, plotted as function of the ionic radii in Å.[57]

coordination, serving both as counterion and hydrogen-bonded link in the cage of a clathrate hydrate. The alkali cations, however, also disturb the water lattice when they move in the interstitial spaces, where the ionic fields contract or expand the lattice, depending on the size of the ion, affecting the local proton configuration in a direction to reduce the proton lattice entropy and the amplitudes of the molecular thermal motions.

It is interesting to apply some of these ideas to the interpretation of the rheological behavior of the aqueous solutions of alkali halides, as was already done by Bernal and Fowler.[60] Figure 23 shows the relative viscosities of the solutions of some alkali metal chlorides as function of concentration. While the addition of NaCl to water gives rise to a monotonic increase of the relative viscosity, KCl displays a viscosity dip below that of pure water at low and intermediate concentrations. The effect is even more pronounced in the cases of RbCl and CsCl.

Since the water lattice disturbance due to the anion is, to a good approximation, the same in all the cases, the observed differences essentially derive

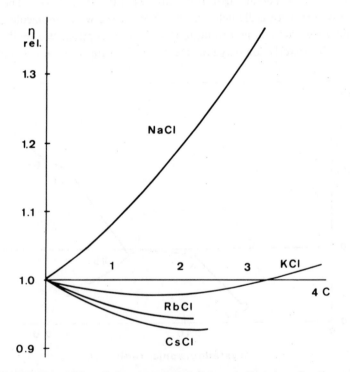

Fig. 23. Relative viscosities of the electrolyte solutions indicated as function of the molar concentration.

from a cation effect on the water lattice. We have seen that Na^+ considerably increases the average strength of the hydrogen bonds in the solution and that K^+ does the same, but to a much lesser degree. However, the different sizes of the ions are expected to influence the intermolecular vibrations of the water and the equilibrium distribution of lattice defects in quite different ways. We may visualize the nature and order of magnitude of the effect by comparing the ionic sizes and the free volume of the interstitial spaces of the ice lattice at $0°$. In this discussion of principles we may disregard the slight expansion of the ice lattice that takes place at the transition point and the rapid increase in the Schottky defect density, which does not invalidate the general arguments. Table IV shows the sizes of the alkali metal ions (column 2) according to Pauling[567] and the residual free volume when the ion occupies an interstitial site in the water lattice (column 3). Negative values imply expansion of the interstitial volume. The ratios of the ionic radii and the minimum radius of the free volume are shown in column 4, giving an idea of the local lattice deformation due to the interstitial ion. The free volume and radius have been derived from the measurements of Megaw[495] and Owston and Lonsdale.[554] The table also includes, in column 5, the ionic mass increase of K, Rb, and Cs relative to the potassium mass, and the corresponding relative viscosity reduction at a constant molarity. We see that the relative viscosity decrease is roughly proportional to the relative change in ionic mass, as expected, since the coupling constant between the two vibrating systems—the water lattice and the dissolved ion—is proportional, to a first approximation, to the attached ionic mass. The coupling constant, on the other hand, determines the nonlocal attenuation of the lattice vibrations that entails a shift in the balance between the lattice and

TABLE IV

Ion	r_i, Å	ΔV, Å^3	r_i/r_{free}	MM_K^{-1}	$\Delta\eta_{\text{rel}}$
Li	0.60	7.28	0.48	—	—
Na	0.95	4.59	0.76	—	—
K	1.33	—1.67	1.06	1	1
Rb	1.48	—5.40	1.18	2.19	2.48
Cs	1.69	—12.04	1.35	3.39	3.22

the interstitial water molecules in a direction to reduce the water density. There are two consequences of this lattice reorganization. First, the general density decrease reduces the viscosity of the system in spite of the added mass of the electrolyte. But this could obviously not happen if the coupling between the dissolved ion, the nearest water neighbors, and the remainder of the water lattice were broken. Second, the reduction of the Schottky defect density reduces the number of broken hydrogen bonds, which facilitates the propagation of dislocations through the lattice and with reduced dislocation blocking the system viscosity decreases. The two cooperative effects—lattice ordering and density decrease—are thus expected to give rise to the observed viscosity reduction of the aqueous solutions of cations possessing a radius larger than the radius of the free interstitial volume in the water lattice. Since a lowering of the temperature increases the lattice order, the effect will increase with decreasing temperature.

From Table IV we conclude that the maximum radial dilatation of the interstitial volume is produced by the cesium ion and attains a value of 35%. The corresponding extension of the nearest hydrogen bonds is, on pure steric grounds, some 0.45 Å, giving rise to a bond length of about 3.22 Å, which is still reasonable, when taken in cooperation with a stabilized lattice.

The viscosity reduction disappears at around 3.3 M KCl, implying that the optimal attenuating efficiency of the cation has been reached around 1.5 M. Above this concentration the Schottky defect density again increases, raising the overall density of the system and accordingly the viscosity. The optimum average distance between the potassium ions is somewhere around 34 Å, between the rubidium ions is around 23 Å, and between the cesium ions is around 21 Å. The molal proton resonance shifts of K, Rb, and Cs are respectively -0.21, -2.28, and -4.63 Hz at 40 MHz resonance frequency. Taking the shift value as a measure of the increase in relative hydrogen bond strength produced by the dissolved cations, we may plot the interionic distance at the viscosity minimum as a function of the induced relative change in hydrogen bond strength. Figure 24 shows that the distance of closest approach between ions at maximum lattice stability decreases with increasing size of the cation and approaches an asymptotic value, given as 18 Å in the graph. This value corresponds to twice the separation of two nonadjacent cavities in the water lattice, implying that no two adjacent cavities may be occupied by ions if the condition of noninterference between primary local lattice distortions is to be satisfied. Obviously, the accommodation of the anions in the system requires the doubling of the noninterference distance for the cations.

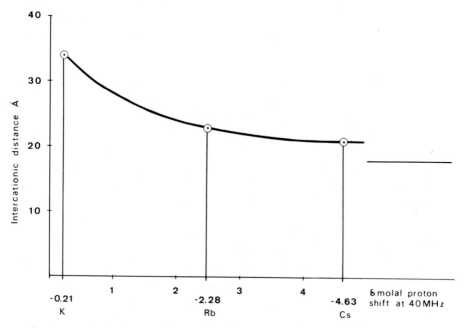

Fig. 24. Noninterference interionic distances in Å as a function of molal proton resonance shifts in aqueous electrolyte solutions (see text).

In order to consider the slightly different behavior of the Li+ and Na+ ions, it is useful to sum up the findings regarding the large alkali ions:

1. The K, Rb, and Cs ions are large enough to replace interstitial water molecules and take over their attenuating effect on the lattice vibrations.

2. The additional lattice attenuation, due to mass increase and steric effects, shifts the balance between the interstitial and lattice water molecules in favor of the latter, increasing bond lifetimes and structural order, reducing the overall density.

3. It is to be noted that the density decreases in spite of the added mass of the electrolyte, which is only possible by way of an extensive long-range structural reorganization.

4. Although the increased strength of the intermolecular hydrogen bonding is principally a consequence of the vibrational attenuation, the entropy-reducing effects of the ionic fields can probably not be disregarded.

5. The attenuation is proportional to the ionic mass, indicating a dynamic coupling of the ionic and water lattice vibrations.

6. The cesium ions almost exhaust the limit of water lattice stabilization due to attenuation effects, the zone of lattice disturbance
approaching the sphere of geometric noninterference.

In the cases of Li^+ and Na^+ ions we see from Table IV that the ionic
sizes do not suffice to attenuate the water lattice vibration by steric effects.
The disturbance due to the anions is rather enhanced by the disturbances
arising from the cation–water interactions, which increase the solution density by augmenting the number of interstitial water molecules. The only
tendency to counteract the density increase derives from the entropy-
reducing effects of the ionic field, which seem to be responsible for part of
the proton resonance shifts and for the lower rate of viscosity increase at
low concentrations. The very small size of the lithium ion, however, leads
to some anomalous effects both in clay–water and biological systems to be
discussed later. The main cause of the proton resonance shifts, indicating a
strengthening of the hydrogen bonding, no doubt derives from attenuation
effects due to the general density increase.

The preceding discussion should, in principle, be applicable to simple
multivalent ions, and to several types of complex ions. In the former case,
however, the intense ionic fields as a rule lead to local destruction of the
lattice order and to the formation of stable ion–water complexes, which
react as independent entities with the water lattice. Some of these ion–water
complexes exist even in the solid state and do not decompose upon solution
of the ionic crystal in water. However, very little is known about the dynamics of the complex–water interaction, since very few investigations have
been carried out at sufficiently low and varied ionic concentrations to be of
immediate relevance to the study of structural details. One of the best
sources of information is to be found in Walrafen's[696] Raman investigation
of the effects of the perchlorate ion on the structure of water, in which he
concludes that (1) the ClO_4^- ion is "structure breaking" (in the sense that
it increases the number of interstitials, i.e., nonbonded water molecules) and
(2) the ClO_4^- ion does not form directed hydrogen bonds with water.

The reason for the second conclusion may, according to Walrafen, be
derived from a rather circumstantial comparison with Raman spectra for
KCl, KBr, and dimethyl sulfoxide solutions in water. The argument may
be summarized as follows. The intermolecular band around 170 cm^{-1} and
the OH oscillator band around 3200 cm^{-1}, both of which we associate with
the four-coordinated species,[254] decrease in intensity on addition of alkali
halides, as expected from the preceding discussion. Since the lattice stabilization induced by the potassium ions diminishes the number of interstitial

molecules—actually observed as an intensity decrease of the 3600 cm^{-1} band—by shifting the equilibrium between the Schottky and Frenkel defects, it follows that the 3400–3500-cm^{-1} bands, ascribed respectively to the three- and two-coordinated species, must increase, as is also observed. Now, the effects produced by the ClO_4^- ions are quite different: the 3200-cm^{-1} band remains constant, the 3400 cm^{-1} band diminishes, and the 3600-cm^{-1} band increases in intensity. There is thus no water lattice stabilization possible in the sense associated with the potassium halides and consequently no attenuation effects due to a coupling between the water lattice and the complex ion, since the sodium ion has no direct attenuating effect on the lattice vibrations. It is significant that the perchlorates of potassium, rubidium, and cesium, which latter ions possess this attenuating ability, also show a poor solubility. The stabilized water lattice cannot accept the misfit of the perchlorate ion. We shall meet cases of similar misfits in the clay–water system, when these ideas are applied in Section 4.4 to the study of the aggregation mechanism.

4.3.3. Water Polymerization at the Hydrophobic Interface

(a) *The Concept of Internal Boundaries in Bulk Water, the Water Cage Formation, and the Thermodynamics of Aqueous Dispersions of Hydrophobic Matter.* In Section 4.3.2(c) we briefly introduced the idea that dissolved or disperse matter may create internal boundaries in an aqueous medium, influencing the structural properties of the latter. In this section we shall consider these boundaries in detail, paying particular attention to the hydrophobic interface.

In a series of beautiful X-ray investigations of clathrate hydrate structures Jeffrey and co-workers[62,63,71,237,487–490,492–494] have clarified the structural organization of water in certain crystalline hydrates, in which water molecules form a cage type lattice[116,568,686] around hydrophobic groups as exemplified in Fig. 25. Klotz[408] suggested that similar water cages may exist around hydrophobic matter in aqueous dispersion. The water "structure" around hydrophobic solutes has been investigated in detail by many techniques—for a full account see Chapter 1. Lindman and co-workers[437,438] have treated the cage formation in aqueous solution as the creation of a water interface defining the boundary between the hydrophobic, disperse molecule and the surrounding water.[252] The structure of this boundary in the liquid aqueous phase was assumed to follow the principles revealed by the X-ray investigations of the solid hydrates. According to these principles, no O–H bond can be directed toward the hydrophobic guest molecule of the cage. This restriction of the permitted molecular orientations

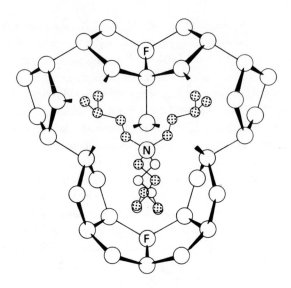

Fig. 25. Tetra *iso* amylammonium fluoride hydrate showing water cage structure (after Feil and Jeffrey[237]).

corresponds to an entropy reduction of the boundary layers, which, together with the general restructuring of these layers, results in a total positive change in the free energy of the system, although the interaction between the hydrophobic phase and the aqueous medium is exothermic, that is, the internal energy of the hydrophobic molecule is lowered in the aqueous medium. The positive change of the free energy, of course, implies that the aqueous phase tends to expel the hydrophobic intruder by forming a defensive boundary of repulsive character separating the two phases. We are, in other words, dealing with a mechanism of phase separation, which may sometimes give the impression that attractive, "apolar" forces exist between the expelled, hydrophobic molecules, and which has led to the unhappy and misleading concept of "hydrophobic bonding" (see Chapter 1). However, the comparative rigidity of the organized water interface, suggested by nmr experiments,* may, as proposed by Klotz,[408] greatly contribute to the conformational stability of an ambiphilic high polymer in aqueous solution.[468] Berendsen[48] has pointed out that the water cages may easily adapt to the normal lattice within a few molecular layers, the bonding flexibility of which provides a smooth transition from the boundary molecules of pentagonal to the normal hexagonal coordination type.

* L. Maijgren, Progress reports, unpublished (1973).

It remains to inquire into the nature of the repulsive fields between the guest molecule and the cage molecules in order to find the connection with the clay–water problems.

The calculated dissociation pressures and heats of formation for some gas hydrates[580] show, in good agreement with experiment, that the xenon hydrate is much more stable than the argon hydrate, indicating that the atom size effects follow the rules expected from a predominance of van der Waals–London forces. Similar conclusions may be drawn from experimental and theoretical studies of a variety of monatomic,[34] diatomic,[7,486,569] and polyatomic[140,491] guest molecules in clathrate hydrates. Davidson in Chapter 3, Volume 2, of this treatise has demonstrated that the predominance of the short-range dispersion forces depends on the mutual cancellation of the dipole fields at the center of the ideal cage. In the real cage small dipolar fields are expected to occur. The quadrupolar and octupolar fields do not cancel and are expected, theoretically, to play an important role in a cage essentially free of dipolar central fields. In practice it turns out that the residual dipolar effects alone are sufficient, as a first approximation, to describe the observations even if the rather oversimplified theoretical model does not permit very detailed conclusions regarding the interactions of the guest molecule with the cage fields. Experimental data, however, indicate that the second moment of the charge distribution of the water molecule in the cage is much less anisotropic than in the gas phase, as may be expected from the nature of the hydrogen bonding and the pseudoisotropic reaction field of the guest molecule. Davidson reports that both the diatomic and the polyatomic guest molecules are displaced from the cage center at low temperatures favoring a van der Waals interaction with the water molecules of the cage. Translational and rotational reorientations, however, occur with high frequency even at low temperatures. The reorientation rate of, for instance, tetrahydrofuran at 90°K is about 1.5×10^{10} Hz and rises with increasing temperature to the order of kT. It appears then that in the cages surrounding guest molecules of the type

$$
\begin{array}{ccc}
H_2C\!-\!\!-\!\!-\!\!-\!\!-CH_2 & & \\
| \qquad\quad | & & H_2C\!-\!\!-\!\!-\!\!-\!\!-CH_2 \\
H_2C \qquad CH_2 & \text{or} & \diagdown \quad \diagup \\
\diagdown \quad \diagup & & O \\
O & &
\end{array}
$$

in which the oxygens are in off-center positions directed toward the cage molecules, there is no other type of bonding than van der Waals interaction

TABLE V

Compound	Melting point, °C	Boiling point, °C	Solubility in H_2O, g/100 ml
Ethylene oxide	-111.3	$+10.7$	∞
Tetrahydrofuran	—	$+64$	∞
Propylene oxide	—	$+35$	65 (30°)
Carbontetrachloride	-22.8	$+76.8$	0.08 (20°)

of weak attractive and strong repulsive fields and especially that, even at low temperatures (90°K), there is no tendency of charge transfer and hydrogen bonding between the cage and the guest molecules, indicating that the polarizing ability of the guest molecule oxygen is insufficient to break up the cage structure. This, of course, is in agreement with the fact that the guest molecule is expelled from the water phase on the formation of an internal interface in the liquid state, from which the crystalline solid is precipitated. Although the guest molecules in question could, at a first sight, be expected to show some hydrophilic properties due to the bridge oxygens, the electronegativity of the oxygen atom in this particular type of bonding seems to be reduced, making the hydrophobic character of the molecule predominate in an aqueous medium. It appears, in fact, that the high solubility of ethylene oxide in water does not depend on hydrogen bonding to the oxygen atom but exclusively on the possibility of water cage formation around the molecule. Table V shows the melting and boiling points of some cage-forming compounds together with their solubilities in water.

The high solubility characteristic of ethylene oxide and tetrahydrofuran is considerably reduced by the symmetry reduction in the propylene oxide which influences the cage stability. The table includes carbon tetrachloride, which is practically insoluble in water, yet is known to form stable cages in the presence of a "help gas" such as H_2S or Cl_2. These latter molecules readily stabilize smaller cages, which, in turn, fit together with and stabilize the larger cages needed for the CCl_4 molecule, which then becomes "soluble."* It is even possible to attain this cooperative cage formation effect

* *Ed. note*: The indications are, however, that H_2S *cannot* "salt in" CCl_4 in the *liquid* aqueous phase. The approach to solubility outlined here will be seen to be a highly individualistic one. Alternative views are discussed in Chapters 1, 3, and 5 of Volume 2 and in Chapter 1 of this volume.

in the case of the weakly soluble iodine molecule used as "help gas." If a solution of iodine is shaken with carbon tetrachloride, the latter molecule readily goes into solution by complex cage formation.[315]

The low tendency of the ether-type oxygen bridge to form hydrogen bonds is by no means restricted to the types of cyclic compounds mentioned above. It is well known that the oxygen bridges joining the glucose rings in cellulose do not participate in the intermolecular hydrogen bonding and in the formation of water bridges between the polysaccharide chains in the so-called amorphous zones of the cellulose lattice. Similarly, in a theoretical study of the hydrogen bond, Fischer and Ehrenberg[241] have investigated the influence on the hydrogen bonding ability of the polarizability of the heteroatom in ethers. They conclude that the availability of the unshared electrons of the heteroatom is of great importance for the water solubility. An involvement of the electrons of the heteroatom through conjugation or displacement, i.e., a decreased polarizability of the electron cloud around the heteroatom, will diminish the attraction of molecules with an exchangeable H atom.

To conclude this survey of the hydrogen bonding properties of the ether-type oxygen bridge, we would like to point out that the surface oxygens of the Hofmann–Endell–Wilm structure form similar bridges between the silicon atoms of certain clay lattices and that these surface oxygens may be expected to display a similar indifference to hydrogen bonding.

4.4. The Aggregation of Clay Monocrystals and Interspersed Water Layers

4.4.1. The Aggregation Mechanism

In Section 3.1.2 we described a general relationship between the concentrations of hydrogen ions and hydrophilic clay crystallites in aqueous dispersion and introduced the phenomenological concept of a critical micelle concentration at which a spontaneous aggregation of the crystallites takes place. In this section we shall attempt a more detailed description of the aggregation process, its generating conditions, and the association mechanism.

In classical colloid chemistry it was assumed that the stability of an aqueous dispersion of minute particles depends exclusively on the electrostatic interaction via the fields surrounding charged particles. The dispersion medium was introduced as a convenient or troublesome, but otherwise uninteresting dielectric carrier or mediator of the electric fields. It is obvious from the preceding discussion that we shall have to desist from this no doubt simplifying view and instead regard the system components—clay, water, and the ions—as an inseparable, structured entity of complex internal interactions.

The existence of a critical clay concentration at which the formation of crystal aggregates begins in the sol may, of course, be understood as an interference phenomenon, due to the increasing contact between the zones of water lattice stabilization around each primary aggregate of clay crystallites. Although there is an unmistakeable change in slope of the concentration curve of Fig. 8 at a clay concentration of about 0.2% by weight, we also note the considerable scattering of the measured proton concentrations observed in the various experiments. A natural clay can, as a rule, only appear as the sol in the form of small polydisperse crystal aggregates that remain stable in most dispersion processes, avoiding destruction or contamination of the natural clay surfaces, and a plethora of interference situations may be expected to occur as the concentration rises in the various experiments.

To simplify the discussion, we neglect, for the moment, ionic effects other than those due to the protolytic hydrogen ions.

We have seen that the extension of the stabilized water zone outside the surface of a hydrophilic clay crystallite can attain an order of magnitude of 100 Å, which is a conservative estimate in the case of a hydrogen montmorillonite. When two such zones approach each other it is to be expected that the intercalated, less stable layers of water will be affected at a certain thickness so that the resulting stabilized zone will exceed the sum of the two primary ones.

There are several consequences of such an event. (1) A repression of protolysis takes place due to the reduction of lattice defects, which serve as sinks in the proton transfer mechanism. (2) As a consequence, the surface charge of the clay decreases, favoring a closer approach between crystallites, which enhances the water lattice stabilization. (3) Since the number of protons that leave the clay per unit time is reduced, the proton concentration in the surrounding liquid will diminish, as revealed by the sudden change in the slope of the concentration curve. (4) Due to the increased ordering of the intercrystalline water, the entropy of the system is reduced and some water is likely to be expelled in the association process. The process in fact represents a phase separation leading to the planar positive tactoids described in Section 3.2.4. (5) Positive tactoid formation entails a reduction of the viscosity of the sol and an increased electrophoretic mobility of the clay phase, in agreement with observation.[510]

The behavior discussed above is characteristic of systems which display enhanced structuring of an aqueous phase serving as the medium of colloidal dispersion. The ideas presented are equally well applicable to the formation of negative tactoids. Indeed, the description given of the association mechanism in clays may serve as a basis for a general theory of hydrogelation.

4.4.2. Effects Giving Rise to Aggregate Dilatation and Concentration

(a) *Electrolyte Effects.* In order to take account of the influence of exchangeable cations other than the protons, already considered, we shall use the montmorillonite–water interaction as a reference system.

Let us suppose that our clay–water system contains sodium as exchangeable ion, and that it is subjected to dialysis by putting an aggregate of crystallites in contact with pure water. Inside the clay aggregate we have a certain amount of adsorbed water molecules and sodium ions that compensate charge defects fixed in the clay crystallites. The water outside the aggregate is initially pure. Let us further suppose that the sodium ions are initially well embedded in the clay lattice, having, as a consequence, no direct steric influence on the water lattice. The enthalpy change due to adsorption of water is negative. So is the change in entropy, and the sign of the change in Gibbs energy depends on the degree of order attained by the transfer of interstitial water molecules to lattice positions. We have seen that "freezing" of the adsorbed water may, in cooperation with the adsorptive fields, lead to positive ΔG and desorption. The water is then expelled from the intercrystalline spaces, which diminish in volume. As the temperature rises, the change in Gibbs energy on adsorption will eventually become negative, measured relative to the Gibbs energy of the outside bulk water and implying that the external chemical potential will exceed the internal one. Water from the outside will in consequence enter the intercrystalline spaces. The ensuing volume increase of the system—the swelling of the clay—accompanying the transport process will proceed until the internal chemical potential has been increased to the external value. At this point the swelling process would come to an end. The mechanism of this equilibration depends, of course, on the reintroduction of water lattice defects as the influence of the adsorptive fields on the new intercrystalline layers of water diminishes. We have, however, seen that the increase of water lattice defects will raise the negative surface charge of the clay crystallites due to facilitated protolysis, and in consequence the probability of sodium migration to the surfaces of the intercrystalline spaces will increase. Once the cations have entered the expanded intercrystalline spaces they will interact with the adsorbed water lattice in the manner already described. In the case of sodium the number of interstitial water molecules will increase, lowering the chemical potential and initiating a swelling of the crystal aggregate. The ions will move along the concentration gradient toward the exterior. This diffusion process will, in turn, disturb the electrostatic balance, which can only be restored by a corresponding dissociation of water molecules according to the scheme

initial state

$$
\begin{array}{c|c}
n\mathrm{Na}^+ & \\
\mathrm{H_2O} & \mathrm{H_2O} \\
ne^- &
\end{array}
$$

final state

$$
\begin{array}{c|c}
(n-x)\mathrm{Na}^+ & x\mathrm{Na}^+ \\
x\mathrm{H}^+ & x\mathrm{OH}^- \\
ne^- & \mathrm{H_2O} \\
\mathrm{H_2O} &
\end{array}
$$

The final state represents a Donnan equilibrium, attained when the internal and external chemical potentials become equal. This may be expressed by equating the internal and external activities of the mobile ions, or

$$
(n-x)x\gamma_i^2 = x^2\gamma_e^2
$$

giving

$$
x = \frac{n}{1 + (\gamma_e/\gamma_i)^2}
$$

the γ's being the respective activity coefficients.

Reversing the process just described by starting from a hydrogen montmorillonitic clay, adding sodium hydroxide, and observing the viscosity of the clay suspension at various stages of neutralization, Mukherjee and Mitra[510] obtained the viscosity and titration curves shown in Fig. 26. Included in this diagram are also the corresponding relationships for a kaolinitic clay suspension. As soon as the titration curve begins to level off there is a sharp rise in the viscosity of the montmorillonitic clay suspension which attains a maximum at the end of the plateau and starts to decrease almost as rapidly as the neutral point is approached and passed. The rise in viscosity corresponds to a gelation process as described in Section 3.2.4 followed by a successive breakup of the structure as the Na^+ concentration increases, associated with contraction and the formation of new aggregates of the sodium clay that will eventually—by a reversal of the diffusion process previously described—end up in the situation where we started.

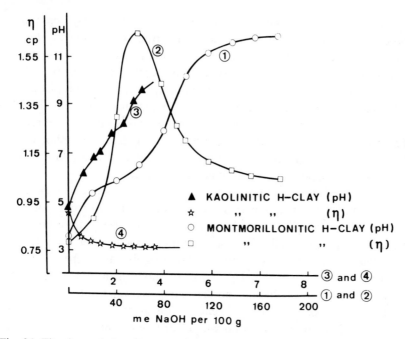

Fig. 26. Titration and viscosity curves for hydrogen montmorillonite and kaolinite clays (after Mukherjee and Mitra[510]).

In the case of the kaolinitic clay no such gelation effects are observed, as expected, since the clay, although partly hydrophilic, is structurally incapable of stabilizing a water lattice.

We have so far paid little attention to the effects of polyvalent ions on the clay–water system. One reason for this is that these ions show, as a rule, pronounced tendencies to form complexes with water that will introduce apparently anomalous features into the clay–water system. It is convenient to compare these anomalies to the structure-making processes already described with a view of calling the reader's attention to some problems in current interpretations of clay research data.

In this discussion of polyvalent ion effects we shall use vermiculite as a reference system, since this mineral has very often been taken as a representative example of the hydrophilic, swelling type of 2 : 1 clays, in spite of the fact that the true vermiculite, fully saturated with magnesium as exchangeable cation, is nonswelling. Barshad[35] characterizes the vermiculite as a magnesium mica, in which the Mg^{2+} ions play the same role played by K^+ ions in micas, showing that an ordinary biotite can be converted into vermiculite by prolonged leaching with $MgCl_2$ solution. On the other hand, a vermiculite, when K saturated, has properties that closely approximate those of ordinary biotite. Now, biotite is a hydrophobic mineral in which the silica layers possess the cis conformation with surface oxygens of the expected weak hydrogen bonding capacity. We know, moreover, from X-ray investigations[310,477] that the cis conformation of the silica layers is retained in the vermiculite and that the oxygen–oxygen distance separating the intercalated water layers from the crystallites is ≥ 2.93 Å. Graham[296] remarks that if it is admitted that the silicate layers are indeed hydrophobic, then it seems possible that they will influence the water structure in a manner similar to hydrophobic ions and molecules in solution.

The boundary conditions of the intercrystalline water in vermiculite, and in some of the intermediary stages of its development from biotite, accordingly seem to differ considerably from those of montmorillonite. In the latter case the stabilization of the water structure derives essentially from the attenuation of the lattice vibrations due to a specific mutual coupling between water and clay. Dissolved monovalent ions interact with the adsorbed water, modifying and deforming the water lattice but without disrupting the structure. There is in effect, little justification for talking about ionic hydration, since there is no way to distinguish between water molecules attached to ions and to the water lattice. They are, indeed, attached to both within the so-called hydration shell, integrated into the complex of clay and water. In fact, all attempts to determine the hydration number of these ions lead, at best, to the normal number of neighbors in the water lattice at the actual, expected state of order. It is a very inconvenient, and sometimes rather dubious, way of assessing the state of lattice order.

The intercrystalline water in vermiculite finds a different type of boundary condition. The structurally disruptive fields of the magnesium ions remove water from the organized lattice of mutually cooperating molecules to form separate, octahedral coordination complexes that may truly be considered as ionic hydration shells. The coupling between the magnesium ion and the surrounding water molecules is essentially dipolar, and the directions of the molecular water dipoles, as determined by the magnesium ion, are unfavorable to hydrogen bonding to a surrounding intercrystalline water lattice. The situation may be compared to the one discussed by Graham[296] citing several investigations[497,669,723] regarding water fixation in natrolite. The orientation of the water molecule in a distorted tetrahedron is such that the protons point toward two surface oxygens while the lone pair electrons are directed toward sodium cations within the cavity. The hydrogen bonds are strained, the configuration being essentially determined by the lone pair interaction with the two sodium ions at distances of 2.40

and 2.38 Å. One hydrogen bond, 2.84 Å, is bent through 11°, and the other, 3.01 Å, through 26° from the coaxial O–H–O directions.

The configuration is similar to the one we have proposed in the discussion of interactions between monovalent cations and the water lattice, which leads to a distortion of the interacting water molecule but does not inhibit its hydrogen bonding capacity to the lattice. In fact, the lattice remains stable, although subjected to the distortions due to the ionic fields.

In the case of the magnesium–water complex no such mutual adaptation and smooth interaction is possible. Compared with the nominal interstitial volume in water (octahedral conformation), the nominal volume of the octahedral ion–water complex is reduced to less than 50% of the former value. The interaction between the complex and a surrounding intercrystalline water phase can only be spurious, because of the limited space between the crystal sheets. Since, on the other hand, the interaction between the intercalated water layers and the surrounding crystal surfaces is expected to be weak and predominantly of dispersion force character, the boundary conditions do not induce significant changes in the normal water structure. The chemical potential stays normal, as expected in a nonswelling system. The introduction, however, of a noncomplexing, defect-producing ion such as Li^+ will initiate an osmotic transport of water into the crystal aggregate which is augmented as the magnesium complexes begin to break up due to outside ionic charge competition for the water shields.

The above observations are in good agreement with the dehydration properties of vermiculite. The number ratio of exchangeable Mg^{2+} ions and intercrystalline water is about 1 : 12,[477] implying that about half of the water molecules are excluded from direct contact with Mg^{2+} ions. They appear as "unbound" water in the dehydration curve.[692] In this connection it is interesting to note that vermiculite saturated with K^+, Rb^+, or Cs^+ ions does not exfoliate upon sudden heating,[35] indicating that the water stabilization effects of these ions leads to efficient expulsion of the lattice-alien Mg– or Ca–water complexes which "explode" at rapidly produced high temperatures, as does the ammonium–water complex. Barshad,[497] however, attributes the inhibition of exfoliation to ion fixation.

We are now in a position to discuss the density measurements of van Olphen et al.[360] previously cited in Section 4.3.2(b). Included in the investigation is an infrared study of the main water band of the two-layer hydrate of the sodium vermiculite used in the density measurement. It shows the characteristic poor resolution of the main water band of a NaCl solution spectrum with, however, a slight enahncement of the high-frequency shoulder normally associated with zero-coordinated water molecules, but probably deriving, at least in part, from the central layer hydroxyls. The band shape is insensitive to changes in the direction of the primary beam, indicating normal water–lattice coordination and average isotropy. The picture is consistent with our previous conclusions regarding the state of intercrystalline water in a magnesium vermiculite, but without an accurate and detailed analysis of the band shape it would be hard to distinguish the sodium vermiculite spectrum from that of a sodium montmorillonite, especially if other parts of the spectrum, including the intermolecular bands, are not included in the analysis.

For the moment the density measurements are more revealing. They confirm, indeed, that the water density beyond a distance of about 5 Å from the vermiculite surface is normal, as expected from a weakly interacting, almost hydrophobic crystal surface. Nuclear magnetic resonance data for a sodium hectorite, which we believe presents the same type of boundary conditions as the sodium vermiculite, give strong support for this interpretation. They will be reported in Section 4.4.3(a).

In Section 4.4.2(a) we proposed a qualitative but detailed description of the swelling mechanism of a sodium montmorillonite starting from the assumption that the Na^+ ions were initially well embedded in the cavities of the silica layers as indicated by the Fourier synthesis of Pezerat and Méring.[578] We predicted a finite swelling of the clay prior to the emergence of the exchangeable cations from the silica layers. The continuation of the swelling process was shown to be dependent on the presence of the mobile cations—not, as is usually presumed, due to the hydration of the cations, but to their influence on the defect balance in the water lattice. The theoretical considerations show, in point of fact, that the swelling mechanism associated with the change in b-axis dimension of sodium montmorillonites observed by Ravina and Low[595] depends on the cation exchange capacity, contrary to the conclusion of these authors. They based their observations on sodium montmorillonite samples derived from six different localities and found that the relation between the free swelling of the clays and the change in b-axis dimension was linear, as shown in Fig. 27, taken from their paper. They also remarked that two of the clays, Otay and Bayard, had almost the same swelling limits but vastly different cation exchange capacities (cf. Table VI) and concluded that free swelling is not related to the cation exchange capacity.

Plotting, however, the free swelling limits, given as g water/g clay, as function of the exchange capacity (Fig. 28), it is clear that the swelling depends on the exchange capacity according to expectation. The erroneous conclusion of Ravina and Low was obviously due to the fact that the two clays

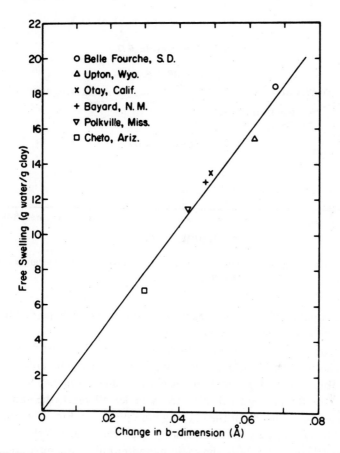

Fig. 27. Relation between free swelling and total change in *b* dimension for six sodium-saturated montmorillonites.[595]

TABLE VI

Clay	Exchange capacity, meq/100 g clay
Belle Fourche, South Dakota	85
Upton, Wyoming	100
Otay, California	52
Bayard, New Mexico	127
Polkville, Mississippi	63
Cheto, Arizona	96

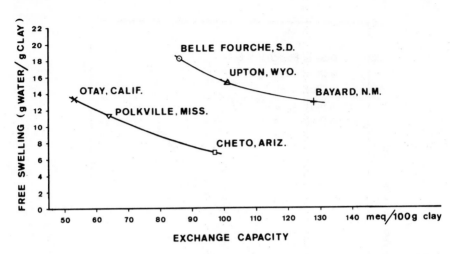

Fig. 28. Free swelling of six montmorillonites as a function of the exchange capacity (after Ravina and Low[595]).

of almost equal swelling limits belong to two different groups of minerals each of which behaves consistently per se.

We propose the following interpretation of Fig. 28: At zero water content the clay samples of group I all have b-axis dimensions around 8.96–8.98 Å, while those of group II all begin at values around 8.94–8.96 Å. According to Ravina and Low,[595] the samples all reach the same final b-axis dimension (Fig. 29) at the swelling limit plotted in Fig. 28 as function of the exchange capacity characteristic of each clay. Table VII shows the mineralogical compositions of the sodium montmorillonite samples used.[141]

It is convenient to begin the discussion with the three last clays (group II) of Table VII, since they show the clearest relation between the mineralogical composition and the physicochemical properties of the clays. We first note that the ratio of the respective charge deficiencies in the tetrahedral and octahedral layers decreases in the order: Belle Fourche 0.72, Upton 0.26, Bayard 0. It is evident that the concentration of counterions in the silica layers dye to the tetrahedral substitutions efficiently contributes to the lattice contraction that relaxes as the layer is depleted of ions in the swelling process. It is satisfactory to note that of the three clays, the Belle Fourche sample displays the most pronounced plateau separating the expulsion of the counterions associated with tetrahedral and octahedral charge deficiencies, respectively, indicating that the counterions are situated at different depths in the holes of the silica layers. But it is equally satisfactory that all three curves of Fig. 29 belonging to group II show a well-defined plateau as expected from the stepwise charging of the surface hydroxyls due to protolytic effects as described in Section 4.4.2(a). In conformity with this interpretation we propose that the initial lattice distortion of the clay in the dry state is produced by the exchangeable ions and determines the initial length of the b axis. The lattice deformation is likely to occur according to various cooperating mechanisms as suggested by Ravina and Low.[595] Relaxation of the contracting fields progresses by ion removal in the swelling process until the lattice is exhausted of exchangeable ions. At this point the swelling stops and the lattice is relieved of the strains caused by the ions. It should be emphasized that the general idea of this mechanism is already to be found in Ravina and Low's paper as one of their proposed alternative interpretations of data, although they discarded the connection between free swelling and cation exchange capacity.

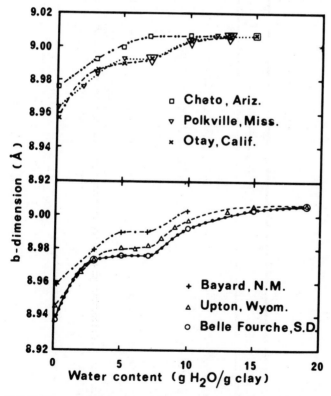

Fig. 29. Relation between b dimension and water content for six sodium montmorillonites (reproducibility of b dimension $= \pm 0.001$ Å at low water content, ± 0.003 Å at high water content) (after Ravina and Low[595]).

TABLE VII

Clay mineral	Mineralogical composition of half unit cell	
Otay, California	$[Al^{3+}_{1.35}F^{3+}_{0.05}Mg^{2+}_{0.60}]^{VI}[Si^{4+}_{3.98}Al^{3+}_{0.02}]^{IV}O_{10}[OH]_2$	$X_{0.62}$
Polkville, Mississippi	$[Al^{3+}_{1.41}Fe^{3+}_{0.13}Fe^{2+}_{0.01}Mg^{2+}_{0.45}]^{VI}[Si^{4+}_{3.93}Al^{3+}_{0.07}]^{IV}O_{10}[OH]_2$	$X_{0.53}$
Cheto, Arizona	$[Al^{3+}_{1.38}Fe^{3+}_{0.09}Mg^{2+}_{0.54}]^{VI}[Si^{4+}_{3.91}Al^{3+}_{0.09}]^{IV}O_{10}[OH]_2$	$X_{0.63}$
Belle Fourche, South Dakota	$[Al^{3+}_{1.60}Fe^{3+}_{0.19}Mg^{2+}_{0.22}]^{VI}[Si^{4+}_{3.84}Al^{3+}_{0.16}]^{IV}O_{10}[OH]_2$	$X_{0.38}$
Upton, Wyoming	$[Al^{3+}_{1.53}Fe^{3+}_{0.19}Fe^{2+}_{0.01}Mg^{2+}_{0.33}]^{VI}[Si^{4+}_{3.91}Al^{3+}_{0.09}]^{IV}[OH]_2$	$X_{0.43}$
Bayard, New Mexico	$[Al^{3+}_{1.46}Fe^{3+}_{0.06}Fe^{2+}_{0.01}Mg^{2+}_{0.47}]^{VI}[Si^{4+}_{4.0}]^{IV}O_{10}[OH]_2$	$X_{0.48}$

In discussing the group II minerals we have slightly oversimplified the lattice distortion problem. It is, indeed, obvious that both the central hydroxyl layer and the silica layers are jointly affected by lattice distortions due to the counterions. This is easily seen if we compare the total charge defect and the charge released as exchangeable cations. The percentages of released charge are: Belle Fourche 83%, Upton 87%, Bayard 98%.

The inhibition of the release of some counterions due to steric and chemical effects obviously runs parallel to the lattice distortions. This effect is much more pronounced in the group I minerals, except that central layer distortions seem to be prevalent in these clays having 31% (Otay), 43% (Polkville), and 57% (Cheto) of the counterions released of those present in the system. It is interesting to note that the ratios of tetrahedral to octahedral charge deficiencies occurring in the above order are 0.03, 0.15, and 0.17, apparently running in a direction to counteract the central lattice distortions.

The data of Ravina and Low[595] thus seem to support the dependence of the free swelling limit on the cation exchange capacity of montmorillonite, in agreement with theoretical expectations and the findings of Calvet and Prost,[98] who conclude from a series of beautifully conceived experiments that limited presence of exchangeable cations also limits the swelling capacity.

(b) *Competitive Adsorption of Neutral Molecules.* Intercrystalline water replacement in clay aggregates may be effected by a large number and types of ambiphilic organic molecules that change the surface properties of the mineral. Reference to some pertinent survey papers were given in the introduction. In this section we shall only be concerned with ethylene glycol[77] and glycerol,[463] which are used for the purpose of mineral identification and artificial expansion of the intercrystalline spaces in clay aggregates to facilitate dispersion and cationic exchange.

For small monovalent exchangeable cations glycerol forms a double layer leading to a basal spacing of 17.7 or 17.8 Å. The corresponding spacings for ehtylene glycol and the same ions on montmorillonite are 16.9–17.1 Å. In both cases the alcohol molecules fit equally well to the surface structure in either the cis or the trans form of the silica lattice. In the former case van der Waals interactions prevail, while in the latter hydrogen bonding determines the interaction with the clay surface. The character of the surface can accordingly not be determined from the observed spacing with conventional X-ray and preparation techniques.

The hydrated Mg vermiculite does not expand beyond about 14.5 Å on glycerol treatment. Expansion to 17–18 Å may or may not occur in clay vermiculites saturated with other divalent ions. For this and other reasons Walker[691] recommended pretreatment with Mg before the identification test with glycerol. The response of vermiculites of different layer charge to glycol treatment varies similarly,[86] which makes glycol less useful for distinguishing vermiculites from montmorillonites.

The rate of penetration of glycerol or glycol into the hydrated intercrystalline spaces depends on the size of the crystal aggregate and on the manner of application. Small aggregates in direct contact with the alcohol saturate in a few seconds at room temperature, while the alcohol uptake from the vapor phase over the liquid heated to a temperature around 100°C for glycerol[86] and around 60°C for glycol may take several hours, and, in quantitative work, is best performed using an electrobalance to assess the water replacement.

As mentioned earlier, complete removal of glycerol from a sodium montmorillonite is equally difficult as the removal of other water replacement agents from cellulose gels or delignified wood in nmr studies of water fixation.

4.4.3. Observations of Phase Transitions of Aggregate Water

(a) *Nuclear Magnetic Resonance.* In 1954 Norrish[534] published a study of the lattice expansion of sodium montmorillonite as a function of the water content. He observed a series of hydration levels above 20 Å basal spacing, which are compared in Table VIII with our theoretical data computed according to the principles set down in Table II. We make use of the same

TABLE VIII

Number of layers	Theoretical water content, g H_2O/g clay	Calculated spacing, Å	Observed spacing according to Norrish, Å
3	0.179	18.73	18.7
14	0.833	39.89	40
16	0.952	43.57	43.8
20	1.190	50.93	49.2
28	1.665	65.65	65.2
32	1.903	73.01	72

designation of adsorbed water layers in both tables, implying that the conventional monolayer is counted as double and given the designation 2, etc.

Norrish obtained his data from a sodium montmorillonite and Table VIII shows clearly that the exchangeable ions do not interfere with the structuring of the adsorbed water layers in the hydrophilic montmorillonite, neither at high nor at low water contents. We shall presently show that the probable small amount of intercrystalline aluminum–water complexes present in a hydrogen montmorillonite does not reveal its presence in nmr spectra while residual intercrystalline ions do so in the hydrogen hectorite spectra.

A freeze-dried hydrogen montmorillonite was subjected to a wide line nmr study of the bandwidth as a function of the temperature in the region surrounding the liquid–solid transition point of the adsorbed water. The water content of the sample was 0.09 g/H_2O/g clay, indicating that the system was polymorphous containing stable two-layer sheets of intercrystalline water leaving about every fourth layer unhydrated.

Figure 30 shows the observed bandwidth as a function of the temperature between $+16$ and $-16°$. The bandwidth is low, only a few decigauss over the whole temperature range, with no indications of proton interactions with unlike spins. A discontinuity appears in the bandwidth curve at the transition point. On the high-temperature side the transition is preceded by a definite increase in bandwidth, starting at the temperature of maximum

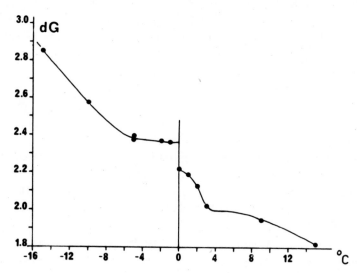

Fig. 30. Proton magnetic resonance bandwidth in decigauss for a freeze-dried hydrogen montmorillonite as a function of temperature. Water content 0.09 g H_2O/g clay.

density of water. It is known that, although the ratio of Schottky and Frenkel defects increases, the number of Schottky defects diminishes continuously with falling temperature until the transition point is reached. At this point a sudden reduction of the number of Schottky defects occurs, associated with lattice expansion and transition to the solid state. As the temperature continues to decrease, the conditions suitable for the formation of tensorial spin coupling[3] rapidly improve without, however, affecting the second moment and hence the bandwidth, which for a while remains constant.

Further decrease of the temperature will continue to increase the dipolar couplings and the bandwidth until a point is reached in the neighborhood of $-60°$ where nmr and dielectric data both suggest the inhibition of spontaneous Frenkel defect formation and consequently the disappearance of proton transfer in the hydrogen bonds. This condition, however, seems to be violated in the temperature region of the transition between hexagonal and cubic ice I which introduces lattice defects giving rise to a limited renewal of proton transfer, as will be shown in the next section.

It is interesting to note that the transition process appears to be quite normal, although the intercrystalline water content is reduced to the first stable minimum of two layers. This, however, is about the amount of residual, "unfreezable" water observed by Anderson and Hoekstra,[10] thus establishing the compatibility of their results with the proposed model.

The case of the hydrogen hectorite presents a different picture. The bandwidth measurements, performed on a sample containing 0.09 g H_2O/g clay, are shown in Fig. 31. It should first be noted that the already low bandwidth of the hydrogen montmorillonite, ascribed to proton transfer involving the hydrogen bonds, is here further reduced by a factor of four or five, indicating a correspondingly shorter lifetime of the proton configuration.

Next, the appearance of discrete peaks of bandwidth enhancement in certain temperature intervals indicates the presence of tensorial spin coupling effects between the resonating protons and other, nonresonating ionic spins. The temperature dependence of these coupling effects shows that the dynamics of the intercrystalline water lattice is involved in the coupling mechanism. The temperatures $(T = hck^{-1}\lambda^{-1})$ at which the two discrete peaks appear correspond to the frequency bands around 187 and 185 cm^{-1}. These are close to the 170 cm^{-1} center of the intermolecular translational water band characteristic of the coordination octahedron of the interstitial volume, normally harboring dissolved ions. The best dynamic fit of ionic masses corresponding to the two peaks are Mg^{2+} at $-3°C$ and Na^+ at $-6°C$, which both appear in the mass spectrum of the sample. We suggest that the magnesium–water complex essentially determines the

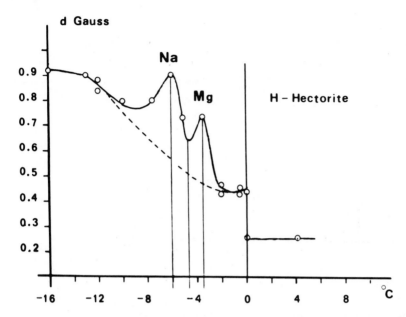

Fig. 31. Proton magnetic resonance bandwidth in decigauss for a freeze-dried hydrogen
hectorite as a function of temperature. Water content 0.09 g H_2O/g clay.

intercrystalline spacing, as in vermiculite, and that intercomplex acidic water of normal density and stabilized by dissolved sodium ions fills up the
remainder of the intercrystalline space. As in vermiculite, we believe that
the magnesium–water complexes only insignificantly influence the structuring of the intercomplex water and this contention is supported by the splitting[376,714] displayed in the wide line spectra of a sodium hectorite at 0.319 g
H_2O/g clay, shown in Fig. 32 at various temperatures. We take the unshifted
central peak to represent the intercomplex water layers containing sodium
ions and spurious, drifting, and reorienting magnesium–water complexes,
while the shifted bands correspond to stationary magnesium–water complexes with internal rotational relaxation and with more or less trapped intercomplex water. At low temperatures the interproton vectors of the water molecules in the stationary complexes are expected to orient parallel to
the hydrophobic crystal surfaces, giving rise to the band shift.[714] The protons belonging to the intercomplex water molecules participate in transfer
processes that give rise to rapid molecular reorientations according to the
schematic mechanism shown in Fig. 33 and previously described. It is significant that the montmorillonite spectrum cannot be resolved at any temperature to produce the spectral band shifts observed for hectorite. This is in

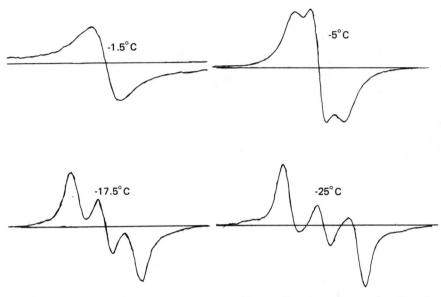

Fig. 32. Successive splitting of the spectral band in a sodium hectorite as a function of decreasing temperature. Water content 0.319 g H_2O/g clay.

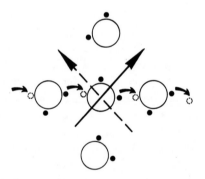

Fig. 33. Schematic representation of proton transfer and dipole reorientation in a four-coordinated water molecule. The collective, simultaneous proton transfer along the hydrogen bonds of ice is postulated to depend essentially on the existence of lattice defects of the Frenkel type, which may serve as sources and sinks for the associated charge displacements. A link in this chain of transfer processes is schematically represented from which it may also be inferred that the formation of a Frenkel defect depends on the breaking of one of the four bonds to the central molecule entailing a 50% a priori probability of creating a proton sink and starting a transfer process. The same argument obtains for the three-coordinated molecule, which is the most important generator of interstitials.

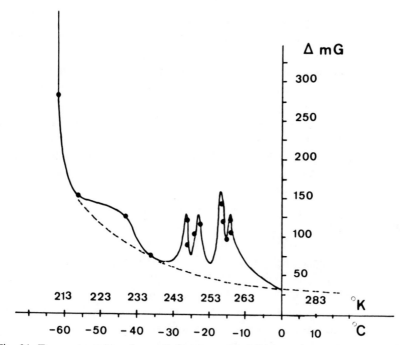

Fig. 34. Temperature dependence of the proton bandwidth of the shifted band in the sodium hectorite spectrum shown in Fig. 32.

fact borne out by the detailed band shape analysis exemplified in Fig. 34 for the shifted band. This figure also demonstrates how powerful a tool the nmr wide line technique really is for the study of hydrogel-forming systems. The sodium and magnesium peaks reappear, but doubled and at shifted temperatures which we ascribe to increased water content and sample polymorphism. The lattice dynamics is different at different layer thicknesses, close to the hydrophobic substrate or screened in a central layer. One obtains in fact a good idea of the situation by considering the six-layer sheet of Table II with the protruding hydroxyls removed, leaving the water to form an interface similar to the free water surface as suggested by Fletcher[243,244] and Watts-Tobin.[700]

(b) *Dielectric Relaxation.* Weiler,[701] in an extensive series of dielectric measurements on hydrated montmorillonite, observed a single low-temperature maximum, independent of water content and exchangeable cation, in the tan δ plot as a function of the temperature, δ being the dielectric loss angle. At about 100 Hz the maximum was situated around $-98°$ and moved toward higher temperatures as the frequency increased.

The measurements were extended by Mamy and Chaussidon,[516] who tried to relate the low-temperature maximum to the structure of the inter-crystalline water. Jacobsson and Forslind[376] took up the idea in an attempt to gain additional information by correlating dielectric and nmr low-temperature data. The results tend to verify the suggestion of Mamy and Chaussidon[516] and may be summarized as follows.

The experiments were performed with the sample temperature rising slowly and linearly with time from $-160°$ to the melting point of ice. The observed temperatures (in $°C$) of the dielectric loss maxima at 1000 Hz are given in Table IX. We believe these maxima to be due to proton transfer resonances.

At about $-60°$ the dielectric loss curve rapidly begins to approach a stationary value while the proton magnetic resonance bandwidth rises more rapidly with decreasing temperature. Both nmr and dielectric data thus indicate inhibition of ionic mobility and strongly impeded proton transfer due to a rapid repression of Frenkel defect formation. However, in the temperature interval between -80 and $-60°$ a transition from cubic to hexagonal ice is expected to occur in pure ice. A lattice disturbance is then expected to shift the transition toward lower temperatures while effects leading to a lattice stabilization will counteract the disturbance.

If we then ascribe the low-temperature loss maximum to the proton transfer resonance made possibile by the reintroduction of Frenkel defects in the intercrystalline water layers subject to the Ic → Ih phase transition, the data of Table IX seem to confirm our predictions.

We note that the hectorite resonances are shifted toward lower temperatures than the corresponding values for montmorillonite, in agreement with our contention that the hectorite water is not stabilized by the substrate whereas the montmorillonite adsorbate is.

We also observe that the resonance temperatures increase in the order of the lattice-stabilizing abilities of the ions as previously shown.

TABLE IX

Clay	H	Na	K
Montmorillonite	-96	-87	-85
Hectorite	-110	-105	-100

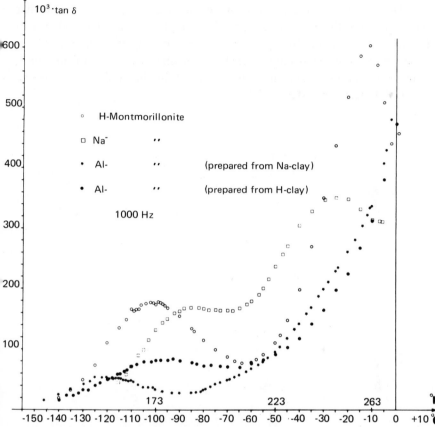

Fig. 35. Dielectric loss angle measured as tan δ for frozen H, Na, and Al montmorillonites at 1000 Hz, plotted against temperature.

In a separate series of experiments it was, however, shown that pure aluminum–water intercrystalline complexes (i.e., in the absence of other cation–water complexes) give rise to a dielectric resonance maximum in montmorillonite at about $-122°$, while the pure sodium clay shows a resonance at about $-87°$. We thus have reason to believe that the conventional way of preparing a hydrogen montmorillonite from a sodium clay by passing the sol through a mixed bed exchange column results in a nonnegligible amount of residual sodium ions. Figure 35 reproduces a set of such measurements on a hydrogen montmorillonite prepared in the manner described. It appears that this observational technique, originating from Chaussidon and his colleagues,[516] might be developed as a valuable tool for assessing the efficiency and suitability of clay cationic exchange processes and the purity of the product. It is, on the other hand, interesting to note that the influence of these structure-disturbing complexes is small at higher temperatures, the coupling to the intercomplex water layers being of little significance. Also, it seems unlikely that any significant amount of an aluminum–water complex could have been present in the sodium clay investigated by Pezerat and Méring.[578]

CHAPTER 5

Adsorption of Water on Well-Characterized Solid Surfaces

A. C. Zettlemoyer, F. J. Micale, and K. Klier

Center for Surface and Coatings Research, Lehigh University
Bethlehem, Pennsylvania

1. INTRODUCTION

There are two aspects to the state of water at solid surfaces. One has to do with the structure and thermodynamic properties of the water in the first layers formed as water is adsorbed from the vapor phase. The other is concerned with the water structure, and usual accompanying ionic arrangement (double layer), at solid/water interfaces. The first subject will be investigated in this summary, although some of the conclusions reached no doubt should spill over into the second field, and vice versa. The first layers of adsorbed water often do not contain ions from the solid (even alkali halides are essentially hydrophobic until ions leave the lattice structure), and this absence of any or many ions is a primary difference in character of the two situations.

Most revealing studies of the nature of adsorbed water at solid surfaces have been conducted with solid powders the specific surface area of which was at least 5 m^2 g^{-1}. With at least this magnitude of interface, there is sufficient sensitivity in the measurements to be described. Whatever other measurements are made, vapor adsorption isotherms should first be monitored; much will be learned from them alone. Nitrogen (or argon) isotherms at the temperature of liquid nitrogen provide the prime tool for measurement of the surface area of the solid.

Then, water vapor isotherms will reveal the fraction of the surface available to adsorbed water, within limits, and the binding energy of water to the surface. With this start, a number of other experimental methods can be applied to learn more about the water behavior and its structure at the surface.

Three methods will be dealt with in some depth. These are infrared absorption measurements of adsorbed water in both the fundamental and overtone regions, heat of immersion measurements, and measurements of dielectric behavior as revealed by the frequency dependence of the dielectric constant and dielectric loss.

We shall see that water at solid surfaces varies in nature from a highly structured form on hydrophilic substrates to a loose, entropic form on more hydrophobic substrates possessing hydrophilic sites. It is likely that these characteristics carry over to the fully immersed state.

2. EXPERIMENTAL APPROACH

2.1. Gas Adsorption

A range of methods has been developed for measuring the adsorption of gases on solid surfaces. The most widely used methods, which employ volumetric and gravimetric techniques, will be discussed in some detail especially with regard to their application for measuring water adsorption isotherms. Some of the lesser used methods, which are generally used for specific applications, include vapor-phase chromatography, radioactive tracer techniques, piezoelectric quartz crystal balance measurements, and ellipsometry.

2.1.1. Volumetric and Gravimetric Techniques of Water Vapor Adsorption

The volumetric method has been the most commonly used technique for measuring gas adsorption isotherms on high surface area solids. The essential features of this technique involve a dosing volume V_d, which is usually adjustable, and a sample volume V_s which contains the adsorbent. The volumes V_d and V_s are first measured accurately, usually by expansion of helium from a standard volume. The adsorbate gas is first introduced into V_d at a known vapor pressure and then allowed to expand into V_s. The sample is allowed to come into equilibrium with the vapor and then the equilibrium vapor pressure is measured. The decrease in pressure from the value given by the ideal gas law is used to calculate the amount of gas adsorbed on the surface of the sample.

A majority of the vapor adsorption data have been measured on "home-made" vacuum apparatus and, as a result, many different configurations can be found. Faith[229] and Orr and Dalla Valle[541] present detailed descriptions of the construction and operation of some of the more popular versions. This equipment, however, is primarily designed for measuring specific surface areas of solids by argon and nitrogen adsorption measurements. Since the adsorption of water vapor, as well as most condensable vapors, presents unique sources of error, an adsorption apparatus which is designed for measuring specific surface area is not ordinarily suitable for measuring water adsorption isotherms. Figure 1 represents a simplified design of what the authors consider to be an "ideal" volumetric adsorption apparatus. The pumping system itself consists of a mechanical vacuum pump, oil diffusion pump, liquid nitrogen cold trap, and an ionization gage for measuring the pressure in the high-vacuum range of 10^{-6} Torr. The adsorption section includes the dosing volumes V_d, V_d', and V_d'', the sample container V_s, and the differential manometer or other suitable pressure measuring devices. Both the sample volume and the dosing volumes should be thermostatted or insulated for maximum temperature stability. The highest sensitivity is obtained when V_s and V_d are as small as possible, usually 10–15 cm³. The

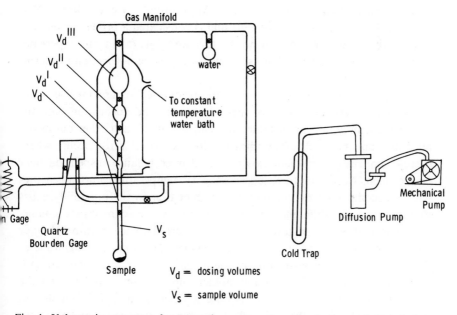

Fig. 1. Volumetric apparatus for measuring water adsorption isotherm. Initial dosing volume V_d should be kept small and approximately the same size as sample volume V_s.

additional dosing volumes are incorporated into the system at high relative pressures or for strong adsorbers.

The most critical component of the adsorption apparatus is the equipment used for measuring the pressure. The ideal requirements for the pressure measuring device are: (1) sensitivity down to 10^{-4} Torr, (2) range up to 25 Torr, (3) low internal volume, and (4) poor water adsorption by the exposed parts of the pressure apparatus. The extent to which these requirements should be met depends upon the water adsorption characteristics of the adsorbent under investigation. The experimental requirements tend to increase as the surface area of the adsorbent decreases. Although a large number of pressure measuring devices exist which meet these requirements to varying degrees, the one instrument which is ideally suited for this particular application is the quartz Bourdon gage (Worden Quartz Co.).

A wide variety of vacuum balances have been utilized for measuring water vapor adsorption isotherms. A survey of the literature reveals that until recently a majority of gravimetric adsorption experiments utilized "homemade" balances which were usually designed for particular applications. Commercial balances for vacuum application are now available with a high degree of reliability and accuracy. A continuing annual series of volumes entitled "Vacuum Microbalance Application" (Plenum Press, New York) has been published since 1961 and is considered a primary reference. Vacuum balance literature will not be reviewed, but rather our experience with a limited number of balances for measuring water adsorption isotherms will be summarized.

Three possible sources of error are recognized in gravimetric water adsorption experiments. Two of these, buoyancy and adsorption onto the balance assembly, are predictable and tend to cancel each other out at relative pressures below 0.5. The third possible source of error occurs whenever a temperature gradient exists along a microbalance suspension fiber or specimen. According to Knudsen's theory, temperature inhomogeneities will be accompanied by pressure inhomogeneities, so the possibility exists that these pressure differences will result in a net force acting on the sample. These normal Knudsen forces are often referred to as thermomolecular flow, TMF, and lead to spurious mass changes. The order of magnitude of these mass changes is a function of the magnitude of the temperature gradient, the geometry of the sample holder and vacuum chamber, and the molecular weight of the gas. For the case of beam balances it is possible, in principle, to eliminate the TMF effects by duplicating the normal Knudsen forces along the counterweight assembly. For the case of single hangdown balances, such as the spring balances, the TMF effects must be measured and the corrections applied to the adsorption data.

One of the more common commercial vacuum balances which has been used for water vapor adsorption measurements is the Cahn RG Electrobalance (Cahn Instrument Company). The popularity of this balance is due to its high reliability, ruggedness, and sensitivity of 0.1 μg with a 1 g load.[97] The balance may be obtained in a vacuum chamber and can be attached to a recorder for continuous monitoring of mass changes. Although the specifications of the Cahn RG meet the demands of high sensitivity necessary for gas adsorption measurements on relatively low surface area solids, i.e., less than 1 m^2 g^{-1}, a number of experimental problems have been experienced which drastically affect the ultimate sensitivity capabilities of this instrument. Although the Cahn RG is essentially a beam balance with symmetric hangdowns for the sample and counterweights, thermomolecular flow effects cannot be eliminated because of the presence of asymmetric heat sources which are generated from the lamp and vacuum tube located within the vacuum chamber. Another unfortunate design feature of the Cahn RG is the large amount of metal within the vacuum chamber. The degassing procedure, which can be facilitated by limited bakeout, tends to be lengthy, depending upon the pumping speed of the vacuum apparatus. A somewhat more serious consideration is that when the balance and sample are sufficiently outgassed the introduction of water vapor results in the sample frequently coming to equilibrium faster than the balance. The sample, therefore, at first adsorbs and then desorbs water as the pressure continues to decrease. It is impossible to measure reliable water adsorption isotherms under these conditions when the isotherms are not completely reversible and they are difficult to measure when they are reversible. Two comparable beam balances, which the authors feel are capable of overcoming these problems to some extent, are the Sartorius Vacuum balance and especially the Wordon Auto-Null Quartz balance, where quartz is the primary material exposed to the water vapor.

A dissertation on vacuum balances for measuring water adsorption isotherms must include the oldest (1926) and simplest balance of them all: the quartz spring. Frequently referred to as the McBain balance,[479] the utility of this instrument has increased as the sensitivity and quality of commercially available springs has increased. Quartz springs with a sensitivity of 0.1 mg mm^{-1} and net sample load of 25 mg have been used extensively in this laboratory for measuring water adsorption isotherms. The cathetometer and optics used to measure the spring extension result in an absolute sensitivity of ± 0.2 μg or a relative sensitivity of ± 8 μg g^{-1}. Although this sensitivity does not compare favorably with the sensitivity of 0.1 μg g^{-1} reported for the Cahn RG Electrobalance, it does have adequate sensitivity for high surface area samples, i.e., for specific surface areas >10 m^2 g^{-1}. The quartz spring balance is sometimes preferred over more sophisticated beam balances for measuring water adsorption isotherms on high surface area samples because of its small sample load which allows for more rapid equilibration with water vapor.

2.1.2. Information from Water Adsorption Measurements

Water adsorption isotherms come in a variety of shapes but most are classified either as type II or type III in the Brunauer[87] classification as given in Fig. 2. The significance of the difference of surface properties between type II and type III is open to question for the adsorption of water on different surfaces. Obviously type III character suggests a hydrophobic surface reluctant to accept water. Often for water on hydrophilic surfaces the knee in the type II curve is not as sharp as for other adsorbates and the "B" point (start of the straight line portion used sometimes for area determinations), if one can be identified, lies at relative pressures well above 0.10.

A variety of models for multilayer adsorption have been developed, [87,302,323,608] but a simple two-parameter adsorption equation due to Bru-

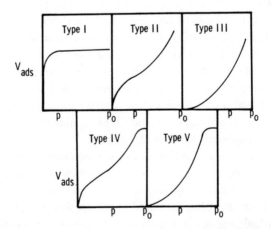

Fig. 2. Five types of adsorption isotherms. Type I is the Langmuir isotherm, which is limited to monolayer coverage. Type II is indicative of strong physical adsorption followed by multilayer absorption. Type III show weak interaction between adsorbate and surface. Types IV and V are considered to reflect capillary condensation, which results in effective reduction of available surface for further adsorption.

nauer et al.[88] remains in wide use. The model is one of multilayers, each of which has its own local equilibrium with the vapor, and only the first layer has an energy of adsorption different from the heat of liquefaction. An equation evolves from this model which, when a plot is made of the BET function, $P/v(P_0 - P)$, versus relative pressure, P/P_0, leads to value of C and v_m, which is the monolayer STP volume. The value of v_m is easily converted into a surface area using a factor depending on the area chosen per adsorbate molecule (as from liquid packing calculated from its density). The BET equation in its linear form is

$$\frac{P}{v(P_0 - P)} = \frac{1}{v_m C} + \frac{C - 1}{v_m C}\frac{P}{P_0} \tag{1}$$

where P is the equilibrium vapor pressure, P_0 is the saturated vapor pressure, v is the volume adsorbed at some pressure P, v_m is the monolayer volume, and C is a constant which may be expressed as

$$C = k\exp[(E_1 - E_2)/RT] \tag{2}$$

where k is a constant which is a function of the net entropy of adsorption, E_1 is the heat of adsorption of the first layer, and E_2 is the heat of liquefaction of the adsorbate and is assumed to be the heat of adsorption for the second and subsequent layers.

The BET equation is widely used in surface area determination with nitrogen frequently used as the adsorbate at the temperature of liquid nitrogen. If the surface is partially hydrophobic, $v_m(H_2O)$ may be smaller than $v_m(N_2)$, giving a measure of the percent surface hydrophilicity. Water adsorption BET plots are not always linear, so sometimes "B" points are taken as a measure of $v_m(H_2O)$. Figure 3 shows a series of water adsorption isotherms calculated from eqn. (1) for different values of the constant C. When the value of C increases, and hence the energy of adsorption E_1 increases, then the point of inflection decreases to lower relative pressures and the isotherms are type II. When the heat of adsorption equals or is less than the heat of liquefaction, i.e., $C \leq 1$, then the isotherms have no inflection and become type III isotherms. In general, the energetics of interaction of the adsorbate with the surface is proportional to the "sharpness" of the inflection point and is inversely proportional to the position of the inflection point on the pressure axis.

The concentration of polar sites evaluated from water adsorption isotherms is frequently used to calculate the degree of surface hydrophilicity. If the cross-sectional area of a water molecule in the adsorbed state is taken as 10 Å² and the B-point monolayer value of adsorption is two molecules/

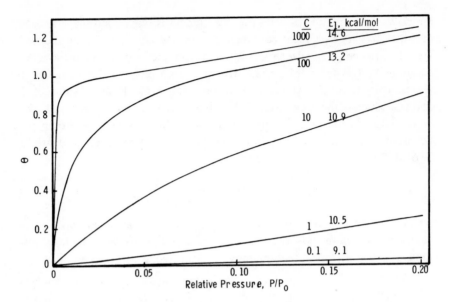

Fig. 3. BET water adsorption isotherms for adsorbate exhibiting different C values, and hence different energetics E_1 of adsorption in the submonolayer region.

100 Å^2, then the surface would be reported to be 20% hydrophilic. Many solution adsorption and dispersion properties of finely divided solids depend on the degree of surface hydrophilicity.

A potentially important aspect of water adsorption isotherms which has received very little attention in the literature is the multilayer region of adsorption which occurs at relative vapor pressures above 0.2. Frenkel,[262] Halsey,[319] and Hill[349] developed an equation for the adsorption of vapors onto solids in the multilayer region:

$$\ln(P_0/P) = K/\theta^3 \qquad (3)$$

where K is a proportionality constant determined by molecular forces, θ is the number of monolayers, and P/P_0 is the relative pressure. The exponent 3 is included in eqn. (3) because the van der Waals forces between an adsorbed atom and the adsorbing surface fall off as the cube of the distance. Halsey[319] chose to generalize eqn. (3) by leaving the exponent of θ unspecified:

$$\ln(P_0/P) = K/\theta^s \qquad (4)$$

Thus s is determined by plotting $\ln \ln(P_0/P)$ against $\ln \theta$. Equation (4) is commonly referred to as the Frenkel–Halsey–Hill (FHH) equation.

The s value is indicative of the type and range of interaction between adsorbent and adsorbate. Large values of s are characteristic of short-range specific interactions, while low values are indicative of forces which are more typically van der Waals and are found in systems where the interaction is nonspecific and longer ranged. The above concepts may be validly applied to water adsorption isotherms only when the adsorption occurs over the entire surface. Frequently the adsorption at high relative pressures occurs primarily on preadsorbed polar sites which can lead to s values which are a function of concentration of sites as well as the energetics of adsorption. The result is that the physical significance of the s value is not clear when a comparison is made of water adsorption on different surfaces.

The authors have recently adopted a different approach for analyzing isotherms at high relative pressures which is based upon the BET approach and assumptions,[88] with the following exceptions. The analysis covers the adsorption isotherm at relative pressures above 0.2 where multilayer adsorption usually predominates. Hence, the monolayer volume must be subtracted from the volume adsorbed and the energetics of adsorption is assumed to be single valued. Application of the BET procedure then yields the following equation:

$$v/v_m = 1/(1 - x) \qquad (5)$$

where

$$x = (a/b)P \exp(Q/RT) = kP$$

with v the volume adsorbed, v_m the apparent monolayer volume, P the equilibrium pressure, Q the heat of adsorption, and $k = (a/b \exp(Q/RT)$, and a and b are the adsorption and desorption coefficients. The desired boundary condition for eq. (5) is that $v \to \infty$ as $P \to P_0$. Although this boundary condition will be satisfied if $x = P/P_0$, the very act of imposing a fixed value for P_0 in eqn. (5) fixes the value for Q if the ratio of a to b is assumed to remain constant. The approach adopted has been to assume that P_0 is variable and is a function of Q [see eqn. (5)] according to the equation

$$P_0 = 1/k = (b/a) \exp(- Q/RT) \qquad (6)$$

Equation (5) may be expressed in its linear form

$$1/v = (1/v_m) - (k/v_m)P \qquad (7)$$

where a plot of $1/v$ as a function of P leads to effective values of v_m and k.

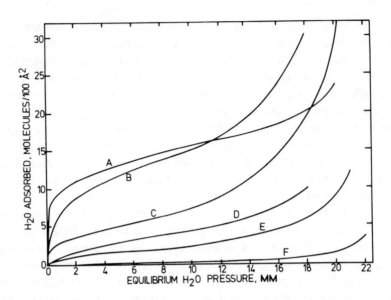

Fig. 4. Water adsorption isotherms at 25° on surfaces which contain different polar site concentrations and varying energetics of adsorption. A, NiO; B, silica (Quso H-40); C, Ni(OH)$_2$; D, silica (Cab-O-Sil HS-5); E, oxidized, graphitized carbon black Elf 4; F, graphitized Elf 4.

The value of the ratio a/b may be calculated from eqn. (6) by substituting the saturation vapor pressure and the heat of vaporization of water for P_0 and Q, respectively, at some temperature T. This calculated value of a/b is then used to calculate Q for different values of k. The assumption is that a/b has the same value as liquid water in the multilayer region which is not waterlike.

Figure 4 presents a series of six water adsorption isotherms at 25°C for surfaces which exhibit different degrees of hydrophilicity. The amount adsorbed is expressed in units of molecules per 100 Å^2 so that the isotherms are normalized with respect to specific surface area. All six isotherms were analyzed according to the B-point method, the BET equation, and eqn. (7). The data from 0.1 to 6 mm water vapor pressure were used for the BET plots and the adsorption data from equilibrium pressures above 6 mm were used to obtain plots according to eqn. (7). The results (Table I) show reasonably good agreement for all three methods insofar as monolayer values are concerned. Since approximately ten molecules/100 Å^2 are required for statistical monolayer coverage, samples A and B are seen to be 100% hydrophilic while the remaining samples exhibit decreasing degrees of hydro-

TABLE I. Water Adsorption Results Calculated from Fig. 4[a]

Sample	BET equation			B Point, v_m	Eq. (7)		
	v_m	C	Q		v_m	k	Q
A NiO	11.2	140	13.4	12	12.5	0.020	10.1
B Quso H-40	8.9	80	13.1	11	9.6	0.037	10.4
C Ni(OH)$_2$	4.3	27	12.4	4.0	4.6	0.042	10.5
D Cab-O-sil HS-5	3.0	10	11.9	2.5	2.6	0.042	10.5
E Oxidized Graphon	1.7	11	11.9	1.2	1.5	0.043	10.5
F Graphon	0.08	12	12.0	0.1	0.19	0.043	10.5

[a] V_m in number of molecules/100 Å^2, Q in kcal mol^{-1}.

philicity. The energy values, which were calculated from the BET theory and Eqn. (7) (columns 4 and 8 of Table I, respectively), represent a qualitative description of the average heat of adsorption over the vapor pressure region used in the calculations. An interesting comparison can be made between samples A and B where the average heat of adsorption for the first layer is greater for sample A and for the second layer is greater for sample B. A comparison of the two water adsorption isotherms in Fig. 4 leads to the same conclusion insofar as sample A adsorbs very little water above 6 mm equilibrium vapor pressure. More information is required, however, for development of a mechanism of adsorption which brings about this phenomenon.

The above discussion has been directed toward an appreciation of what information may be obtained concerning the hydrophilic properties of surfaces from water adsorption isotherms measured at one temperature. An appreciation of the concentration and relative average energetics of adsorption over the first monolayer and in the multilayer region was seen as a possibility. An accurate description of the energy site distribution, which cannot be deduced from single isotherms, may be obtained from multitemperature water adsorption isotherms. Application of the Clausius–Clapeyron equation[4] to multitemperature adsorption at surfaces,

$$\left[\frac{d(\ln P)}{d^1/T} \right]_{n_a} = \frac{q_{st}}{R} \tag{8}$$

where n_a is the water coverage and q_{st} is the isosteric, or differential, heat of adsorption, predicts that a plot of $\ln P$ vs. $1/T$ at constant coverage will yield a straight line with a slope proportional to the isosteric heat of adsorption. Two requirements are that the adsorption process is completely reversible and that q_{st} is constant over the temperature range investigated. The latter requirement is not absolutely necessary in the sense that if q_{st} is not constant with temperature, then the Clausius–Clapeyron plot will result in a curve of changing slope. Examples of Clausius–Clapeyron plots obtained from water adsorption isotherms on TiO_2 measured over the temperature range 25°–200°[727] are presented in Fig. 5. The unique feature of these results is the appearance of two distinct slopes at low coverages, i.e., below 0.2 of a monolayer. The higher slopes at temperatures above 120° have been attributed to chemisorption (reversible at these temperatures). The isosteric heats of adsorption as a function of coverage, calculated from Fig. 5, are presented in Fig. 6.

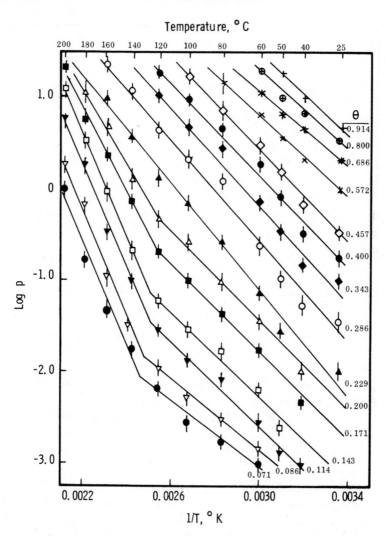

Fig. 5. Clausius–Clapeyron plots of water adsorbed on TiO₂ in the temperature range 25–200° and at fractional monolayer coverages of 0.071–0.914. A change in slope, indicative of a different mechanism and energy of adsorption, is observed at about 120°.

2.2. Infrared Spectroscopy

2.2.1. *Transmission and Reflectance Techniques*

Three methods of sample preparation have been employed to examine infrared spectra of adsorbed species such as water on solids. At first, the solid powder was sprinkled on a transmitting window, but this arrangement

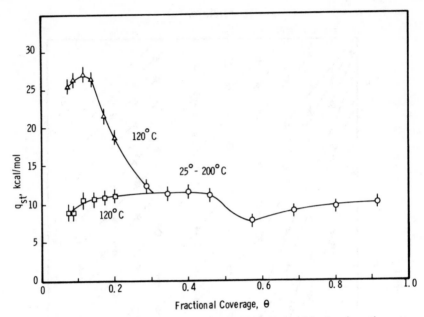

Fig. 6. Isosteric heats of adsorption of water on TiO_2. Reversible chemisorption occurs above 120° and at fractional coverages below $\theta = 0.3$.

was not very satisfactory because of excessive light scattering and other problems. Milling into hydrocarbon or fluorocarbon oils was the next method employed. Although excellent transmission was obtained and the preparation was rapid, the oil affected the surface to various degrees. Both of these methods are useful only to probe strongly bound species or chemisorbed complexes.

The third method of sample preparation depends upon the pressing of a disk of the powder and this preparative method is most used for spectroscopic investigations of water at surfaces; a transmitted beam is examined with these disks. This procedure will be discussed next. Then, the newer method of reflection spectroscopy will be described. The latter approach avoids some of the difficulties of the disk and transmission method.

(a) *Transmission Spectroscopy.* An early approach[398] was to press the powder to be investigated with potassium bromide with or without added solvent at about 1400 kg cm^{-2} in a suitable die to form an optically acceptable disk. Not only could the amount of adsorbed water not be varied at will in this method, but some surface groups could ion exchange with the KBr. It has now been shown for oxides and for a variety of catalysts that a more advantageous method is to press the disks in the absence of the potassium bromide. Best results are often achieved if the powder is not evacuated and is not completely dry. The disk is so waferlike that one speaks of thickness in terms of milligrams per square centimeter, say 4 mg cm^{-2}. To avoid sticking to the die, the disk is sometimes pressed between tissue paper or thin sheets of polymeric film. Then the paper or possible organic contaminant is removed by burning[317] as reported by Hair in his useful survey. The O–H stretch band v on oxides is usually followed in the fundamental region at about 3740 cm^{-1}, or nearby if the hydroxyl groups interact. McDonald[484] was apparently the first one to identify surface hydroxyls (on silicas) in this region.

The disk technique gives rise to several problems in the infrared investigation. For one thing, the pressing operation may cause alterations in the surfaces including particle–particle cross-links and surface annealing dehydration. Then, too, interparticle condensation may be enhanced when vapor molecules are admitted to the pressed disk.

Ingenious cell designs have been developed for supporting the disk and for allowing vapors to be admitted[215,636] at both low and high temperatures. Both volumetric adsorption measurements and gravimetric measurements [571,620] have been made simultaneously with the spectroscopic measurements in cells built for the purpose.

Perhaps the most serious problem in transmittance spectroscopy is the heating effect caused by the adsorbed radiation. This effect is particularly serious with the undispersed beam, where the temperature can rise as much as 50°. If monochromatic beams are directed at the disk, the heating effect is much reduced.

(b) *Reflectance Spectroscopy*. There is a spectral region, however, where the overtone band ($2v$) for the O–H vibration occurs in the neighborhood of the combination band ($v + \delta$) for H_2O. Thus, the surface O–H behavior and the adsorbed water as it is admitted can be monitored at the same time by reflectance spectroscopy.

The arrangement for holding the powder sample is shown in Fig. 7. The depth of the powder layer is usually several millimeters, sufficient for it to be considered infinite, so that R_∞ (ratio of reflected to incident intensity) is recorded at each wavelength. (In an earlier arrangement[407] undispersed diffuse illumination was used.) Temperature control can be achieved to ±0.02 deg and is abetted by using monochromatic radiation. The port to the sample bed is connected to a volumetric water adsorption apparatus so that exposure to water vapor and, consequently, water isotherms can be measured simultaneously with the recording of the spectra.

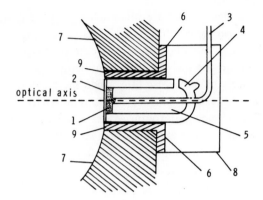

Fig. 7. A vacuum cell for reflectance spectra measurements at various temperatures: 1, sample; 2, quartz window; 3, gas inlet and vacuum; 4, vacuum jacket; 5, vessel for thermostatting liquid; 6, metal holder; 7, integrating sphere; 8, light shielding; 9, Araldite joints.

Spectral data are recorded by a digital readout system in intervals of 1 nm on the wavelength scale, processed by computer, and plotted in suitable coordinates.

2.2.2. Analysis of Reflectance IR Spectra

Spectra are analyzed in the first place according to the Schuster–Kubelka–Munk (SKM) theory[405,413] which gives the ratio of the absorption (K) to the scattering (S) coefficients as

$$K/S = (1 - R_\infty)^2/2R_\infty = F(R_\infty) \tag{9}$$

where R_∞ is the reflectance measured as the ratio of the total intensities of the light reflected from the sample and from the standard. At each wavelength the 5-mm-thick cell was tested to show that the semiinfinite thickness requirement is satisfied. For the range of spectral bands studied for powdered silicas, as explained below, their scattering coefficients are constant with changing wavelength, but not with particle size. Thus, $F(R_\infty)$ represents the absorption spectrum except for a multiplicative constant $1/S$, and log $F(R_\infty)$ represents the logarithm of the absorption coefficient except for an additive constant $-\log S$.

An overtone and a combination band appear nearby as stated previously. The overtone OH vibration for Si–OH (2ν) occurs at 7300 cm^{-1}, while the broad band at about 7150 cm^{-1} represents the $\nu_1 + \nu_3$ band of molecular water with some $2\nu_3$ and $2\nu_1$ admixtures. The combination band at 5300 cm^{-1} represents both deformation and vibration, H_2O $(\nu + \delta)$ or $(\nu_3 + \nu_2)$ band, and can therefore only be due to molecular water. Thus, in this region the addition of water to a surface containing hydroxyls can be followed at $\sim$5300 cm^{-1}, while the titration of the surface hydroxyls can be followed at 7300 cm^{-1}. The latter band is actually deflected to lower frequency due to the effect of the adsorbed water on the vibrations of the O–H bond. The magnitudes of the band center shifts permit the nature of the adsorbate–adsorption center interaction to be determined, while the band shapes reveal the character of the rotational motion of the molecules.

A word might be added here about the attenuated total reflection (ATR) mode in which the IR beam is internally multiply reflected from two surfaces of a large ($\sim$5 cm) crystal. Water has been studied at a sapphire (Al_2O_3) surface[318] in this manner in the fundamental region. Application to the 2ν and $\nu + \delta$ regions and line shape analysis might well prove a useful extension of the IR results reported here.

2.3. Heats of Wetting of Solids

Heat-of-immersion calorimetry provides one of the most fruitful avenues to pursue for studying solid/vapor interactions. At first, immersion of a solid into a liquid may appear to be a surprising approach to the investigation of the interaction of a solid with a vapor. But the method is applicable when direct measurements with the vapor are difficult or virtually impossible. Furthermore, the thermodynamics of vapor adsorption and of the immersion process are directly related. To illustrate the approach, let us examine the study of the heterogeneity of site energies for a particular adsorbate in a qualitative way.

Suppose the energetics of hydrophilic sites toward water are to be mapped. Then the bare solid followed by sequentially increasing, partially coated samples can be immersed in liquid water in a calorimeter. It can readily be seen that the first added increment of vapor will adsorb on the most active sites, reducing the heat evolved accordingly. The second added increment will cover the next-most-active sites, and so on. The heats of immersion will fall with each step and so the site energetics will be revealed. The coverage can be assayed by measuring adsorption isotherms by volumetric or gravimetric methods. Of course, the site–energy distribution developed is for the particular adsorbate employed; another adsorbate may produce an entirely different heat of immersion vs. coverage curve.

Several advantages of the heat-of-immersion technique emerge. If adsorption from the vapor phase is an activated and slow process, then the energies may still be obtained with precision after the equilibrium or final adsorbed state is attained. The method is suitable also for low vapor pressure liquids and for measurements at quite low coverages. It is noteworthy that the heats of interaction can be established even if the adsorption is irreversible; chemisorption indeed usually produces higher heat evolutions than physical adsorption. When entropy changes are evaluated from one reversible isotherm and heats of immersion, the results are usually more precise than those calculated from multitemperature isotherms.

2.3.1. Experimental Techniques for Heats of Immersion

The various types of calorimeters and some of the difficulties in the measurement and interpretation of heats of immersion have been discussed by Zettlemoyer in an earlier review.[726] Calorimeters consist essentially of two general types: adiabatic and isothermal. The adiabatic calorimeter, which is the more commonly used type for surface chemistry application, is constructed so that the heat of interaction is confined to the sample chamber by thermally insulating it from the environment. The isothermal calorimeter is constructed so that the calorimeter chamber is in thermal contact with its environment. This discussion will be restricted to the application of adiabatic calorimeters.

Certain important factors that have to be taken into account in the design and construction of a calorimeter include:

1. A steady and sufficient stirring rate to ensure good mixing with the minimum of heat generation.
2. Accurate measurement of temperature changes.
3. An accurate measurement of the electrical energy when calibrating the heat capacity of the calorimeter.
4. Attainment and maintenance of a steady change in temperature with time.
5. A reproducible method of sample bulb breaking.

Simple calorimeters have been designed to fulfill these requirements. A thermistor calorimeter developed in this laboratory is shown in Fig. 8. A Mueller bridge and an electronic galvanometer allow the temperature to be measured to a few ten-thousandths of a degree. A calorimeter of this type can be used to measure heat effects of the order of 0.01 cal and is adequate for the study of solids with surface area greater than 5 m^2 g^{-1}. Thermistor calorimeters are in wide use because of their ease of construction and operation.

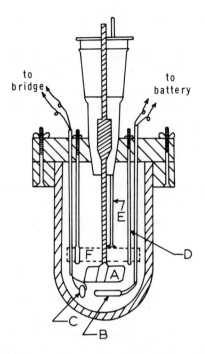

to
bridge

to
battery

E

F

A

D

C

B

Fig. 8. Thermistor calorimeter for measurement of heat of immersion. A, Stirrer; B, heater; C, thermistor; D, sample holder; E, breaking rod; F, sample tube. A sensitive resistance bridge such as the Mueller bridge and a galvanometer of sensitivity of the order of 10^{-10} A allows the determination of heats of the order of 0.01 cal. The calorimeter is placed inside an air thermostat. Evolution of heat up to about 10 min can be followed using a simple arrangement of this type. Submarine-type calorimeters enclosed in large water thermostats are needed to follow heat effects over longer periods. (Courtesy of American Chemical Society.)

Microcalorimeters are needed in the case of solids with low surface areas. Twin differential microcalorimeters have been described by Berghausen et al.,[56] Makrides and Hackermann,[469] and Whalen,[705] Calorimeters of this type are capable of measuring heat effects of the order of 0.001 cal and can be employed in the immersion wetting of solids of surface areas as low as 0.1 m^2 g^{-1}. Berghausen has shown that slow heat evolution (up to 2000 min) can be followed with differential adiabatic calorimeters.

One of the important sources of uncertainty in immersion calorimetry is the heat evolved during bulb breaking plus accessory events. Contributions could arise from the mechanical energy introduced during the fracturing of the bulb, residual strain in the glass, the $p \, \Delta V$ work as the residual volume from the bulb is filled with the immersional liquids, and the vaporization of liquid to fill the additional space in the calorimeter. The magnitude of the correction depends on the liquid and the type of bulb. Guderjahn *et al.*[311] determined the heat of bulb breaking for three types of bulbs: (1) metal bulbs with a thin brass foil at the bottom, (2) carefully preparad glass bulbs supplied by a commercial source, and (3) bulbs blown in their laboratory. In the case of the first two types of bulbs, the heat of bulb breaking could be accounted for by the equation:

$$\Delta H_B = pv + 0.03 \pm 0.07 \text{ J} \qquad (10)$$

There was considerable scatter in the case of bulbs blown in their laboratory. Since the general practice is to use bulbs blown by simple techniques, it is necessary to obtain an average correction factor from separate experiments.

2.3.2. Analysis of Heat of Immersion Data

Heat is evolved when a solid is immersed in a liquid and the enthalpy change is always negative. The total heat of immersion ΔH_I, the heat evolved on the immersion of a clean, outgassed solid into a liquid, is the most straightforward quantity. Since specific surface areas of most solids can be established with reasonable certainty, the heat of immersion may be put on a unit area basis:

$$\Delta H_I/\Sigma = h_{SL} - h_{S^0} = h_{I(SL)} \simeq e_{I(SL)} = e_{SL} - e_{S^0} \qquad (11)$$

where $h_{I(SL)}$ is the heat of immersion per unit area of the solid, ΔH_I is the total heat of immersion per gram of solid, Σ is the surface area of the solid per gram, h_{SL} is the enthalpy of the solid/liquid interface, h_{S^0} is the enthalpy of the solid/vacuum interface, $e_{I(SL)}$ is the change in internal energy for the immersion process, e_{SL} is the internal energy of the solid/liquid interface, and e_{S^0} is the internal energy of the solid/vacuum interface. Since there is very little pressure–volume work during the immersion process, the enthalpy change is essentially the same as the change in internal energy. The total heat-of-immersion values are themselves useful in comparing different solid surfaces and in assessing their polarity. Variation of heat of immersion with activation temperature has been of great value in studying the nature of bound water on oxide surfaces.

The immersion wetting can be carried out after precovering the solid surface with a known amount of the adsorbate from the vapor phase. This process is depicted in path 2 of Fig. 9. Path 1 represents the immersion

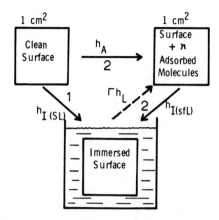

Fig. 9. Diagrammatic representation of the relation between adsorption and immersion processes (after Harkins). In path 1 the clean surface is immersed in the liquid. In path 2 the surface is first covered by N_A molecules of the adsorbate from the gas phase and subsequently immersed in the liquid. The relationship between the two processes is discussed in the text. Following this method the integral and differential heats of adsorption and site energy distribution on a solid surface can be determined.

of the bare solid. The heat of immersion can then be related to the heat of adsorption of the same molecular species. To make the second path equivalent to the first path, the molecules for adsorption have to be evaporated from the liquid. Then the enthalpy changes for the two paths can be equated:

$$h_{I(SL)} = - h_{A(SV)} - \Gamma h_L - h_{I(SfL)} \simeq e_a - e_L \qquad (12)$$

where $h_{A(SV)}$ is the integral heat of adsorption up to the coverage obtained at the vapor pressure represented by V, $\Gamma = N_A/\sum$ represents the molecules/cm² adsorbed during the precoverage, h_L is the heat of liquefaction per molecule of the vapor of the wetting liquid, and $h_{I(SfL)}$ is the heat liberated on immersion of the precovered solid. The negative signs are included to emphasize the exothermic character of the various heat effects. On rearranging eqn. (12), the integral heat of adsorption is given by

$$h_{A(SV)} = h_{I(SL)} - h_{I(SfL)} - \Gamma h_L \qquad (13)$$

Equation (13) clearly illustrates the relationship between the immersion and adsorption processes.

When the solid is precovered to the extent that the film is liquidlike (at least in terms of enthalpy), then $h_{I(SfL)}$ becomes h_{LV°, the enthalpy of the liquid surface per unit area. Sometimes the situation develops when only one monolayer has been preadsorbed, as for water on titania and water on alumina. In other cases, additional layers are required. The net integral heat of adsorption is then obtained from the equation

$$h_{A(SV)} = h_{I(SL)} - h_{LV^\circ} - \Gamma h_L \simeq e_a - e_L \qquad (14)$$

Equation (14) has been used in some instances to calculate net integral heats of adsorption. It is not difficult to determine $h_{I(SfL)}$ experimentally and employ eqn. (13) to calculate the net integral heats of adsorption. The use of eqn. (14) may sometimes yield erroneous values of $h_{A(SV)}$, especially in the case of organic liquids interacting with polar surfaces.

Application of the Clapeyron–Clausius equation to multitemperature vapor adsorption isotherms was previously seen to yield the isosteric heats of adsorption q_{st}. Either differential, q_d, or integral heats of adsorption can be obtained from vapor-phase calorimetry. The isosteric and the differential heats differ by a pressure–volume work term. Ideally, they are related by

$$q_{st} = q_d + RT \qquad (15)$$

The difference is often within experimental error, and so the two heats are generally treated as equivalent.

Immersion calorimetry yields integral heats, since all the interaction up to the precoverage is included. The integral heat and the isosteric heat are related by

$$h_{I(SL)} - h_{I(SfL)} = \int_0 q_{st}\, d\Gamma + \Gamma h_L = e_A{}' - e_L \qquad (16)$$

Equation (16) relates the measured heat effects to the molar energies of the adsorbate/solid system and that for the liquid. If perturbations of the solid surface may be neglected, $e_A{}'$ can be regarded as the molar energy of the adsorbate itself. Adsorption indeed implies that the solid surface is perturbed but the energy involved in the perturbation is likely to be small in the case of physical adsorption.

Chemisorption, on the other hand, implies severe surface perturbations and may lead to largely irreversible adsorption isotherms. Heats of adsorption cannot be calculated from isotherms that are not reversible. However, the direct determination of heats of immersion as a function of precoverage allows reasonably accurate evaluations of heats of adsorption from eqn. (13), even when rates of attainment of equilibrium are very slow.

The term heat of immersion isotherm is employed to describe the variation of heat of immersion at a given temperature with coverage. Experimental evidence for five different types of heat of immersion isotherms have been reported in the literature and are given in Fig. 10. The curves depicted in Figs. 10a and 10b are the most common heat of immersion isotherms found for polar solids in water, and also for polar and nonpolar solids in organic liquids. The linear decrease of heat of immersion with coverage (Fig. 10a) is typical of a homogeneous surface with localized adsorption where lateral interactions set in only at high coverages. The exponential decrease of heat of immersion with coverage shown in Fig. 10b is typical of heterogeneous polar surfaces. Many examples of oxide surfaces which exhibit this behavior may be found in the literature. Extremely nonpolar hydrophobic surfaces, for example, Graphon, give the heat of immersion isotherm shown in Fig. 10c. The interaction between the surface and water is weak here, and water adsorption occurs primarily in isolated patches followed by clustering at high relative pressures.

The type of immersion isotherm shown in Fig. 10d is usually found with clay–water systems, where a variety of processes can occur upon the incorporation of water into the solid. Swelling, ion exchange, and changes due to chemical, thermal, or mechanical modification of clays and other minerals can be followed by means of heat of immersion measurements in water.

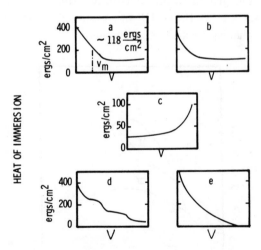

Fig. 10. Known types of heat of immersion isotherms, i.e., heat of immersion vs. coverage at a constant temperature. (a) Homogeneous surfaces. (b) Heterogeneous surfaces, e.g., water on most oxide surfaces. (c) Wetting of a hydrophobic surface with a few hydrophilic sites, e.g., water on Graphon. (d) Swelling of clay or stratified mineral with internal surface area, e.g., water on Wyoming bentonite. The adsorbate penetrates through platelets of the mineral at definite relative pressures. (e) Gradual filling of pores, making a large part of the surface inaccessible for the wetting liquid, e.g., benzene on graphitized black. (Courtesy of American Chemical Society.)

The isotherm in Fig. 10e is generally found in the case of porous solids and possibly organic fibers. The enthalpy of immersion progressively decreases and reaches a value lower than the enthalpy of the liquid surface due to the filling of pores which gradually reduces the effective surface area.

Heat of immersion experiments represent a useful technique for investigating surface properties as a function of pretreatment, e.g., activation temperature. A surface interacting with water in a simple manner should show a continuous increase in heat of immersion with increasing activation temperature, and any peculiarities in the interaction with water will be reflected in the variation of heat of immersion with the activation temperature. The general types of heat of immersion curves observed in the case of oxides interacting with water are depicted in Fig. 11. Silica surfaces generally show a variation of heat of immersion with activation temperature corresponding to curve I. Curve II is characteristic of alumina and thoria, and curve III is usually obtained with hydrous titanium dioxide. The initial, relatively flat portion in curve I is due to removal of physisorbed water; the large increase in the heat values is due to rehydration of dehydrated silanol groups; and the decrease in the heat of immersion at higher temperature is due to stabilization of siloxane groups so that rehydration does not take place in relatively short times. Curves II and III may essentially be interpreted as variations of the mechanisms which occur in curve I.

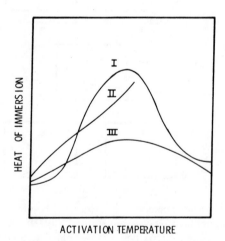

ACTIVATION TEMPERATURE

Fig. 11. General types of variation of heat of immersion with activation temperature. Silica, alumina, and hydrated titania surfaces exhibit variations of types I, II, and III, respectively. The temperature scale covers a range of 25–600°. The absolute values of heats of immersion vary from sample to sample, and the curves do not necessarily indicate the magnitude of the heat of immersion. (Courtesy of American Chemical Society.)

2.4. Dielectric Constant Measurements on Adsorbed Water

The use of dielectric relaxation measurements has led to significant advances in understanding the orientations and motions of molecules in the liquid and solid states. Thus, it should be expected that the application of these same methods to adsorbed water films might also be worthwhile. The theoretical background required for studies of surface phases is the same as that involved in the study of homogeneous phases, with the one important difference of the added factor of heterogeneity, which poses the additional problem of differentiating the effects of the adsorbed film from the composite system. Nevertheless, the fact that the adsorption of a polar gas onto a solid in a dielectric cell results in an increase in the electric capacitance is in itself an important observation which can lead to significant conclusions.

The dielectric method essentially involves a study of the dielectric constant, the dielectric relaxation, and the dielectric loss of a system. The dielectric constant is defined simply as the ratio of the observed capacitance of the system between the plates of a condenser to the capacitance of a vacuum between the plates. Dielectric relaxation may be defined as a delayed response in the ability of polar molecules, which possess permanent dipoles, to respond to an alternating esternal field. This dipole orientation is expected to be slow for very large polar molecules or clusters of molecules, for media of high viscosity, or for alternating currents of very high frequency. The rotation of dipolar molecules in the presence of an ac field encounters quasi-frictional resistance which results in dissipation of part of the energy of the applied field as heat. The energy dissipation is referred to as dielectric loss. Relationships between dielectric constant, dielectric relaxation, dielectric loss, and frequency of the ac field have been proposed according to a number of theories. Summaries of these theories have been presented by McIntosh[485] and McCafferty.[480]

2.4.1. *Experimental Techniques*

A number of experimental methods have been described by McIntosh.[485] The experimental approach outlined here has been utilized by McCafferty[480] to investigate the dielectric properties of water adsorbed on iron oxide. The capacitance cell (Fig. 12) consists essentially of two concentric stainless steel cylinders, the annular space of which is filled with the powder sample. The metal cylinders are 10 cm long and the separation between inner and outer cylinders is approximately 2 mm. This electrode assembly is fitted into a Pyrex glass cell bakeable to about 300° and attached

To Adsorption Apparatus

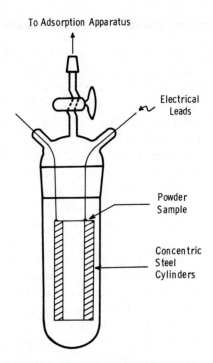

Electrical
Leads

Powder
Sample

Concentric
Steel
Cylinders

Fig. 12. Sample cell for dielectric studies. Powder sample is contained between concentric stainless steel electrodes. Adsorption apparatus allows for high vacuum and source of water vapor.

to a conventional adsorption apparatus which is capable of operating at 10^{-6} Torr and of exposing the sample to water vapor at known pressures.

Most dielectric measurements employ alternating current methods which are generally of two types, depending on the frequency range employed. For the frequency range between 10^{-2} and 10^5 Hz, bridge methods may be used. For the MHz range, however, inductance effects become increasingly important, so that a reasonance circuit is required. The experimental arrangement described here depends upon the alternating current bridge method to study the dielectric properties in the frequency range from 70 Hz to 300 kHz. Measurements are made by the substitution method[89] with a commercial Schering bridge consisting of a Rhode and Schwartz type SRM generator with type UBM detector and a General Radio type 716-C capacitance bridge with 716-P4 guard circuit.

Point-by-point permittivity measurements in the frequency domain, as described above, can be very accurate but are expensive and time-consuming. A method has recently been proposed by Fellner-Feldegg[238] which involves making the measurement not in the frequency domain but in the time domain, using a pulse which, simultaneously, contains all the frequencies of interest. The instrument is referred to as a time domain reflectometer and consists of a pulse generator which produces a fast rise time step, a sampler which transforms a high-frequency signal into a lower frequency output, and an oscilloscope or high-speed recording device. Fellner-Feldegg[238] has used this method to measure the dielectric properties of different alkyl alcohols over a wide temperature range in the frequency range of 1 MHz to 6 GHz. The results were in agreement with the dielectric properties obtained by frequency domain measurements. The main advantage of the time domain method is that the same information can be obtained over a wide frequency range in only a fraction of a second. Over the last two years the method has been considerably developed both in its instrumentation and the theoretical treatment of the data,[445,652] and it has been successfully applied to various problems associated with water–solute interactions[260,658] (see also Chapter 6 in Volume 4 of this treatise).

2.4.2. Analysis of the State of Water on the Surface

The most efficient approach to illustrate the information which can be obtained about the adsorption process by dielectric measurements is to show a concrete set of results. McCafferty's dielectric measurements[481] of water adsorbed on α-Fe_2O_3 has been chosen for this application. Gas adsorption isotherms and dielectric measurements as a function of coverage were measured on α-Fe_2O_3 at the temperatures of 5, 15, 25, and 35°, and in the frequency range of 70 Hz to 300 kHz. Dielectric isotherms for the adsorption of water vapor at 15° on α-Fe_2O_3 are shown in Fig. 13. The real part of the dielectric constant ε' increased with relative pressure for each fixed frequency. The effect of increasing the frequency at a fixed coverage was to reduce the dielectric constant in accord with theory.

The dielectric constants are characteristic of the combined system, adsorbate + solid. The dielectric properties of the adsorbate alone can be separated from the effective values for the combined solid + gas interface according to the method of McCowan and McIntosh.[483] Nevertheless, consideration of the apparent values for the combined system yields valuable information about the solid + gas interface.

Figure 14 shows the dielectric constant as a function of coverage for a fixed frequency of 100 Hz. Coverages were taken from the adsorption isotherms and the BET monolayer volume and refer to the number of physically adsorbed layers beyond the underlying hydroxyl layer. Figure 14 shows that the first layer of physically adsorbed water does not cause any change in

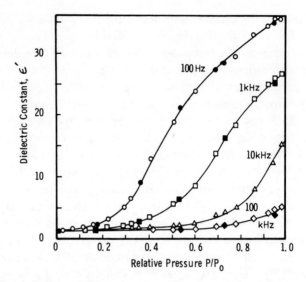

Fig. 13. Dielectric isotherms of water vapor at 5° on α-Fe_2O_3 outgassed at 25°. Solid points indicate desorption, which shows the system to be reversible. The dielectric constant remains constant over all frequency ranges below a relative pressure of 0.1.

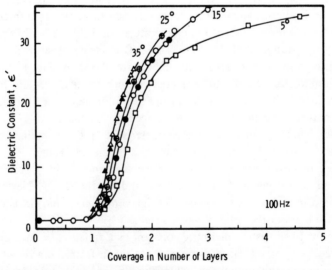

Fig. 14. Dielectric constant at 100 Hz as a function of coverage for various temperatures (solid points indicate desorption). The dielectric constant remains constant only in sub-monolayer region where the adsorbed water is not able to respond to ac signal.

dielectric constant. This behavior is attributed to the existence of a mono-layer in which the adsorbed species are held rigidly to the solid surface at fixed sites. Such molecules are unable to orient with the alternating field and hence cannot contribute to the capacitance of the system.

The dielectric constant rises sharply with the start of the second layer. This increase is due to increased ability of the adsorbed water molecules to respond to the ac field, so that these multilayers must be more mobile than the first layer. After the formation of approximately three layers, the relative change in ε' levels off. This decrease in the growth of ε' indicates that the adsorbed film has again become more ordered. A possible cause is increased hydrogen bonding as the adsorbate approaches the bulk liquid or bulk "icelike" state.

The temperature effect in Fig. 14 is consistent with the above model. Within the first layer, water molecules are bound so strongly that a temperature increase of 30° has no effect on the ability of the molecules to orient in the electric field. At increased coverages, however, where the adsorbed film is mobile, thermal excitation is able to overcome the order produced by strong local fields and thus aids orientation of the adsorbate in the applied field, producing an increase in the dielectric constant. Additional curves similar to Fig. 14 were obtained for frequencies between 100 Hz and 300 kHz although the trend was less pronounced for higher frequencies.

The dielectric isotherm measurements allow for calculation of the characteristic frequencies as a function of coverage. Figure 15 gives these results for all four isotherms measured at 5, 15, 25, and 35°. With the buildup of several layers of water it is expected that the properties of the adsorbate should approach those of bulk water or bulk ice, depending on the degree of hydrogen bonding in the adsorbed water film. If a hydrogen-bonded "icelike" solid could persist above 0°, it would exhibit characteristic frequencies of the order of 10 kHz at room temperature, while the characteristic frequency of water at the same temperature is of the order of 10^{10} Hz. Figure 15 shows that the characteristic frequencies reach or approach 10 kHz when about three layers of water have been adsorbed on the surface. These results were interpreted to mean that the adsorbate has developed stronger hydrogen bonding than liquid water, and adsorbate molecules are oriented to some extent like those in ice. The water dipoles will no doubt be oriented with the negative end toward the surface, i.e., the oxygen of the water doubly hydrogen bonded to two surface hydroxyls, thus explaining the immobility of the first layer. This concept will be further developed in Section 2. Subsequent layers will be oriented in somewhat the same direc-

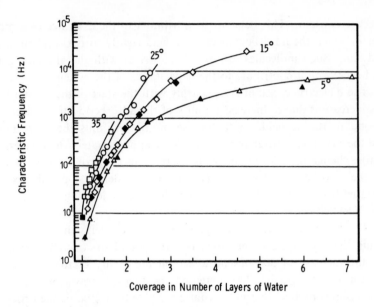

Fig. 15. Characteristic frequency of adsorbed water vapor on α-Fe_2O_3 as a function of coverage (solid points indicate desorption). The extrapolated value of the characteristic frequency of ice at 25° is 10 kHz, which indicates that the adsorbed water is icelike in resonance frequency up to three monolayers.

tion. Therefore, while the resonance frequencies suggest an "icelike" structure, the adsorbed water is not quite ice. In accord with this thesis, α-Fe_2O_3 is not a particularly good ice nucleant.

3. INORGANIC OXIDES

High surface area silicas have been mostly studied by infrared techniques so attention will be given to such studies. These amorphous silicas, often used as adsorbents and catalyst supports, transmit IR radiation down to about 1450 cm^{-1}. Like all metal oxides, they possess hydroxyl groups on the surface unless they have been heated to temperatures above 400°. In this latter property of remaining dehydroxylated and hydrophobic even in moist air, silica surfaces differ from those of most oxides such as Al_2O_3, Fe_2O_3, and ThO_2; one speaks of "hard" versus "soft" hydroxyls. The population of surface hydroxyls on silicas has been reported to be from 4.2 to 10 per 100 Å^2. A value nearer 5 is more probable, and after heat treatment at 800° there is about one OH per 100 Å^2. There are also internal hydroxyls often present due to defect sites and these may be spectroscopically accessible but inaccessible for reactions such as deuteration. In addition, the surface roughness on an atomic scale allows some silicon ions to possess two hydroxyls (geminal) while a majority possess one hydroxyl (vicinal). Flame-hydrolyzed silicas, such as Cab-O-Sil made from $SiCl_4$, often possess one-quarter the surface concentration of a fully populated wet-precipitated silica. Table II lists the surface concentrations of a variety of oxides as measured by several methods, as indicated. When the surface hydroxyl concentration is low, as in partially hydrophobed silicas or titanias or possibly germanias, we shall see that reflection IR studies show clearly that the BET method depending on water isotherms gives much too low values.

Emphasis will now be placed on the interaction of water with silica surfaces. The detailed studies of the nature of silica surfaces by the IR technique have been summarized by Hair.[317] Young showed clearly[722] that the amount of water physically adsorbed on a silica was directly related to the number of hydroxyls present on the surface so that it was reasonable to conclude that the water interacted with these groups. IR spectroscopy has provided evidence as to how the interaction occurs.

There is ample evidence that water prefers to adsorb on two neighboring hydroxyls when such sites are available. Benesi and Jones[47] found that exposure of a silica evacuated at 400° to D_2O yielded little increase in the intensity of the isolated OD band, but considerable broadening occurred. Its integrated intensity was more than doubled, suggesting reaction to form neighbor Si–OD groups and subsequent adsorption with two of them rather than adsorption on the remaining OH groups. It is certainly not unexpected that neighboring OH or OD groups would be preferred sites for the adsorption of water because of the double hydrogen bonding that can occur. Kiselev and co-workers[400] seem to have been the first to propose that the oxygen of the water molecule is hydrogen-bonded to the two OH hydrogens. We shall see that there is good evidence for this position of the first layer of adsorbed water molecules on a fully hydroxylated surface.

The overtone region, 4000–8000 cm^{-1}, was first studied by Anderson and Wickersheim[13] by the disk method. For a 750° dehydrated silica a sharp band at 7326 cm^{-1} was identified with the first overtone of the fundamental OH stretching vibration peak at 3740 cm^{-1}. The overtone peak gradually decreased as the titration with water vapor was effected at 25° until it finally disappeared. The sharpness of the overtone peak at the outset and its gradual disappearance indicate that the water molecules were adsorbing onto isolated hydroxyls. Therefore, single hydroxyls will take up water, too, as McDonald[484] had claimed in his classic 1958 paper. It should be mentioned, however, that infrared spectroscopy alone does not provide sufficient

Table II. Total Concentration of Surface Hydroxyl Groups

Oxide	Number of OH groups/100 Å^2	Investigator	Method
ZnO	6.8–7.5	Nagao[513]	BET
SiO$_2$	4.8	Davydov et al.[146]	D$_2$O exchange
	4.2	Bermudez[95]	—
	4.6	de Boer and Vleeskens	Weight loss
TiO$_2$ (anatase)	6.2	Jurinak[390]	BET
(anatase)	4.9	Boehm[69]	D$_2$O exchange
(rutile)	8.5	Hallabaugh and Chessick[356]	D$_2$O exchange
(rutile)	2.7–7.1^a	Boehm[69]	D$_2$O exchange
α-Fe$_2$O$_3$	4.6–7.9^a	Morimoto et al.[509]	BET
	5.5	McCafferty and Zettlemoyer[482]	BET
α-Al$_2$O$_3$	15	Morimoto et al.[509]	Volume water lost upon heating
γ-Al$_2$O$_3$	10	Morimoto et al.[509]	Volume water lost upon heating
η-Al$_2$O$_3$	4.8	Boehm[69]	CH$_3$MgI (CH$_4$ evolution)

^{a} For different samples.

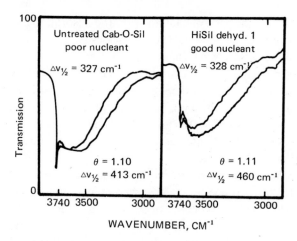

Fig. 16. Infrared spectra obtained from pressed disks of Cab-O-Sil and dehydrated HiSil under high vacuum, $\theta = 0$ and $\theta = 1.1$ monolayers. Of particular significance is the half-bandwidth $\Delta \nu_{1/2}$ between 3700 and 3200 cm⁻¹ for HiSil, which indicates hydrogen-bonded, clustered surface hydroxyls The Cab-O-Sil shows isolated hydroxyls (3740-cm⁻¹ peak).

evidence for a single hydroxyl or multiple hydroxyl bonding of water, and should always be accompanied by measurements of adsorption isotherms which help to determine the chemical equivalence of OH : H_2O.

Certain other bands also were detected in this overtone region. One is a combination band of OH stretching and deformation at ~ 4550 cm⁻¹. Another was found at 4420 cm⁻¹ which Benesi and Jones attributed to a newly formed OH (called OH_2 as distinguished from original OH_1) as modified by adsorbed water.

An interesting case where the disk IR technique helped to differentiate among several partially hydrophobed silicas will be described.[725] We had reported that partially hydrophobed silicas had improved ice nucleating ability over the fully hydroxylated parents. Yet Cab-O-Sil (and other source samples) with the apparent required lower population of OH sites performed poorly in cloud chamber tests. The IR spectra in the fundamental region revealed a significant difference.

Typical spectra of "poor" and "good" nucleants are depicted in Fig. 16. For $\theta = 0$ (evacuated), it is clear that the "poor" nucleant possesses a greater number of single hydroxyls. Furthermore, the 3740-cm⁻¹ peak was not diminished (not visible due to overlap of the spectra) by adsorption of water, $\theta = 1.1$ (which stands for a nominal 1.1 water molecules per OH—probably diminished somewhat by the heating effect). In contrast, the "good" nucleant shows a weak single OH peak which was diminished by water adsorption on the basis of combined IR and adsorption measurements. It was concluded that the good nucleant possesses patchy or near-neighboor hydroxyls around which water molecules cluster more readily and then form ice at appropriately low temperatures. The fundamental IR region has one significant disadvantage in the studies of water on surfaces completely or partially covered with hydroxyls, namely that the stretching fundamentals of molecular water and the OH groups overlap and that the deformation[87] band of molecular water at 1640 cm^{-1} is weak and overlaps with lattice vibrations of the silicas.

These problems can be overcome by monitoring the spectra of water and of the OH groups in the overtone and combination band region, earlier denoted as NIR (5000–10000 cm^{-1}, or 2 to 1 μm). Luck[453] has presented spectra of bulk water in that region for a wide range of temperatures, states of association, and electrolytic solutions. The combination band $\nu + \delta$ is five times more intense than the stretching overtone 2ν, and none of these overlaps with any lattice vibrations of the silicas.[407] Further advantage of the studies in the NIR region is that diffuse reflectance technique can be employed, avoiding the necessity of working with pellets and collecting all light scattered from the silica powdered specimens so that more reliable band shapes are obtained than in the transmission techniques, where part of the scattered light is lost to the detector. At first we worked with undispersed incident IR beams but then we changed to dispersed beams to reduce the temperature effect to about ±0.1°C. Examples of the spectra obtained for one "good" nucleant are plotted in Fig. 17. The numbers on the spectra correspond with coverages numbered in the same order in Fig. 18; number 1 corresponds to the vacuum-degassed substrate containing no molecular water.

The band at 7300 cm^{-1} is the OH stretch (2ν) of the Si–OH groups. It is quite clear that this peak progressively diminishes as the amount of water adsorbed (or the relative pressure) is increased. Since the silica contains about 1% of impurities, largely NaCl, it is interesting to learn that the hydroxyls still provide the only adsorption sites. As the OH peak diminishes, the broad band at 7100–7130 cm^{-1} develops. This band is due to the 2ν vibrations of molecular water and the 2ν mode of the silanols as perturbed by the adsorbed water. The magnitude of the shift for the silanols, at least 100 cm^{-1}, is close to that observed for a donor hydrogen bond in water dimers isolated in solid nitrogen.[670] On the other hand, the molecular water band H_2O ($\nu + \delta$) is shifted only about 20–30 cm^{-1} to lower frequencies with respect to monomeric water, indicating that water is the acceptor of a hydrogen bond. The surface complex is thus one in which the

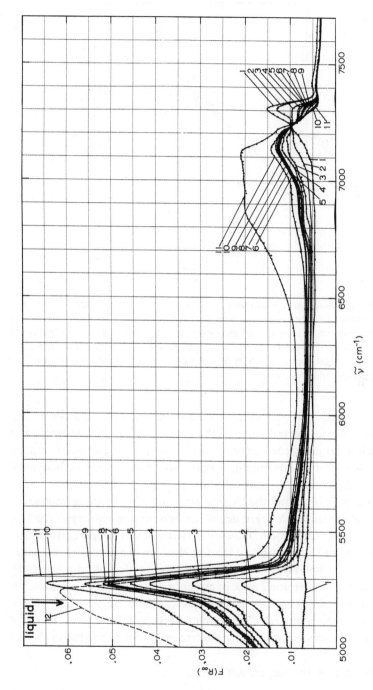

Fig. 17. Near-infrared reflectance spectra at different water coverages where $F(R_\infty)$ is the Kubelka–Munk function. Curve 1 is for the bare surface and additional numbers from 2 to 11 are for increasing concentrations of surface water as shown in Fig. 18.

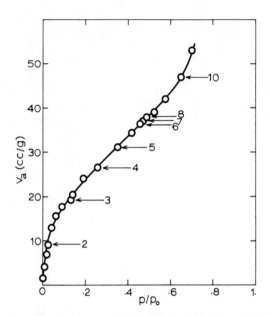

Fig. 18. Water adsorption isotherm on HiSil doped with sodium chloride. The numbers 2 to 10 correspond to the water coverages used for the infrared results in Fig. 17. At point 3 the isotherm is indicative of monolayer coverage, but the analysis in Fig. 19 (or Fig. 17) shows that the hydroxyls are not all titrated at the apparent "*B* point"; clustering of water molecules with "first down" water molecules occurs before all surface hydroxyls possess adsorbed water.

water is attached "oxygen down" to the silanol groups. Analysis of isotherms and of the spectral changes shows that on partially hydrophobed silicas each water is attached on one hydroxyl, and this is consistent with the relatively low heat of adsorption, 6 kcal mol⁻¹.* On highly hydroxylated surfaces such as on iron oxide the initial heat of adsorption is higher (10–12 kcal mol⁻¹)[482] and the likeliest structure of the surface complex is one in which water sits "oxygen down" on *two* surface hydroxyls. The lack of orientational (or rotational) mobility in electric field strongly corroborates this structure, and is revealed by the fact that the characteristic resonance frequency of the first layers is seven decades below that of bulk water.[482]

An especially interesting piece of evidence demonstrating that the water dipole is oriented toward the surface was provided by the study of the conductance of thin wafers of germanium (GeO and GeOH covered); the *n* conductivity increased with water adsorption, presumably due to the de-

---

* D. R. Bassett, E. A. Boucher, and A. C. Zettlemoyer.

crease in surface-trapped charges, and also the adsorbed water was held less firmly with increasing n conductivity of the wafers.[373] The adsorption heat effect of 10–12 kcal mol^{-1} strongly suggests double hydrogen bonding. It is appealing that on a fully hydroxylated surface this arrangement should develop for the first molecules adsorbed.

An example can be given of how the quantitative relations between adsorbed amounts and infrared intensities can reveal not only the structure of the adsorbed complexes but also their stability range and the onset of clustering of water molecules into a multilayer. Figure 19 demonstrates the intensities, represented by the SKM function of the Si–OH (2ν band and the H_2O $\nu + \delta$ band, during the course of water adsorption. As the adsorbed amount increases, the intensity of the H_2O $\nu + \delta$ molecular band increases and that of the free SiOH 2ν band at 7300 cm^{-1} decreases.

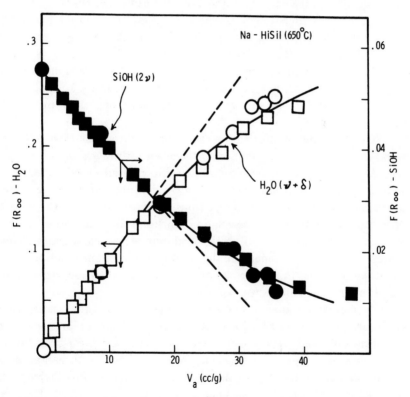

Fig. 19. Dependence of intensity $F(R_\infty)$ of SiOH (2ν) bands at maximum (7294 cm^{-1}) and water molecule ($\nu + \delta$) bands at maximum (5291 cm^{-1}) as a function of adsorbed water. The $F(R_\infty)$ function for OH continues to diminish long after 20 cm^3 g^{-1} is adsorbed (B point or BET value).

A very interesting observation can be made on comparing Figs. 18 and 19. Departure from linearity of the two plots in Fig. 19 sets in at just about the "B point" or BET monolayer value, point 3 on the isotherm, Fig. 18. But importantly, the decrease in the Si–OH 2ν band continues. In other words, water molecules adsorb on previously adsorbed water molecules before all the Si–OH groups are titrated. Clustering obviously occurs before almost half the Si–OH sites have gained a water molecule. This clustering is common to adsorption on partially or almost entirely hydrophobic surfaces, including partially hydrophobed oxides, graphitic surfaces, and low surface energy organic polymers (oxygenated or surfactant sites). The characteristic nature of this cluster adsorption has been summarized.[725] The isosteric heat of adsorption is often close to or less than the heat of liquefaction in the first layers and it gradually rises toward the heat of liquefaction. Correspondingly, the entropy of the first adsorbed molecules is of the order of that for a two-dimensional gas molecule adsorption and the heat of immersion of the substrate into water *rises* as a function of precoverage with adsorbed water, as indicated previously. The easily reversible isotherms also *rise* with increasing temperature. In addition, the reflection IR results tell us that the apparent BET monolayers represent, in the case of weakly bound adsorbates, a far smaller number of polar sites than are actually there.

One other kind of information can be derived from the IR bands. That is the correlated rotational motion of the adsorbed water molecules as perturbed by the solid surface and/or by neighboring adsorbed molecules as studied by Klier.[406] These observations can contribute to the development of mechanistic concepts of heterogeneous nucleation of water vapor to ice or of bulk water to ice. The correlation function is the real part of the normalized Fourier transform of an IR band onto a time basis and it characterizes the time development of the rotational–vibrational motion of the molecules. Changes in the correlation motion induced by the surface cause the phase transformations to take place.

The correlation functions of water on the surfaces of three heat-treated HiSils are presented in Fig. 20, where Na-HiSil (650°) is representative of the silicas discussed earlier. The correlation motion exhibits "beats" which are a consequence of a low-frequency rotovibrational perturbation of the motion of a "selected" molecule, representing the ensemble average, by other molecules in its neighborhood or by a proximal surface. The beats occur only a few times during the rotational period which permits the interpretation that they correspond to temporary formation and breakage of hydrogen bonds. Their amplitude tells us how many molecules on the aver-

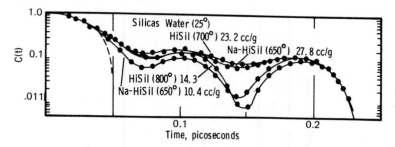

Fig. 20. The dipole correlation function of water adsorbed on silicas: (a) coverages close to a monolayer and less than a monolayer; (b) saturated multilayers. The physical meaning of the correlation function $C(t)$ represented on the ordinate axes is that of the projection of molecular dipole at time t onto the molecular dipole at time zero. Abscissa shows the time in picoseconds. The "beats" on the time development of the correlation function are low-frequency vibrational–rotational perturbations effected by the surface and by other water molecules onto the "average" molecule whose motion is followed. It is seen that at low coverages the bets are slow and pronounced, at monolayer coverages they become damped, and in multilayers their frequency increases to a value close to that of bulk liquid water.[406]

age undergo such process, and their frequency reveals how often the bonds with the neighbors or with a surface are formed. Correspondingly, the highest frequencies are observed in bulk water, next in water clustered in multilayers, and lowest in monolayers due to the lower number of neighbor molecules. The average angular perturbations $\{ \Delta \omega_{\theta \varphi} \} = \Delta \omega + \Delta \omega_r$ are given in Table III along with the band shifts with respect to a common reference.

The method deriving the correlation functions from infrared line shapes is particularly attractive for following phase transformations such as freezing of water at surfaces, since during these transformations the rotational motion is largely arrested and dramatic changes occur in the correlation motion. The "beats" are transformed into true vibrations of the solid lattice, corresponding to the transformation of perturbed rotation into a rocking motion of the water molecules about fixed lattice positions.

Finally with regard to the IR technique, it should be mentioned that some success has been achieved[318] in studying adsorption of water on single crystals by internal reflectance spectroscopy (ATR). The spectra obtained on α-Al_2O_3, for example, possess very broad bands. Normally a large bandwidth corresponds to a fast rotational relaxation but in this particular case the adsorbed water would have to undergo a faster rotational motion than that on the silica surfaces. It is quite possible, however, that there are different adsorption sites on the surfaces of α-Al_2O_3 crystals, giving rise to different mechanisms for water adsorption on one single crystal plane.

TABLE III. Average Perturbation Frequencies and Band Shifts of Water on Silicas and in Zeolite Cavities

	HiSil (800°)	HiSil (700°)	Na-HiSil (650°)[b]		Type A Zeolite	
	Saturated[c]	Saturated[c]	Saturated[c]	Saturated[b]	Saturated[b]	Residual
Surface coverage by water[a], cm³ g⁻¹	14.3	23.2	10.4	27.8	250, 250[a]	1
$\Delta\omega_v$, cm⁻¹	−30	−30	−30	−50	−150, −100	−150, −100
$\langle\Delta\omega_{\theta\varphi}\rangle$, cm⁻¹	−380, −420 −450, −480	−380, −420 −450, −480	−380, −420 −450, −480	−380, −420 −450, −480	—	—

[a] Temperature of adsorption was 25° in all experiments, except that in the last column, for which it was −20°.

[b] Na-HiSil (650°) was the 1:2 Na-HiSil specimen of Ref. 407.

[c] Saturated samples are those exposed to relative humidity close to 100%. No adsorbed amount could be measured, but it is estimated from the band intensities to correspond to the equivalent of 4–10 monolayers.

The fundamental region only has been studied thus far, so that it remains to be seen what the overtone region will reveal and whether smoother surfaces will indicate more definitive results.

4. HYDROPHOBIC SURFACES

Hydrophobic surfaces may be characterized by a number of experimental approaches, none of which is absolutely definitive. The classical definition of a hydrophobic surface is one on which water does not spread, hence there is a contact angle of water with the surface always measured through the water drops.[725] From practical considerations, which will not be pursued here, a surface is sometimes defined as being hydrophobic when the contact angle of water is greater than 90°. Since the contact angle of a liquid on a surface is a measure of the degree of interaction of the two substances, i.e., large contact angle is the result of a low degree of interaction and a low contact angle is the result of a high degree of interaction, then experimental methods which measure this degree of interaction may be used to define hydrophobicity.

Water adsorption isotherm measurements are most commonly used to determine surface hydrophobicity of high surface area samples. When a water adsorption isotherm exhibits a type III character (see Fig. 2) then that surface is conceded to be hydrophobic. The convex character of the isotherm toward the pressure axis is due to the nonspecific nature of adsorption of polar water molecules on the surface. The problem is that very few surfaces exhibit complete type III character. Water adsorption isotherms on graphitized Elf 4 and polystyrene latexes which have been measured in our laboratory exhibit pseudo type III character in that inflections normally characteristic of type II isotherms were observable at low relative pressures and at low coverages of one site per 1000 Å^2. The question actually reduces to an estimate of the degree of hydrophobicity, where the assumption may be made that no surface is 100% hydrophobic. An example of surfaces which exhibit different degrees of hydrophobicity is presented in Fig. 21 for water adsorption isotherms on carbon black Elf 4, graphitized Elf 4, and the graphitized Elf 4 which had been oxidized. Although, as mentioned above, graphitized Elf 4 appears to yield the classical type III isotherm, a small inflection was detectable on an expanded scale at relative pressures below 0.1 and at a coverage of 0.1 water molecule/100 Å^2. The oxidized graphitized Elf 4 water adsorption isotherm exhibits a type II character with a B point at one water molecule/100 Å^2, while the carbon black Elf 4 isotherm does not fall into any of the types presented in Fig. 2 but does achieve monolayer coverage at a relative pressure of 0.5. The conclusions are that graphitized Elf 4 is 99% hydrophobic; oxidized graphitized Elf 4 is 90% hydrophobic; and carbon black Elf 4 is not hydrophobic at all.

Heats of immersion in water and isosteric heats of water adsorption have also been used to define hydrophobic surfaces. Figure 9c shows a typical heat of immersion curve for Graphon in water as a function of water coverage. The integral heat of adsorption at low water coverage is less than the heat of liquefaction and rises with increasing water coverage. Apparently the degree of interaction of water with itself is greater than the degree of interaction of water with the Graphon surface. Isosteric heats of water adsorption on carbon black Elf 4 and oxidized graphitized Elf 4 were calculated from multitemperature water adsorption isotherms. The results (Fig. 22) show that both surfaces have a low concentration, one site/100 Å², of hydrophilic sites in a hydrophilic and hydrophobic matrix for Elf 4 and oxidized graphitized Elf 4, respectively. The conclusion to be drawn from these results is that Elf 4 is essentially hydrophilic and the oxidized graphitized form of Elf 4 is hydrophobic with a low concentration of hydrophilic sites.

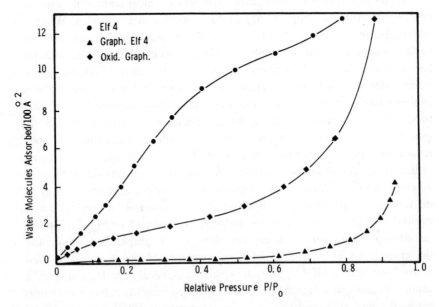

Fig. 21. Water adsorption at 25° on carbon surfaces. Graphitized Elf 4 exhibits a type III isotherm, while its oxidized form is type II. The Elf 4 isotherm indicates that this surface is relatively hydrophilic.

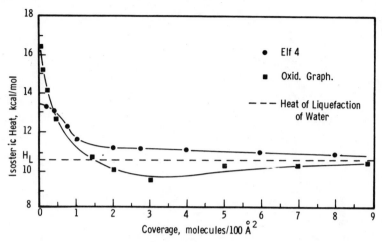

Fig. 22. Isosteric heat of adsorption of water on carbon black Elf 4 and oxidized, graphitized Elf 4. Although both surfaces appear to have a high-energy polar site concentration of one site/100 Å², the amorphous Elf 4 matrix is hydrophilic in that the isosteric heat of adsorption is always greater than the heat of liquefaction.

The different silicas which are commercially available represent a series of surfaces which are difficult to describe unequivocaly in terms of hydrophobicity. For example, B-point monolayer values obtained from water adsorption isotherms show that Cab-O-Sil has a polar site concentration of two sites/100 Å² compared with ten sites/100 Å² for HiSil silica. Cab-O-Sil, therefore, is frequently regarded as hydrophobic even though the isosteric heats of water adsorption are always greater than the heat of liquefaction of water. It is likely that the adsorption of water on Cab-O-Sil above the B point is due to clustering and hence a higher energy of interaction develops, as opposed to water adsorption on oxidized, graphitized Elf 4, where clustering does not occur.

Many methods have been proposed for hydrophobing surfaces, some of which have been discussed earlier in this chapter. The most common method which potentially has a wide application is methylation of OH sites present on oxide surfaces. The simplest approach, but not necessarily the most efficient, is to chemisorb low-chain alcohols, especially methanol, at elevated temperatures. Hexamethyldisilazone, HMDS, has been utilized to hydrophobe the surfaces of glass tubes and water adsorption isotherm results on HMDS-treated Cab-O-Sil show that the polar site concentration has been reduced by a factor of five with the result that this Cab-O-Sil could not easily be dispersed in water. The HMDS-treated tubes also show a definite increase in the water contact angle.

Specific Interactions of Water with Biopolymers

Herman J. C. Berendsen

Laboratory of Physical Chemistry, The University of Groningen
Zernikelaan, Groningen, The Netherlands

1. INTRODUCTION

1.1. General Aspects of Hydration

There is a widespread confusion about the terms "hydration" and "bound water" in relation to biopolymers, both in solution and in disperse systems. The amount of "bound water" is very much dependent on the experimental techniques and on the definitions and computational procedures used to determine its magnitude. If such definitions and procedures are not given, the term "bound water" should preferably not be used at all. Among the definitions that have been used, implicitly or explicitly, are the following:

(a) The water that is bound in equilibrium at a certain relative humidity and temperature, or the amount of water that is retained after a prescribed drying procedure.

(b) The water that fills the first adsorbed monolayer, in connection with a certain theoretical interpretation of the adsorption isotherm (usually the BET theory).

(c) The water that is not available for the solvation of solutes, depending on the type of solute.

(d) The water that does not freeze at a sharp transition temperature, e.g., the normal freezing point. This amount may depend on the technique used to determine unfrozen water.

(e) The water that gives no orientational contribution to the dielectric constant at high frequencies, and hence is not free to rotate with respect to the macromolecule to which it is attached. Such definitions depend on the frequency under consideration.

(f) The water that moves with the macromolecule in sedimentation, diffusion, or viscosity experiments, increasing the hydrodynamic size of the macromolecule.

(g) The water that shows a slower rate of rotation as measured by magnetic resonance techniques.

(h) The water that shows a slower rate of self-diffusion.

(i) The water that is shown by X-ray or neutron diffraction to occupy regular molecular positions with respect to a macromolecule in a crystal.

(k) The water that deviates from the normal liquid in density, as determined by light scattering or small-angle X-ray diffraction of macromolecular solutions.

(l) The water that can be shown by infrared or raman spectroscopy to be engaged in hydrogen bonding to a macromolecule.

Each of these "definitions" yields different results because they all have bearing on different aspects of the properties of water. The following aspects of hydration should be distinguished: *thermodynamic aspects*, relating to enthalpies and entropies of interaction between macromolecules and water and to influences on the activity coefficients of solutes (cases a–d); *dynamic aspects*, ranging from details of molecular motion to hydrodynamic properties (cases e–h); and *structural aspects*, relating to the average positions and orientations of water molecules with respect to each other and to the macromolecules (cases i–l).

No clear-cut relations between these different aspects of hydration can be given in general. Hence a full description of hydration should comprise all aspects: A description on the basis of one aspect or experimental method only is necessarily incomplete. Theoretically, knowledge of the structural aspects, together with a detailed knowledge of all molecular interactions, provides all that is necessary to derive both the thermodynamic and the dynamic properties of water bound to macromolecules. The actual derivation of these properties is at present beyond the practical possibility of realization because of the great complexities involved. Among the theoretical approaches the computational simulation procedure known as "molecular dynamics" is at present rapidly developing; in the future this method will undoubtedly contribute to the understanding and prediction of the properties of bound water and is expected to provide a link between structural, dynamic, and thermodynamic properties. The successful simulation of the time-dependent behavior of a large number of water molecules in the liquid state by Rahman and Stillinger[590,650] has already shown that complicated systems can be simulated by the method of molecular dynamics and there is no reason to doubt that such methods can be extended to water bound to macromolecules. While such approaches are not yet available, simplified theoretical models can be used to predict the properties of macromolecular hydration, albeit at the expense of reliability.

1.2. Specific versus General Hydration

In many instances it is possible to indicate certain specific sites at the surface or in the interior of macromolecules where water molecules are bound. It is the purpose of this review to emphasize such sites and their significance for the structure and properties of the macromolecule. The distinction between "specific" and "general" hydration will be made principally on a structural basis, according to definition (i) of the previous section. However, as will become evident later, there are strong indications that specific hydration exists also in cases where structural data are not available and that the binding of water molecules at well-defined sites at biological macromolecules is a general phenomenon with at least structural and probably also functional significance. Such water molecules are distinguishable from other water molecules by a higher binding energy, an appreciably lower rotational freedom, and an extended lifetime at the binding site. The distinction between specific and general hydration can only be made in a meaningful way if the specifically bound molecules differ appreciably from the other molecules, e.g., by order-of-magnitude longer reorientation and residence times. Any type of hydration will influence dynamic and static properties to some extent and normally a range of values for motional rates and binding energies will be observed. Only when a number of molecules can be clearly separated out from the range of values will it be meaningful to speak of specific hydration.

2. THERMODYNAMIC PROPERTIES OF SORBED WATER

2.1. Introduction

The experimental determination of adsorption isotherms in principle gives the thermodynamic properties of water adsorbed onto macromolecules. In several instances adsorption data have been analyzed for this purpose. As will be shown in this section, the results obtained are critically dependent on the theory used and unless the adsorption data satisfy a clear-cut theoretical case, a detailed interpretation in terms of specific hydration is not justified.

Adsorption isotherms, if determined under equilibrium conditions, give the number of moles of water n adsorbed on a given amount of adsorbent at a given temperature T as the function of water vapor pressure p. Hence the thermodynamic potential μ of the vapor is an independent variable, given with sufficient accuracy by

$$\mu = \mu_L + RT \ln(p/p_0) \tag{1}$$

where μ_L is the thermodynamic potential of the liquid at the temperature under consideration and p_0 is the saturation vapor pressure of water at that temperature. Thus one determines $n(\mu, T)$ experimentally and any interpretation in terms of the properties of bound molecules will depend on further assumptions concerning models of hydration. The isotherms of Langmuir[421] (monolayer), Freundlich[267] (empirical relation), Brunauer et al.[88] (BET, multilayer), and Bradley[76] (multilayer) are well known. Most theoretical treatments are based on irrelevant kinetic arguments or are partly or fully empirical. The best theoretically justified treatment of adsorption processes is that of Guggenheim,[312] based on the use of grand partition functions. Guggenheim's approach has been applied by Grigera and Berendsen to the hydration of collagen.[305] Since this treatment is suitable for a large number of special cases and provides a great deal of insight, it is outlined briefly below. For further details and proofs the reader is referred to the original literature.[312]

2.2. Guggenheim's Treatment

We first consider an adsorbent with a given number of separate, not necessarily equivalent, adsorption sites. The ordinary partition function

$$q_i = \sum_r \exp(-\varepsilon_r/kT) \tag{2}$$

is the quantity describing the thermodynamic properties of a molecule at the ith site.

Somewhat loosely, one may define a "standard thermodynamic potential" for the ith site:

$$\mu_i^\circ = -RT \ln q_i \tag{3}$$

where "standard" refers to full occupation of the site, and compare such μ_i° values with the thermodynamic potential of a reference state such as the liquid.

One then proceeds to calculate the grand partition function Ξ if the adsorbent is in equilibrium with vapor at the thermodynamic potential μ and absolute activity λ:

$$\lambda = \exp(\mu/RT) \tag{4}$$

and derives the number of adsorbed molecules from the relation

$$n = \lambda\, \partial(\ln \Xi)/\partial\lambda \tag{5}$$

which yields the theoretical adsorption isotherms.

For a given number of sites N_s the grand partition function is given by

$$\Xi = \prod_{i=1}^{N_s}(1 + \lambda q_i) \tag{6}$$

For a single species of adsorption site this leads to the classical Langmuir adsorption. For two species of sites

$$\Xi = (1 + \lambda q_1)^{N_1}(1 + \lambda q_2)^{N_2} \tag{7}$$

the adsorption isotherm resembles that of Freundlich, as was shown by Guggenheim.

In the case of multilayer adsorption a site can be occupied not only by one molecule with partition function q_i, but also by a second layer molecule q_2^i, a third layer molecule q_3^i, etc. This leads to

$$\Xi = \prod_{i=1}^{N_s}(1 + \lambda q_i + \lambda^2 q_i q_2^i + \lambda^3 q_i q_2^i q_3^i + \cdots) \tag{8}$$

for the general case. In the special case of one primary site q_1 and under the simplifying assumption that all multilayer molecules have the same partition function q, eqn. (8) reduces to

$$\Xi = (1 + \lambda q_1 + \lambda^2 q_1 q + \lambda^3 q_1 q^2 + \cdots)^{N_s}$$
$$= [1 + \lambda q_1/(1 - \lambda q)]^{N_s} \tag{9}$$

This can be easily extended to many other special cases.

From eqns. (5) and (9) one now derives the adsorption isotherm

$$n = N_s \lambda q_1/[(1 - \lambda q)(1 - \lambda q + \lambda q_1)] \tag{10}$$

This is equivalent to the BET isotherm, but under the assumption that the multilayer molecules are thermodynamically equivalent to the liquid, i.e.,

$$q = \exp(-\mu_L/RT) \tag{11}$$

In that case, $\lambda q = p/p_0$, and eqn. (10) reduces to the familiar BET isotherm

$$n = \frac{N_s c(p/p_0)}{[1 - (p/p_0)][1 + (c - 1)(p/p_0)]} \tag{12}$$

where $c = q_1/q$ has the significance of an equilibrium constant between the primary adsorption sites and the liquid. In the notation of eqn. (3)

$$\mu_1^\circ = \mu_L - RT \ln c \tag{13}$$

It is generally found that the BET isotherm does not provide a suitable expression for experimental water adsorption data on biopolymers at relative humidities above 35–50%. This discrepancy is due to the simplifying assumption that the partition function of the multilayer molecules is equal to that of the liquid. This assumption is not justified for biopolymers for two reasons. The partition functions of the multilayer molecules will cover a range of values. Only the outer layers will approach liquid properties and even then this approach will be incomplete for cross-linked polymers with a limited degree of swelling. In the thermodynamic potential of water a contribution is incorporated from the change in polymer interactions as a function of water content. As hydration proceeds polymer–polymer interactions become less favorable (salt bridges, van der Waals contacts, and hydrogen bonds are broken or replaced by polymer–water interactions), leading to a positive contribution to the thermodynamic potential of water. Swelling pressure also leads to a positive contribution. As was suggested by Guggenheim[312] and shown to be valid for collagen,[305] a good fit to experimental data can be expected if the multilayer molecules are all assumed to have the same partition function q not equal to that of the liquid. Equation (12) now becomes

$$n = \frac{N_s c(p/p^*)}{[1 - (p/p^*)][1 + (c - 1)(p/p^*)]} \tag{14}$$

where p^* is a measure for the thermodynamic properties of the average multilayer molecule:

$$\mu^\circ(\text{multilayer}) = \mu_L + RT \ln(p^*/p_0) \tag{15}$$

The positive contributions to μ° are dominant, implying that $p^* > p_0$. The constant c has the same meaning as before: $c = q_1/q$, which now implies

$$\mu_1^\circ = \mu_L - RT \ln(cp_0/p^*) \tag{16}$$

2.3. Specific Hydration from Sorption Data

Despite its limited applicability, BET theory is often used for the interpretation of sorption data for biopolymers. Considering the accuracy of experimental data and the difficulties involved in achieving a real equilibrium, Guggenheim's treatment [eqn. (14)] with three adjustable parameters will in practice suffice to obtain a good fit to experimental curves. It is easy to invent more elaborate models, but where these contain four or more adjustable parameters, extension of the Guggenheim model is neither meaningful nor justified. At the same time this means that if experimental data fit a theory with one type of binding site, such a fit does not necessarily prove the existence of only one type of binding site. With more types of specific binding sites an equally perfect fit to experimental data will be obtained. Data on primary binding sites on biopolymers from sorption isotherms are therefore of limited value and reliability unless additional knowledge about the number or properties of specific binding sites is available. The observed fit of experimental data to any particular theoretical isotherm does not imply that the model on which that isotherm is based applies to the substance under consideration. For example, the fit of the hydration data on living cells to a Bradley isotherm[440] cannot be trusted as evidence that the water in living cells exists in the form of organized polarized multilayers.

It is of interest to investigate whether the data obtained on the primary adsorption sites through a BET treatment at low relative humidities are significantly different from the data obtained from a full Guggenheim treatment. From a BET plot of the reciprocal of $n[(p_0/p) - 1]$ versus p/p_0 one obtains values for N_s and c according to eqn. (12). A calculation shows that the real values of N_s an c from eqn. (14) relate to N_s' and c' from the intercept and slope of a BET plot as

$$N_s = N_s'\left[1 - \frac{(p^* - p_0)}{p^*c'}\right]^{-1} \tag{17}$$

and

$$c = 1 + (p^*/p_0)(c' - 1) \tag{18}$$

Where p^*/p_0 is between one and two and c' is fairly large ($\sim$10) it follows that $N_s \approx N_s'$ and $c \approx (p^*/p_0)c'$. The latter equality means that eqn. (16) becomes

$$\mu_1^{\circ} = \mu_L - RT \ln c' \tag{19}$$

which is equivalent to eqn. (13). Thus it follows that the usual BET treatment for low humidities yields correct values for the number of primary adsorption sites and the standard thermodynamic potential of these sites.

Enthalpy data can be obtained either from calorimetric measurements or from the temperature dependence of sorption isotherms, through the relation of Clausius–Clapeyron, e.g., in the form

$$\varDelta h = h(\text{ads}) - h(\text{liq}) = - RT^2 \frac{p_0}{p} \frac{[\partial n/\partial T]_{(p/p_0)}}{[\partial n/\partial (p/p_0)]_T} \tag{20}$$

The enthalpy of vaporization can also be obtained by observing the pressure dependence of the temperature of the endothermic DTA peak due to bound water in the range 105–130°.[363] With this method one should be aware of possible major conformational changes at temperatures below that range. If available, isothermal calorimetry should be preferred.

These values relate to the change in enthalpy when 1 mol of water is bound under the pressure and temperature of the measurement. The enthalpy of the specific adsorption sites can be obtained from the temperature dependence of μ_1° through an analysis of the isotherms.

For biopolymers that undergo a major conformational change at low relative humidities the interpretation of sorption data in terms of specific hydration is highly questionable. Indeed, such polymers may have specific binding sites at low humidities, but these may be structurally unrelated to specific binding sites of the conformational state at higher humidities. In such cases sorption data provide little knowledge on specific water binding to the native state polymer. One may fear that this caution should be applied to most biopolymers, except such structurally stable ones as collagen, keratin, silk fibroin, cellulose, chitin, etc.

2.4. Applications to Biopolymers

Sorption studies have been performed on several biopolymers. A number of examples are listed in Table I, where the number of primary sites according to a BET treatment is given together with thermodynamic data where available. For most biopolymers the primary hydration amounts to 0.2–0.5 mol of water/100 g of polymer. Several guesses have been made as

TABLE I. Water Sorption Data for Biopolymers

Biopolymer	Ref.	Method[a]	BET sites, mol H_2O[b]	Thermodynamic data, kcal mol^{-1} H_2O	Remarks
DNA	231	AI	2[c]	$\Delta\mu^\circ$(liq-ads) = 2	h
DNA	363	DTA PD	— —	Δh(vap-ads) = 11.9 (wet)	i
Sol. RNA and polyriboadenyl	230	AI	1.9[d]	$\Delta\mu^\circ$(liq-ads) = 1.8	Hysteresis
V-amylose	528	VP	—	Δh(vap-hydr) = 10.4	—
Poly(ala,gly)	81	AI	—	$\Delta\mu^\circ$(liq-ads) = 3.4 $\Delta\mu^\circ$(coop) = 0.5	j
Poly(glu,lys)	81	AI	—	For COOH and NH_2 $\Delta\mu^\circ$(liq-ads) = 3.1 $\Delta\mu^\circ$(coop) = 1.0 for COO$^-$NA$^+$, NH$_3$$^+Br^-$ 3.8 and 0.7, resp.	j
Poly gly I	622	AI	0.18	Δh(vap-liq) = 3[d]	—
Poly gly II	622	AI	0.23	Δh(vap-liq) = 3[d]	—
Poly gly	585	DTA area	—	Δh(vap-ads) = 11.4[e]	i
Poly glu	585	DTA area	—	Δh(vap-ads) = 12.1[f]	i
Poly pro	585	DTA area	—	Δh(vap-ads) = 12.6[g]	i
B.s.a.	363	DTA PD	—	Δh(vap-ads) = 10.4 (wet)	i
B.s.a., collagen, β-lactoglobulin, casein	58	DSC	—	Δh(vap-ads) = 12 above 1 mol/100 g = 10 below 1 mol/100 g	See Fig. 1[i]
Hemoglobin	80	AI	0.320	—	Hysteresis
Methem. A	397	AI	0.317	—	Hysteresis[k]
Methem. S			0.322	—	k
Metmyoglobin			0.350	—	k
Collagen	305	AI	—	$\Delta\mu^\circ$(liq-ads) = 1.4	—
Collagen	91	AI	0.529	$\Delta\mu^\circ$(liq-ads) = 1.75	—

TABLE I. (Continued)

Biopolymer	Ref.	Method[a]	BET sites, mol H_2O[b]	Thermodynamic data, kcal mol^{-1} H_2O	Remarks
Gelatin	91		0.485	$\Delta\mu°$(liq-ads) = 1.73	—
Elastin	91		0.345	$\Delta\mu°$(liq-ads) = 1.50	—
Silk	91		0.226	$\Delta\mu°$(liq-ads) = 1.55	—
Wool	91		0.366	$\Delta\mu°$(liq-ads) = 1.46	—
Serum albumin	91		0.374	$\Delta\mu°$(liq-ads) = 1.47	—
Egg albumin	91		0.342	$\Delta\mu°$(liq-ads) = 1.45	—
β-Lactoglobulin	91		0.370	$\Delta\mu°$(liq-ads) = 1.30	—
Pseudoglobulins	91		0.397	$\Delta\mu°$(liq-ads) = 1.51	—

[a] AI = adsorption isotherm; DTA = differential thermal analysis; PD = pressure dependence; VP = vapor pressure; and DSC = differential scanning calorimetry.
[b] Per 100 g, unless noted otherwise.
[c] Per nucleotide.
[d] At 0.17 mol/100 g.
[e] At 0.29 mol/100 g.
[f] At 1.37 mol.
[g] At 0.61 mol.
[h] Structural changes occur that are not reflected in water binding.
[i] DTA and DSC methods measure total heat of adsorption in a range 80–130°. For comparison, Δh(vap-liq) = 9.7 kcal mol^{-1} at 100° for water.
[j] Sorption data were interpreted with a cooperative binding model. The binding free enthalpy for a site next to an already occupied site is larger by $\Delta\mu°$ (coop) than that for an isolated site.
[k] For D_2O the number of sites is, respectively, 0.314, 0.317, and 0.350.

to the nature of these primary binding sites, such as a monomolecular adsorption on polar sites (arg, his, lys, asp, gly, cys, met, ser, thr, tyr, trp), and multimolecular adsorption on specific sites. Also peptide backbone hydration of proteins has been suggested as primary adsorption. Such guesses are entirely hypothetical and arise from assumptions only. As will become more evident in Section 3, it is likely that biopolymers (specifically proteins) will exhibit specific binding sites where water molecules will bind to two, three, or four hydrophilic molecular groups simultaneously. The existence of such binding sites depends on specific geometric arrangements in the macromolecules. Any hydrogen-bonding donor or acceptor may serve as hydrophilic group. If more than two bonds are made to one water molecule, evidently no more than two groups may be of the donor or of the acceptor type.

Commonly occurring donors include: peptide–NH, $-NH_2$, $-NH_3^+$, –OH, –COOH, his–H^+, –SH (weak), trp NH.

Acceptors are: peptide CO, $-NH_2$, –OH, –COOH, $-COO^-$, his, $-S^-$ and –SH (weak), trp N, esters, ethers, ketones, phosphate groups.

Probably sites that provide only single hydrogen bonds to water molecules do not serve as primary sites of any reasonable free energy of binding that would become visible in a BET isotherm. This is due to the fact that already in the liquid almost two hydrogen bonds per molecule exist. While in addition the standard entropy of primary sites (at full occupation) is lower than that of the liquid, at least two "good" H bonds have to be present in order for the standard thermodynamic potential of the binding site to be lower than the potential of the liquid. Thus, depending on the geometry of hydrophilic groups in the neighborhood, a specific group may or may not form a primary binding site. Unfortunately, very few results have been published on the existence of such specific multiple binding sites. Weak bonds such as to cysteine or methionine hardly count as hydrogen bonds in this respect; they can only serve as weak additional stabilizing interactions.

In Table I some available data on binding enthalpies, generally with respect to the vapor state, are given. Different methods are not exactly comparable. Enthalpies from the temperature dependence of sorption data yield differential heats, $h = \partial H/\partial n$, while DTA or DSC measurements yield integral heats of adsorption for all the water present. These latter methods, moreover, yield the enthalpies of adsorption at the temperature of the observed water-dependent endothermic peak, which frequently occurs in the range of 80–130°.

In general the binding enthalpies at low coverage are between 1 and 3 kcal mol^{-1} below the liquid-state enthalpy. This is in the range which one may expect, indicating two to three hydrogen bonds per primary site. The fact that binding enthalpies with respect to the liquid are close to the fusion enthalpy of ice obviously implies nothing about possible icelikeness of bound water.

In Fig. 1 data of Berlin et al.[58] for a number of proteins have been plotted. A remarkable decrease of the binding enthalpy at low coverage is observed. If not an artifact of the method, this could indicate (as Berlin et al. suggest) that below 1 mol H_2O/100 g protein conformational changes occur in these proteins, changing the availability of suitable H_2O binding sites. It was shown by Aviram and Schejter[19] that lyophilized cytochrome c undergoes a conformational change (involving a change in spin state)

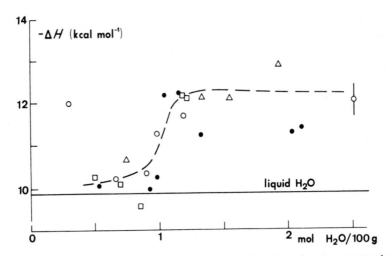

Fig. 1. Adsorption enthalpies for several proteins as a function of water content, plotted from data of Berlin et al.[58] (●) casein, (○) bovine serum albumin, (□) β-lactoglobulin, (△) collagen.

when dissolved, indicating a nonnative conformation in the lyophilized state. Bull[91] already remarked that lyophilized egg albumin is not as hydrophilic as the unlyophilized material, a difference that seems to be irreversible. Collagen shows the X-ray diffraction characteristics of the native three fold helix only when the relative humidity exceeds 10%,[221] a value at which the primary sites are about half filled.[80] The water sorption properties may not always reflect structural changes under dehydration: the conformation of DNA is critically dependent on humidity[142] but the adsorption isotherm is perfectly regular,[231] indicating the binding of two water molecules per nucleotide.

Also, the observation of adsorption/desorption hysteresis points to slow conformational changes (possibly even cross-linking) accompanying a change in hydration. Such conformational changes, although not applying to all biopolymers (e.g., polyglycine seems to be stable[622]), may be the rule rather than the exception. This should once more caution us against unjustified molecular interpretations of sorption data.

2.5. Freezing Behavior

The freezing behavior of hydrated biopolymers, as well as of tissue and cells, is generally observed to be anomalous in the sense that only part of the available water shows a phase transition even down to very low temperatures. In partially dehydrated material often no phase transition is observed at all. Moreover, a considerable hysteresis of freeze/thaw behavior is common.

An interesting study on DNA was made by Falk et al.,[232] which possibly may serve as an example of a general behavior. These authors studied the infrared absorption of water in hydrated DNA by a difference technique, clearly distinguishing ice and liquid water. It was observed that an inner layer of about ten water molecules per nucleotide is incapable of crystallization, even when the surrounding layer crystallizes to ice I. The "bound" water showed infrared spectra (of the stretching modes) indistinguishable from liquid water.

In collagen[151,503,505] a phase transition can be easily observed through magnetic resonance, since an observed splitting of the proton resonance is inversely proportional to the amount of exchangeable water present. In partially dehydrated samples, e.g., containing 2–3 mol of water/100 g collagen, no phase transition is seen. The rotational mobility of the water molecules simply slows down with decreasing temperature until the resonances become unobservable. Even at $-50°$ the water molecules retain a considerable motional freedom.

Berlin et al.[58] measured the heat of fusion of solvent water by DSC. They found that about 0.5 g water/g protein does not participate in the phase transition down to $-70°$. The authors consider this as supporting the view that such water of hydration already exists in an ice configuration at room temperature and hence does not freeze at lower temperatures. In view of the known dynamic properties of protein hydration (see Section 4.5), this explanation is almost certainly incorrect. In fact, this water of hydration is likely to be bound is such geometric configurations that it is incapable of ice formation without extensive rearrangements which are thermodynamically unfavorable.

In disperse systems, in which water may occur in isolated pockets of molecular size, one does not expect sharp transition temperatures. For example, the liquid → ice transition of 20 water molecules, disregarding all surface effects, covers for 10–90% completion a range of more than $20°$. Surface effects may shift the range of freezing temperatures. Thus phase transitions in water may occur that do not have the appearance of proper freezing transitions.

A comparison with sorption isotherms shows that the observation of roughly 0.5 g unfreezable water/g protein is entirely reasonable. No peculiar properties of hydrates at low temperatures need be invoked that are not already to be found in partially dried material. If one assumes that the hydrate structure in a sorption experiment is in principle equivalent to the hydrate structure in frozen conditions at the same thermodynamic potential, the freezing temperature T can be found from the following relation:

$$\int_{273}^{T} [s(\text{ads}) - s(\text{ice})]\, dT = RT \ln(p/p_0) \tag{21}$$

where $s(\text{ads})$ is the partial molal entropy of adsorbed water for a given water content and p/p_0 is the relative humidity at $0°$ in equilibrium with this water content. Or, roughly

$$T_m \sim \frac{RT \ln(p/p_0)}{s(\text{ads}) - s(\text{ice})} \tag{22}$$

The adsorption entropy can be derived from the temperature dependence of the adsorption isotherms. For a typical example, if we plot Bull's data[91] for collagen as the amount adsorbed versus $\mu - \mu_L = RT \ln(p/p_0)$, we find from the temperature dependence that at low water contents the adsorption entropy is close to that of ice. For higher water contents (above about 0.6 mol H_2O/100 g) the adsorption entropy is about 1 cal $K^{-1} mol^{-1}$ below the liquid entropy. For a water content of 2 mol H_2O/100 g one then expects freezing to begin at -25 to $-30°$ and extend slowly to lower temperatures, reaching -80 to $-100°$ for a water content of 1 mol/100 g. Experimental observation of such behavior would not be interpreted as freezing. Molecules bound with the entropy of ice or less would of course genuinely not freeze at all.

Hysteresis in freezing/melting is to be expected for disperse biopolymers. For water in small pockets or channels nucleation will not easily occur, giving rise to considerable supercooling. Once crystals begin to form, water from neighboring areas diffuses to the growing crystals. If the freezing process has not been extremely fast, crystals of reasonable size are formed which show a normal melting behavior. The melting curves are closer to thermodynamic equilibrium than the freezing curves.[505]

The "nonfreezing" water represents a larger quantity than the primary BET hydration. It is likely that any identifiable, specifically bound water belongs to the nonfreezing fraction, but that in addition more loosely bound molecules are involved.

3. STRUCTURAL PROPERTIES OF SORBED WATER

Structural data on hydration are obtained by X-ray and neutron diffraction methods and, to a limited extent, by nuclear magnetic resonance (nmr) and infrared dichroism. In macromolecular solutions, in the presence of paramagnetic ions or groups, the positions of water molecules relative to the paramagnetic centers may in favorable cases be determined by nmr. In single crystals in the solid state the orientations of water molecules can be determined both by proton and deuteron nmr (of H_2O and D_2O, respectively). For simple hydrates very useful results have been obtained, but for biopolymers the applications are very limited. In the case of proton resonance a prohibitive overlap with macromolecular resonances occurs, while in the case of deuteron resonance large single crystals are required and the exchange of water molecules complicates the resonances. In oriented liquid crystalline systems, such as can be prepared from dispersions of certain polypeptides,[615] DNA,[611] and lipid mixtures[190,335] or occur naturally as oriented fibrous systems such as collagen,[503] nmr measurements are quite feasible. If not complicated by rapid proton exchange processes, such measurements yield the average anisotropy of the orientation of water molecules with respect to the symmetry axis of the system. From proton resonance one obtains the average of $3\cos^2\theta - 1$, where θ is the angle between the symmetry axis and the vector connecting the two protons of a water molecule. From deuteron resonance in D_2O one obtains the average of $3\cos^2\theta'$ -1, where θ' is the angle between the symmetry axis and the direction of the O–D bonds. From both types of data the average orientation can be determined.

Similar information on the average orientation of water molecules in fibrous systems can be obtained from infrared dichroism of the vibrational bands of water. The fundamental vibrations of the water molecules are not very suitable for such investigations because of the strong overlap with macromolecular vibrational bands. The stretch frequencies occur as a band around 3450 cm^{-1}; this band is generally too broad[75,109] to allow resolution of the symmetric and antisymmetric stretch modes. Since the two modes have perpendicular transition moments, no straightforward structural information can be obtained from dichroism of this band. The deformation mode near 1600 cm^{-1} is strongly overlapped by amide-I absorptions in proteins.

Although very little explored, it seems that a combination band at 5150 cm^{-1} is relatively free from overlap with macromolecular absorptions and is very suitable for studies of water orientation.[261] The band is due to a combination of the antisymmetric stretch and the deformation mode of the water molecule; its transition moment is therefore parallel to the vector connecting the two protons. Thus we find for a fibrous system that

$$\overline{\tfrac{1}{2}(3\cos^2\theta - 1)} = (A_\| - A_\perp)/(A_\| + 2A_\perp)$$

where $A_\parallel$ and $A_\perp$ are the optical densities with the fibers parallel and perpendicular to the electric field, respectively. The angle θ has the same meaning as in the case of proton dipole splitting. Hence similar information is obtained with two important differences: (a) nonexchangeable molecules also contribute and (b) the sign of the average angular factor is obtained as well, in distinction with the observation of proton splitting. For collagen, infrared results[261] are in fair agreement with nmr results.[503]

Infrared difference studies between hydrated and dry bovine serum albumin[94] have shown the presence of water molecules that adsorb around 3300 cm^{-1}, having an IR spectrum similar to that of ice. Of course no structural inference can be drawn from this observation, but such absorption may well be expected from specifically bound hydrogen-bonded water molecules in the protein interior.

Detailed structural information on water associated with biopolymers can be obtained by X-ray diffraction of single crystals. For this purpose high-resolution data (~ 2 Å) are required and a considerable effort has to be made to interpret electron density maps in terms of water molecules because in large polymers single oxygen atoms hardly stand out from the background noise. On the surface of a biopolymer motional freedom of the water molecules and side chains, as well as statistical occupation of sites, renders interpretation even more difficult. This, combined with the fact that the determination of the arrangement of water molecules is generally not the prime concern of the protein X-ray crystallographer, is probably the reason why little attention has been paid in the literature on X-ray biopolymer structures to the hydration structures.

In the case of collagen, Ramachandran and Chandrasekharan[591] have proposed a very specific hydration structure on the basis of model building and X-ray diffraction. In this structure (Fig. 2) two water molecules are specifically bound to the backbone of the macromolecule for every three amino acid residues. The water molecules make two hydrogen bonds to two different protein chains of the collagen threefold helix, thus stabilizing the tertiary structure of the macromolecular array. Also in polypeptides that can serve as a model for collagen structure, such water molecules are likely to exist.[721] A careful analysis of the nmr data on oriented hydrated collagen[305] has shown that Ramachandran's model is entirely consistent with the experimental data.[503]

In carboxypeptidase A, Quiocho and Lipscomb[588] found ten water molecules trapped inside the protein molecule. Two make four, seven make three, and one makes two hydrogen bonds each. Some of the hydrophilic side chains situated inside the molecule are bound to occluded water mol-

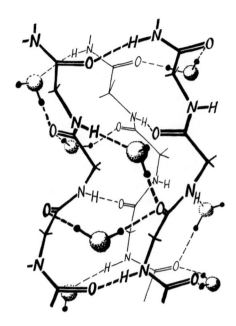

Fig. 2. Specific hydration structure proposed for collagen by Ramachandran and Chandrasekharan.[591] The drawing is based on a photograph in Ref. 591.

ecules. A water molecule is also coordinated to the zinc ion found in the active site of carboxypeptidase A; this molecule is displaced by the binding of substrates.[525] In addition, water molecules are found in the active site pocket. In the surrounding liquid, distinct densities for about 50% of the water molecules were observed,* but no clear hydration structure did emerge. The occupation of sites is apparently not the same in all unit cells, because apparent water–water distances were often observed to be too short.

Carbonic anhydrase C[436] represents a case where water molecules are involved in the catalytic process. A zinc ion is situated at the active site (bound to three histidines) which carries water as its fourth ligand. In the process of catalysis this water molecule is displaced. Liljas et al.[436] have interpreted the electron density in the cleft of the active site as representing nine water molecules in a configuration characteristic of ice I (see Fig. 3).

In ribonuclease S two water molecules have been reported.[719] One of these is hydrogen-bonded to His-119, an amino acid that is implicated in the enzymatic mechanism.

* W. N. Lipscomb, personal communication.

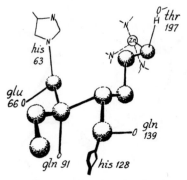

Fig. 3. The structure of water in the active site of carbonic anhydrase C, as proposed by Liljas *et al.*[436] Protons are not shown.

A thorough investigation of the electron density map of papain[197] by Drenth and Jansonius has shown many structural details of the specific binding of water molecules to the macromolecule.* About 40 water molecules are resolved by X-ray diffraction. A large fraction of these molecules is found inside the macromolecule, some entirely isolated from the surrounding solvent, some bound to other water molecules in chains connecting to water at the surface of the molecule. Other molecules occupy surface positions and are presumably also bound to molecules in the liquid. Most molecules engage in two or three hydrogen bonds to the macromolecule or to other water molecules; in some cases even four bonds are made. Most abundant are hydrogen bonds to N–H and $C=O$ groups of the peptide backbone of the protein. Almost equally abundant are water–water bonds. Less frequent are bonds to side chains (Lys NH_3^+, Tyr OH, Glu COO^-, Trp NH, Ser OH, Thr OH, Gln amide CO, Gln amide NH_2). In Figs. 4–7 some examples of structural relations are given. Binding of water molecules to adjacent peptide groups on a protein chain frequently occurs, sometimes with one molecule (Fig. 4), and sometimes with two (Fig. 5) or three molecules (Fig. 6). In many instances water molecules connect different portions of the protein chain (Fig. 4 and 7). An interesting pocket of water occurs inside the molecule where two lysine and one glutamic acid residue are situated without any other contact to the outside than through a hydrogen-bonded chain of a few water molecules (Fig. 8). These specifically bound water molecules clearly fulfill a structural role, stabilizing the native struc-

* The author is indebted to his colleagues J. Drenth and J. N. Jansonius of the Department of Structural Chemistry of the University of Groningen for providing the opportunity to study their molecular model of papain.

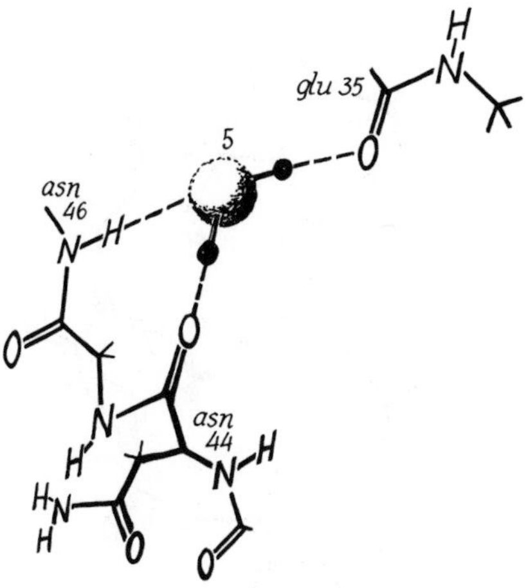

Fig. 4. Example of the arrangement of a specifically bound water molecule in papain, as detected by X-ray diffraction. Figures 4–7 are in approximately correct perspective, but not drawn to scale. Proton positions in Figs. 4–7 indicate in several cases only one of the possible configurations.

ture. They are particularly important for internal hydrophilic groups, where the presence of water molecules provides for a favorable interaction energy.

It is likely that such water molecules will be found in many more biopolymers if high-resolution electron density maps are carefully studied. In metalloproteins water is often found as a ligand of the metal. For example, in subtilisin Novo[355] an alkali-metal binding site has been found in the vicinity of which three water molecules are visible, one being a ligand of the ion.

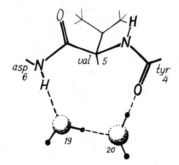

Fig. 5. Example of a specific hydration structure in papain, showing a two-molecule water bridge between adjacent peptide groups.

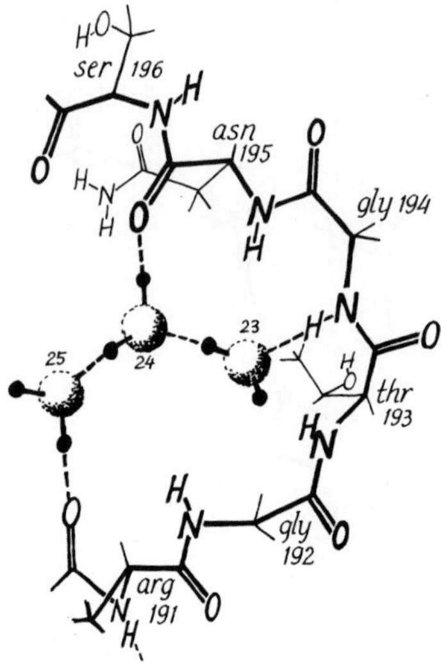

Fig. 6. Example of a three-molecule water chain in papain, forming bridges between backbone peptide units on the peptide chain.

An X-ray study of a hemicellulose, β,D(1–4′)-xylan hydrate,[529] has revealed an important structural role of water. A helical chain of water molecules was found to provide hydrogen bonds to oxygen atoms and hydroxyl groups of the polysaccharide molecules. Each water molecule is engaged in four hydrogen bonds. The macromolecules are arranged in a three-fold helical conformation. It is only because of the favorable hydration

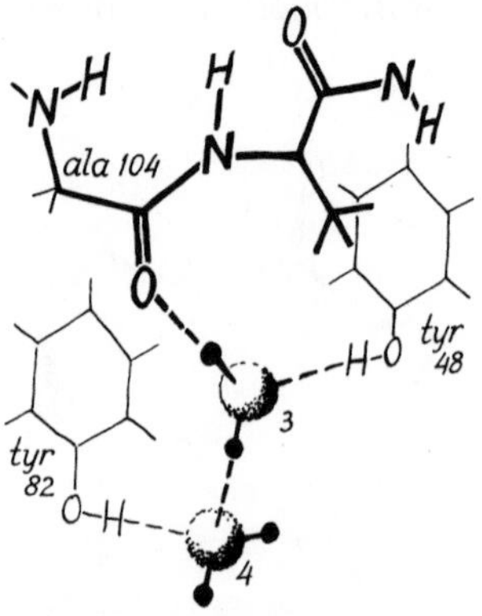

Fig. 7. Example of water molecules in the interior of papain, binding to three sequentially unrelated parts of the protein and involving bonds to tyrosine.

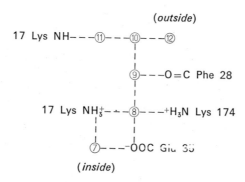

Fig. 8. Arrangement of water molecules in a hydrophilic cleft inside the papain molecule that contains two lysines and one glutamic acid. The water chain communicates on one side with the outside solvent. Only the topology, not the structure, is represented.

structure that a threefold helix is preferred; energy calculations have shown that otherwise a twofold screw helix would be more favorable. In a marine algal polysaccharide, poly-α-L-glucuronic acid,[18] a water molecule provides four hydrogen bonds, linking three polysaccharide chains together. In fact, all macromolecular oxygens (if not involved in intramolecular bonding) participate in water-mediated interchain bonding. It is clear that in these cases water molecules provide an essential contribution to conformational stability.

No experimental evidence has been found for more extensive specific hydration structures around the macromolecules, extending several angstroms into the surrounding liquid. Thus the remarkably frequent occurrence of repeating distances of 4.74 Å or multiples thereof in biopolymers[50,51] remains of mysterious significance. This distance is the expected second neighbor distance in regular structures that could occur in water and that could conceivably be stabilized by substrates providing hydrogen bonds in a fitting pattern. The nucleation of ice is related to this property, and many steroid hormones have been found to be excellent ice nucleators.[276] In (poly) saccharides such structural fits, which appear to depend critically on configuration, occur as well (Chapter 1, Volume 2). Nevertheless, evidence for "structured" water as a result of these structural similarities is nonexistent. In a study of DNA hydration[504] no relation between the degree of orientation of water molecules and the occurrence of a multiple of 4.74 Å repeat was found when the A, B, and C forms of DNA were compared. At present, early proposals of extensive icelike structure of biopolymer hydration[375,656] are primarily of historical value.

4. DYNAMIC PROPERTIES OF SORBED WATER

4.1. Types of Motion Occurring in Hydrates

Dynamic aspects of hydration are specified by the type and rate of motion of the water molecules. Three types of motion should be distinguished: proton exchange processes, translational motions, and rotational or librational motions.

Proton exchange processes limit the lifetime of a water molecule as a unit. In solution, and also in macromolecular systems studied thus far, the lifetime of a water molecule at room temperature is generally of the order of milliseconds. Since rotational and translational lifetimes in macromolecular hydrates are as a rule several orders of magnitude shorter, it is meaningful to consider the H_2O molecule as a separate entity for the purpose of rotational and translational processes. This is not a trivial matter at all: If hydrates would behave like doped ice with fast proton exchange and slow molecular motion, a situation could occur in which the lifetime of a water molecule as an entity is of the same order as the translational lifetime of an oxygen nucleus. The terms translation and rotation of water molecules would then lose their usual significance.

Translational motions are characterized by the velocity autocorrelation function

$$K_v(\tau) = \langle \mathbf{v}(t) \; \bar{\phi} \; \mathbf{v}(t + \tau) \rangle \tag{23}$$

where $\mathbf{v}$ is the velocity of the molecular center of gravity, and averages are taken over the time t or over the ensemble of molecules. The velocity autocorrelation function is related to the self-diffusion coefficient D by

$$D = \frac{1}{3} \int_0^\infty K_v(\tau) \, d\tau \tag{24}$$

The diffusion coefficient can be measured by pulsed nmr techniques in combination with the application of a magnetic field gradient. Related, but not equivalent, to the velocity autocorrelation function is the time correlation of intermolecular distances. Proton relaxation times contain a contribution from the fluctuation of certain functions of the intermolecular distances. The difference with autocorrelation functions results from cooperative motion (as in clusters). In principle the most complete description of molecular motion can be obtained through the measurement of the change in energy of scattered cold neutrons, which yields after proper Fourier transformation the time-dependent pair distribution function $G(r, t)$. This function describes the probability that a molecule is found at position r and time t, given the presence of a molecule at the origin at time zero. Cold neutron scattering has been applied extensively to liquids, but not to hydrated biopolymers.

Rotational motions cannot be described by single correlation functions. One reason is that the water molecule has three different moments of inertia about its three molecular axes. Another reason is that one can define correlation functions of more than one type of goniometric function of the orientation angles. Which function is appropriate depends on the experimental technique employed to gain information on the rotational motion. The useful techniques are measurements of dielectric relaxation and proton, deuteron, and 17-oxygen nmr. The band shapes of infrared and Raman lines contain information on rotational motions, although in hydrated biopolymers complications arise that limit the applicability of these methods.

In addition to these types of molecular motion one may in hydrated biopolymers distinguish various classes of bound water molecules by their different behavior on a time scale which is long compared to the characteristic times for individual molecular motions. Thus, a certain number of molecules may be trapped in binding sites with a residence time much longer than the rotational times of molecules outside these sites. It then depends on the time resolution of the measuring method whether such molecules are seen as a separate class. Generally speaking dielectric relaxation and frequency-dependent nmr relaxation can distinguish such classes, while single frequency nmr does not.

In disperse biopolymer systems complications may (and often will) arise from the heterogeneity of the system. This applies to both dielectric and nmr measurements. This aspect will be treated in the following sections.

4.2. Dielectric Relaxation

The dielectric constant is a complex quantity $\varepsilon = \varepsilon' - i\varepsilon''$, of which the real part ε' represents the ratio of the in-phase component of the dielectric displacement $\mathbf{D}$ to the electric field $\mathbf{E}$, and the imaginary part ε'' represents the 90° out-of phase component of D. ε' and ε'' correspond to the lossless and the dissipative parts, respectively, of the response to an oscillating field and both are functions of the applied frequency.[271,348] The interesting contribution to the dielectric constant from the viewpoint of hydration comes from the dipole moments of the water molecules. The frequency dependence of this contribution to ε' is directly related to the

Fourier transform of the correlation function $\langle m_z(t)m_z(t + \tau)\rangle$ of the dipole moment in the field direction m of a large assembly of molecules. If the dielectric behavior is found to be of the Debye type:

$$\varepsilon(\omega) - \varepsilon(\infty) = [\varepsilon_0 - \varepsilon(\infty)][1 + i\omega\tau_d]^{-1} \qquad (25)$$

this correlation function is exponential

$$\langle m_z(t)m_z(t + \tau)\rangle = \langle m_z^2\rangle \exp(-\tau/\tau_d) \qquad (26)$$

and the Cole–Cole plot of ε'' versus ε' follows a semicircle. $\varepsilon(\infty)$ is the high-frequency limit of the dielectric constant above the frequency range of dipolar relaxation and ε_0 is the limiting value of ε below that range. τ_d is the dielectric relaxation time, which equals 9.6×10^{-12} sec for liquid water at $20°$, 2.2×10^{-5} sec for ice at $0°$, and which will have intermediate values for hydrated biopolymers. This dielectric relaxation time is not equal to the correlation time of the molecular dipole moment, because in water (and other associated liquids) the dipole moment of any molecule is related to the dipole moments of its neighbors. Hence the time dependence of the dipole moment of an assembly of molecules is not simply determined by the time dependence of the statistical sum of individual molecular dipole moments, but is also dependent on molecular interactions and on the rate of propagation of the influence of molecular orientations to neighboring molecules. The exact description of this process depends on a detailed molecular model. It is not at all sure that the correlation function of the motion of the molecular dipole moment is an exponential function of time, even if the macroscopic dielectric relaxation function is found to be exponential.

The correlation time τ_m of the molecular dipole moment is defined in the case of exponential relaxation time by

$$\langle \cos\theta(t)\cos\theta(t + \tau)\rangle = \langle \cos^2\theta\rangle \exp(-\tau/\tau_m) \qquad (27)$$

where θ is the angle between the molecular dipole moment and a z axis. No general relation between τ_m and τ_d exists, although Glarum[284] derived the useful formula

$$\tau_m = 2\varepsilon_0[3\varepsilon_0 + 3(\infty)]^{-1}\tau_d \qquad (28)$$

More serious than the uncertainty in the derivation of τ_m from the observed dielectric relaxation are the complications arising from sample heterogeneity. In disperse systems of biopolymers one is often confronted with a structure consisting of macromolecules with a low conductivity, surrounded by an aqueous layer containing mobile ions with a relatively high conductivity. In such structures a dispersion of the dielectric constant will be observed that resembles a dispersion due to dipolar relaxation, but that is totally unrelated to such a relaxation. These *Maxwell–Wagner effects*[690] have to be carefully identified and preferably excluded before dielectric relaxation data can be interpreted correctly.[306] The following consideration of an oversimplified model serves to provide a feeling for the cause of the Maxwell–Wagner dispersion.

Consider a system consisting of alternate layers of macromolecules with dielectric constant ε_1 and liquid with conductivity σ. Let us apply a field perpendicular to the layers. The dielectric displacement D will be continuous perpendicular to the layers. In the macromolecular layer $D = \varepsilon_1 E_1$; in the liquid layer the current density $J = i\omega D = \sigma E_2$. Thus $E_1 = D/\varepsilon_1$ and $E_2 = i\omega D/\sigma$. If both layers have equal thickness, the macroscopically applied field E is equal to the average of E_1 and E_2:

$$E = \tfrac{1}{2}D[(1/\varepsilon_1) + (i\omega/\sigma)] \tag{29}$$

and the observed $\varepsilon = D/E$ is given by

$$\varepsilon = \frac{2\varepsilon_1}{1 + i\omega(\varepsilon_1/\sigma)} \tag{30}$$

Hence a normal Debye dispersion is observed with a relaxation time ε_1/σ. For $\varepsilon_1 = 2\varepsilon_0$ (ε_0 being the permittivity of vacuum: 8.854×10^{-12} F m^{-1}) and $\sigma = 0.015 \ \Omega^{-1} \ m^{-1}$ (as for 1 mF KCl solution) the "relaxation" time is 10^{-9} sec. The dispersion centers around 100 MHz in this case, but in most practical cases Maxwell–Wagner dispersion is found in the 100 kHz to 10 MHz range. The effect is so strong as to obscure dispersions due to dipolar relaxation.

Other complicating dispersions arise from electrode polarization and surface charges of polyelectrolytes at low frequencies. Water relaxation will be found at relatively high frequencies (100 MHz to 10 GHz). This used to be an experimentally extremely difficult frequency range. Since the development of time domain reflectometry (TDR),[238] sometimes referred to as time domain spectroscopy (TDS),[653] measurements in this range are quite feasible, including applications to biopolymers.

4.3. Nuclear Magnetic Relaxation

With *proton magnetic resonance* one measures the rate at which the magnetic coupling between protons is modulated by molecular motions.

Since the magnetic moment of the proton is large and the abundance of other magnetic nuclei (^{17}O and ^2D) is small, the proton–proton dipole interactions are the dominant factor determining the relaxation times of proton spins and the linewidth of the proton resonance. This statement is not true if paramagnetic centers such as Cu^{2+}, Fe^{3+}, and Mn^{2+}, spin label molecules, or magnetic impurities are present. We shall not consider this special case; the explicit use of proton relaxation in the presence of paramagnetic centers has been reviewed by Cohn.[132] The exact calculation of relaxation processes caused by dipolar interactions requires a quantum mechanical treatment that can be found in various textbooks.[3,6,100] A simplified treatment that nevertheless shows the essential features of the interaction is based on the notion of local magnetic fields produced by neighboring protons at the site of one proton. Such fields modify the resonance of this proton, and since the local fields rapidly fluctuate with the molecular motion, they provide a possibility for the exchange of energy of the proton both with neighboring protons and with the modes of molecular motion (the "lattice"). Thus, by both of these processes the lifetime of a proton in a particular spin state is limited, producing a broadening of the resonance as a result of Heisenberg's principle $\Delta E\, \Delta t \geq \frac{1}{2}h$. The characteristic lifetime of a spin state is called the *spin–spin relaxation time* or transverse relaxation time T_2 because it is equal to the correlation time of the average transverse component of the proton magnetic moment (perpendicular to the external field), which is the component measured in a resonance experiment. The lifetime-determined line shape is Lorentzian, i.e., proportional to $[1 + (\omega - \omega_0)^2 T_2^2]^{-1}$. The exchange of energy with the lattice alone is characterized by a different relaxation time T_1, the *spin–lattice relaxation time*. T_1 is also referred to as the longitudinal relaxation time because it is equal to the time constant with which the average z component of the proton magnetic moment (in the direction of the external field) adjusts itself to its thermal equilibrium value. Under special experimental circumstances one can measure yet another relaxation time,[449] $T_{1\varrho}$, which for our purpose is practically equal to T_2.

The local field in the z direction produced by a proton with magnetic moment μ (1.41×10^{-23} erg G^{-1}) at a distance r and in a direction at an angle θ with respect to the external magnetic field is given by

$$H_{\text{loc}} = (\mu/r^3)(3 \cos^2 \theta - 1) \tag{31}$$

Quantum considerations modify this expression by a factor 3/2 if the interaction between two equivalent nuclei is considered. The nearest neighbor to each proton is the other proton at the same water molecule. The local field can then be oriented up or down depending on the spin state of the neighboring proton, thus causing a splitting of the resonance. The local field fluctuates because of rotational motions that modify the intramolecular interaction by changing θ and translational motions that modify intermolecular interactions by changing both r and θ.

If the rotation of the water molecule is isotropic, the angular term in eqn. (31) averages to zero. In anisotropic samples, however, the rotation is often also anisotropic and results in a residual average intramolecular interaction. If not modified by fast proton exchange processes, this results in a splitting of the observed resonance. In DNA[504] and collagen[503] line splittings in the range of 0.1–1 G have been observed. This represents a considerable reduction compared to the value of 21 G for the maximal splitting in a rigid water molecule with interproton distance of 1.59 Å.

The relaxation times are determined by certain Fourier components of the fluctuations of the local field. Both T_1^{-1} and T_2^{-1} are proportional to the mean square fluctuation H_{loc}^2, which in the absence of motion is equal to the second moment of the absorption curve. The spin–lattice relaxation is determined mainly by the Fourier component at the resonance frequency. For T_2, however, the Fourier component at zero frequency is important as well. Under the usual assumption that the fluctuations have an exponential time decay with a correlation time τ_c, the following formulas apply for a proton pair[100]:

$$T_1^{-1} = (6/5)\gamma^2\mu^2 r^{-6}\tau_c[(1 + \omega_0^2\tau_c^2)^{-1} + 4(1 + 4\omega_0^2\tau_c^2)^{-1}] \tag{32}$$

$$T_2^{-1} = (9/5)\gamma^2\mu^2 r^{-6}\tau_c[1 + (5/3)(1 + \omega_0^2\tau_c^2)^{-1} + (2/3)(1 + 4\omega_0^2\tau_c^2)^{-1}] \tag{33}$$

Here γ is the gyromagnetic ratio of the proton (2.675×10^4 sec^{-1} G^{-1}) and ω_0 is the resonance angular frequency. The behavior of T_1 and T_2 for an isolated rotating water molecule is given in Fig. 9 for several resonance frequencies as a function of the rotational correlation time.

For water in both the liquid and the solid states the rotational and translational lifetimes are about equal.[50,313] This is a result of the strongly directional intermolecular forces in water and is expected to apply as well to accessible water molecules in hydrated biopolymers. The intramolecular contribution to the total interaction in the condensed state can then be estimated from the ratio of the second moment for a single water molecule ($1.8\mu^2 r^{-6} = 22$ G^2) to the second moment of ice[418] (37 G^2). This means that 60% of T_1^{-1} and T_2^{-1} are due to intramolecular contributions, in agreement with conclusions of Krynicki[417] based on quite different arguments. Hence

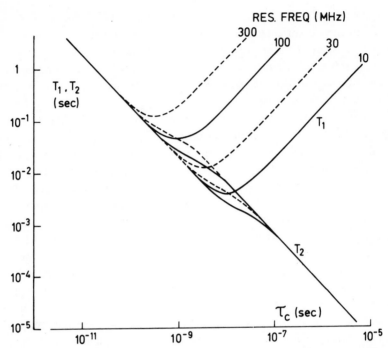

Fig. 9. The spin–lattice and spin–spin relaxation times for a rotating water molecule as a function of the rotational correlation time τ_c, for various values of the resonance frequency.

the curves in Fig. 9 are expected to shift to lower values by a factor of 1.7 due to intermolecular contributions to the relaxation. In biopolymers with low water content these intermolecular interactions may be of less relative importance because the approach to external protons is not as close as in liquid. In collagen it was concluded from isotope substitution measurements that the linewidth was mainly determined by intramolecular interactions.[503]

The interpretation of T_1 and T_2 in hydrated biopolymers is by no means as simple as Fig. 9 suggests. The reasons are: distribution of correlation times, exchange with specific binding sites, proton exchange processes, spin coupling to macromolecular protons, spin–rotation interaction, and heterogeneity on a macromolecular scale.

A *distribution of correlation times* (corresponding to nonexponential correlation functions of the spin interactions) leads to a frequency dependence of T_1 and T_2 that differs from that of Fig. 9. With the assumption of a lognormal distribution of the form

$$P(\tau)\,d\tau = B\pi^{-1\,2}\exp(-B^2Z^2)\,dZ \qquad (34)$$

where $Z = \ln(\tau/\tau^*)$, B is a constant, and τ^* is the median correlation time $\exp\langle\ln\tau\rangle_{\mathrm{av}}$, complete functions for T_1 and T_2 have been derived by Resing.[599] Lynch et al.[461] have published useful formulas and graphs for temperature-dependent correlation times. The application of such formulas is not without danger: By the introduction of adjustable parameters it represents a way of curve fitting that by itself does not justify the particular form chosen for the distribution of correlation times. Only an extensive nmr investigation over a wide frequency range enables us to determine the shape of the distribution function. Experimental limitations prevent investigations at frequencies >300 MHz, while at least part of the water relaxation is commonly found above that frequency. The additional determination of ultrahigh-frequency dielectric relaxation, possible with TDR techniques, may prove to be essential for a correct interpretation of nmr data. In the low-frequency range special instrumentation is required[449,598] and measurements are not easily performed.

If the distribution of correlation times extends to times longer than 10^{-5}–10^{-6} sec, part of the resonance is broadened to (or approaching) the "rigid lattice," i.e., motionless, resonance. In such cases a narrow line is observed, superimposed on a broad one. The broad water resonance may not be distinguishable from that of polymer protons. The intensity ratio of the two lines is temperature dependent and the resonance may be interpreted as an apparent phase transition, while no real phase transition exists. This effect is clearly observed in hydrated zeolites and has been excellently reviewed by Resing.[600]

If *specific binding sites* exist, a two-state distribution may be a more faithful representation of the real distribution of correlation times than a log-normal distribution as given in eqn. (34). The approximation may be made that a single type of specific binding site exists with a residence time τ_b of a water molecule in that site, while all other water molecules have a much shorter correlation time τ_f, approaching the correlation time in the free liquid. Let p represent the fraction of bound water molecules. The correlation time of the molecular motion of the water molecule at the binding sites may be shorter than τ_b if a rotational motion still exists in the bound state. In solutions that may be generally the case; in hydrated solid systems the backbone mobility will be so low that the correlation time of the interactions of the water molecule will be limited by the residence time τ_b. The total correlation function of the molecular motion will now simply have the form $p\exp(-\tau/\tau_b) + (1-p)\exp(-\tau/\tau_f)$. For $\tau_b > 10^{-5}$ sec the bound fraction will show its rigid lattice width and the fraction $1 - p$ will behave according to eqns. (32) and (33), which with short τ_f leads to equal and frequency-independent values of T_1 and T_2. For the more interesting case $\tau_b < 10^{-6}$ sec both T_1^{-1} and T_2^{-1} will be weighted averages of the inverse relaxation times of the bound state (with correlation time τ_b) and the "free" state (with correlation time τ_f). This leads to values of T_1 much larger than T_2. In fact, T_1 may in practice be determined by other processes as well, such as interaction with macromolecular protons. For the important case that $\tau_b \gg \omega_0^{-1}$ and $\tau_f \ll \omega_0^{-1}$, eqns. (32) and (33) yield the following simple forms:

$$T_1^{-1} = p(12/5)(A/\omega_0^2\tau_b) + (1 - p)6A\tau_f \tag{35}$$

$$T_2^{-1} = p(9/5)A\tau_b + (1 - p)6A\tau_f \tag{36}$$

where

$$A = \gamma^2\mu^2r^{-6}(8.7 \times 10^9 \text{ sec}^{-2}) \tag{37}$$

The strength of interaction A may be modified because of intermolecular contributions as discussed before. Since $\tau_b \gg \tau_f$, the spin–spin relaxation time will be determined by the bound molecules even if the fraction bound is very small. However, the free molecules do contribute strongly to the spin–lattice relaxation.

Spin coupling to macromolecular protons occurs with correlation times comparable to the residence time at binding sites. The distance of nearest approach to intramolecular protons will not be smaller than 2–3 Å, as compared to 1.59 Å in the water molecule. The direct contribution to the relaxation times from this coupling will not exceed 10–20% of the intramolecular contribution, and thus this coupling may not seem significant. However, the phenomenon of *spin diffusion* in the macromolecule may influence the spin–lattice relaxation to a significant extent in cases where $T_2 \ll T_1$ for the water molecules.

Spin diffusion is the exchange of energy within the system of coupled spins. The rate at which a proton exchanges its average energy with surrounding protons is determined by T_2^{-1}. If $T_2 \ll T_1$, the energy exchanges much more rapidly within the system of spins than with the thermal sink of the lattice. The consequence of this is that the T_1's of all coupled protons tend to become equal even if the contributions of individual protons are different. Spin diffusion between mobile water molecules is of no importance for the simple reason that molecular diffusion is much faster than spin diffusion because $\tau_c \ll T_2$. In macromolecules, where generally $T_2 \ll T_1$, spin diffusion is of major importance. If for the water of hydration $T_1 \gg T_2$, there will be exchange of energy between the macromolecular protons and the water protons through spin coupling. The effect will be an averaging of the spin–lattice relaxation of the water protons and the macromolecular protons. The extent of this effect is not easily evaluated: Experimentally T_1 of the macromolecular protons should be determined for a sample hydrated with D_2O as well as H_2O to estimate a correction to the measured T_1 of water. Thus the use of T_1 for the interpretation of water relaxation data in terms of the dynamics of the water molecules requires more caution than does the use of T_2.

If the sample is *heterogeneous*, the interpretation of T_2 is no longer straightforward and a contribution to the line broadening may result that does not reflect the local dynamic properties of the water molecules. This effect, which is similar to the pronounced heterogeneity broadening in the sodium resonance,[52] occurs as well in proton and deuteron resonance. As already discussed in connection with eqn. (31), oriented biopolymers may give rise to splittings of the water resonance line in the range of 0.1–1 G separation when the axis of anisotropy is oriented along the external field. It is clear that such material in a randomly oriented sample yields a superposition spectrum for all orientations, provided that the molecular ordering is still present on a microscopic scale. If diffusion of water molecules occurs rapidly enough between regions of different orientation, the superposition spectrum is modified into a line of Lorentzian shape. We can derive that the relaxation rate of this line is given by

$$T_2^{-1} = (1/20)\gamma^2(\Delta H)^2\tau_c \tag{38}$$

where ΔH is the maximum line splitting observed for the completely aligned material and τ_c is the rotational correlation time determined by the diffusion process. Here τ_c measures the time in which water molecules must have diffused to a region with appreciably different orientation through diffusion. We shall define a distance a by the equation

$$a^2 = 6D\tau_c \tag{39}$$

where D is the diffusion constant of water. The distance a is the *structural correlation distance* of the heterogeneous sample and represents roughly the distance over which molecular ordering persists. From eqns. (38) and (39) we find

$$T_2^{-1} = \gamma^2(\Delta H)^2 a^2/120D \tag{40}$$

In Table II the T_2 values are given for two values of ΔH and various structural correlation distances, assuming a diffusion coefficient $D = 10^{-5}$

TABLE II. Spin–Spin Relaxation Time T_2 of Water Protons Due to Sample
Heterogeneity[a]

	T_2, sec	
a, nm	$\Delta H = 0.1$ G	$\Delta H = 1$ G
10 (100 Å)	1.7×10^2	1.7
10^2	1.7	1.7×10^{-2}
10^3 (1 μm)	1.7×10^{-2}	1.7×10^{-4}, $\sim$H
10^4	H	H

[a] a is the structural correlation distance. ΔH is the maximum line splitting for oriented material. H ("heterogeneous") indicates that the line shape is not Lorentzian, but approaches
the superposition spectrum. A diffusion constant of 10^{-5} cm² sec⁻¹ is assumed.

cm² sec⁻¹, close to that of liquid water. For lower diffusion coefficients the
line broadening increases.

It is clear that in fibrous macromolecules the effect of sample heterogeneity on T_2 may be very severe, leading to T_2 values in the millisecond
range. If there is any suspicion of local ordering over several hundred
angstroms or more, the usual interpretations of T_2 cannot be relied upon.
Complete clarity can only be obtained if the diffusion coefficient of water
is measured and the material can be compared in states of different microscopic ordering, including preferably that of complete alignment. If that
is not possible, the temperature dependence of T_2 should be compared to
the temperature dependence of D. It is not expected that ΔH will be strongly
temperature dependent, and the correlation time of water molecules in binding sites is expected to cause a stronger temperature dependence of T_2^{-1}
than that of the diffusion coefficient.

The effect of heterogeneity is the addition of low-frequency components
of the dipole (or quadrupole) interactions due to diffusional motion through
ordered regions. An elegant technique exists by which such low-frequency
components can be detected,[715-717] and the effects of heterogeneity can be
evaluated. With a normal spin-echo experiment one applies a 90° pulse
(rotating the magnetization in a rotating coordinate system over 90°),
followed after an interval τ by a 180° pulse. At a time τ after the second pulse
an echo occurs, the amplitude of which is an exponential function of the
interval τ with characteristic time T_2. The effect of the 180° pulse is to

reverse the influence of chemical shifts and magnetic inhomogeneities, but it does not change dipolar or quadrupolar interactions. Hence such interactions exert their full effect on the echo amplitude after a 90°–180° sequence, also if the interactions are slow or stationary. If, as Woessner et al.[716,717] have demonstrated, two 90° pulses are applied, an echo occurs as well, but its amplitude is not influenced by stationary dipolar or quadrupolar interactions. This is due to the fact that the 90° pulse has, roughly speaking, the effect of reversing the dipolar or quadrupolar interaction. Thus, by the 90°–90° method, another relaxation time T_3 is obtained to which slow dipolar interactions do not contribute. The cancellation is only effective for those low-frequency components for which $\omega\tau \ll 1$, where τ is the 90°–90° pulse interval. Dipolar or quadrupolar relaxation caused by higher frequency components do influence the echo amplitude. By this method Woessner and Snowden[715] have measured T_3 for D_2O in solutions of a high molecular weight bacterial polysaccharide. By also measuring T_2 and the diffusion coefficient D, these authors were able to derive the local ordering influence of the polysaccharide on the water as well as the domain size. The latter turned out to be 15 μm for a polymer concentration of 0.1 g cm^{-3}, which is quite long range.

The effect of submicroscopic heterogeneity on T_1 is not nearly as strong as it is on T_2. This is because the diffusion through ordered regions adds a long time-scale correlation to the molecular correlation function, which produces low-frequency components of the spin interaction. The diffusion process does not appreciably alter the Fourier component at the resonance frequency and thus leaves T_1 unchanged.

Another heterogeneity effect may occur in samples where water behaves differently in different regions of the sample. An example is cellular material, where water in intracellular and extracellular space may have different properties. Membrane material and any polymer material with regions of different degrees of crystallinity or structure are suspect from this point of view. There may be exchange of water molecules between the different regions at such a rate that the relaxation times are affected. Relaxation experiments should be carefully analyzed for the presence of more than one relaxation time. Consider the simplified case of two types of water molecules with spin–spin relaxation times T_{2A} and T_{2B} and with average lifetimes in the two states τ_A and τ_B, respectively, without difference in chemical shift of the two species. It can be shown, e.g., by the theory of Swift and Connick,[655] that the resulting relaxation is the sum of two exponentials (equivalent to a superposition of two Lorentzian lines). Both the relaxation times and the amplitudes of the two components are functions of the exchange rate between the two species. Hence in such cases one will observe two resonance components, but their amplitude ratio is not a direct measure of the relative amounts of the two species of water molecules, and therefore a direct interpretation of the two resonance components in terms of two species is deceptive.

For the general case the relationships are quite involved. For the important case where the bound water with relaxation time T_{2B} represents a small fraction p of the total water, the observed linewidth of the major resonance component is given by

$$T_2^{-1} = T_{2A}^{-1} + p(T_{2B} + \tau_B)^{-1} \tag{41}$$

This formula applies as well to homogeneous systems where a fraction of the water is bound with an average lifetime τ_B to a site where it retains a certain degree of mobility, as in macromolecular solutions.

Relaxation by *quadrupole coupling* will be the dominant mechanism in the case of ^{17}O and ^2H resonances. Since the electric field gradients are intramolecular, the relaxation is determined by the rotational correlation time. Although inhomogeneity may cause complications, the interpretation of deuteron relaxation times will generally be more straightforward than that of proton relaxation times. Equations relating T_1 and T_2 to the rotational correlation times can be found in Abragam's textbook.[3]

4.4. Diffusion

The coefficient of self-diffusion of water D can be measured conveniently by nmr relaxation techniques,[99] using pulsed field gradients. An overall restricted mobility of the water molecules is expected to result in a decreased diffusion coefficient. If we assume, as before, that the correlation times for translational and rotational motions are approximately equal, and if we further assume that all water molecules behave similarly, both T_2 and D should be influenced to the same degree with respect to their values for the liquid state. Thus, a T_2 of 3 msec would imply a D of 2.5×10^{-8} cm^2 sec^{-1} under these assumptions (the corresponding liquid-state values are 3.2 sec and 2.5×10^{-5} cm^2 sec^{-1} at 25°.)

However, it is often found in macromolecular solutions or disperse systems that D is quite large[103,699] and still close to the liquid value, while T_2 is severely reduced. A specific binding model offers an entirely satisfactory explanation for this apparent discrepancy. Assume that a fraction p of the water molecules is specifically bound with a residence time τ_b at a binding site. T_2 is now determined by eqn. (36). The reduction of T_2 compared to T_{2w} of liquid water is then given by

$$T_{2w}/T_2 \approx p\tau_b/\tau_w \qquad (42)$$

where τ_w is the correlation time in the liquid (2.6×10^{-12} sec at $25°$). For $p = 10\%$ and $\tau_b = 10^{-7}$ sec the reduction in T_2 is more than three orders of magnitude. The self-diffusion coefficient is only decreased by 10%, however, as a result of the specific binding. This can be shown as follows: A water molecule spends an average time of τ_b at a binding site, and an average time $\tau_f = \tau_b(1-p)/p$ between binding sites while diffusing with a diffusion coefficient D. In that time τ_f the mean square displacement $\overline{r^2}$ is given by

$$\overline{r^2} = 6D\tau_f = 6D\tau_b(1-p)/p \qquad (43)$$

Because the molecule is trapped during τ_b, $\overline{r^2}$ is its mean square displacement during a time $\tau_b + \tau_f$. Hence its real diffusion coefficient D' is given by

$$D' = \overline{r^2}/6(\tau_b + \tau_f) \qquad (44)$$

From eqns. (43) and (44) we derive

$$D' = (1-p)D \qquad (45)$$

We may conclude that a large coefficient of self-diffusion combined with a restricted mobility as measured by nmr can be explained by specific binding models. This explanation is indeed the most likely one.

4.5. Applications to Biopolymers

A large number of publications on the use of magnetic resonance for the study of biomolecular hydrates has appeared, where the cautions mentioned in the previous sections have been applied to a greater or lesser extent. A review by Walter and Hope[699] has covered pre-1970 studies on macromolecular solutions, adsorbed water systems, and living cells. Several aspects of the influence on water by biological macromolecules have been challengingly reviewed by Tait and Franks.[657]

Investigations on fibrous collagen[106,107,152,277,503] have shown that the reorientation of water molecules is anisotropic. Such anisotropies have also been found for DNA,[504] rayon,[150] and keratin.[460] Several explanations have been put forward, of which the simplest is the specific hydration mentioned in Section 3. The average rotational correlation time of the water molecules was found to be of the order of 10^{-7} sec. However, dielectric relaxation measurements[306] with time domain reflectometry (TDR) methods have shown a dielectric relaxation frequency of the bulk water in hydrated collagen of 1 GHz. This means that the short nmr relaxation time is determined by a small bound fraction, while the greater part of the water is rotating at a rate one order of magnitude slower than the liquid. This observed behavior applies to a partly hydrated system containing about 3 mol water/100 g protein.

TDR measurements by Harvey[324] on partly hydrated lysozyme over a wide frequency range have shown a Debye dispersion at about 0.25 GHz and a second dispersion around 10 GHz. Both dispersions have been attributed by the author to water. The 10-GHz lysozyme dispersion should be compared to the 1–GHz collagen dispersion. Also in protein solutions (bovine serum albumin) indications are that a hydration layer of 1–3 mol H_2O/100 g exhibits a dispersion in the 250–2000 MHz range.[301]

It is a general observation[46,67,127,136,151,154,232,324,419,420,503,505,654] that in protein solutions, hydrated disperse biological macromolecules, and tissues some water remains unfrozen down to the lowest observable temperatures, since a narrow line is observed which is characteristic of mobile water molecules. The amount of this water (a few mol/100 g) and its mobile properties suggest that this water should not be identified with specific hydration but with the hydration responsible for the dielectric absorption in the 0.1–10 GHz range. Apparently this water does not have the ability to freeze, although its properties are close to those of the liquid state. The influence of a macromolecular surface may thus extend to one or two molecular diameters from the surface. There is no evidence for long-range influences.[14,123]*

Frequency-dependent measurements[410] of the spin–lattice relaxation times of a solution of apotransferrin have shown that not more than 13 water molecules are irrotationally bound to the protein molecule. The residence time of these water molecules at the protein surface was found to be $<3 \times 10^{-5}$ sec. The authors remark that this amount of hydration is only 2% of the water to be expected in the first hydration shell. It seems clear, however, that this water should be regarded as specific hydration, while "the first hydration shell" contains water with rotational rates several orders

---

* See also Chapter 2 for a discussion of the properties of water in capillaries and micropores.

of magnitude faster than the protein molecule itself and not differing from the bulk liquid by more than a factor of ten.

Of the large number of measurements carried out on various tissues, the experiments on muscle[44,79,104,133,136,154,239,329,330,547,548] and tropomyosin[136] deserve special attention. Freezing experiments[46] indicate that 20% of the cell water is incapable of freezing and stays mobile at least down to −80°. D_2O spectra in frozen aqueous dispersions of tropomyosin[136] show temperature-dependent spectra that indicate the existence of a number of different types of D_2O. The ice phase seems to consist of two different types (probably with different hydrogen bonding strength), while the interesting mobile phase shows a complex behavior with a transition in the range −20 to −40°. At ordinary temperatures T_1 is found to be much longer than T_2, while T_2 increases under contraction.[79] The diffusion coefficient of water in muscle[104,239] is half that of liquid water. These observations fit into the pattern expected for an inhomogeneous system with a specific hydration of exchanging, strongly bound water molecules and a weak non-immobilizing influence on a large number of water molecules.

5. CONCLUSIONS

A general picture of the hydration of biological macromolecules emerges from the observations. A generalization necessarily means a simplification and any particular system will deviate in detail from the following description. It is likely that the general characteristics for disperse or "solid" systems apply just as well to solutions and to heterogeneous wet systems such as biological cells.

5.1. Specific Hydration

Biological macromolecules possess several specific binding sites for water molecules, which provide a favorable hydrogen-bonding pattern with at least two, but often three or four, bonds to each water molecule. Very often such specifically bound water molecules also form hydrogen bonds between each other. The specific sites can be situated internally or on the surface of the macromolecule. The free energy of such sites is about 1.5–2 kcal mol^{-1} below that of the liquid state; their enthalpy is 1–3 kcal mol^{-1} below the liquid enthalpy. The lifetime of a molecule is of the order of 10^{-6} sec at a specific site on the macromolecular surface (for internal isolated sites no data are available, but lifetimes in the range of seconds are plausible). In solutions such water molecules are irrotationally bound to the macromolecule, although rotations about the symmetry axis of the water molecules are likely to occur. The number of specific binding sites is 0.2–0.5 mol water/100 g of macromolecule, at least for proteins, thus amounting to a few percent specific hydration.

5.2. Nonspecific Hydration

In addition, an amount of about 3 mol water/100 g macromolecule (50%) is affected by the macromolecule in the sense that this water is not able to freeze and shows a lower rate of rotation than molecules in the bulk liquid state. Characteristic dielectric relaxation times of this water are in the range $2-50 \times 10^{-11}$ sec. This correspond to rotational correlation times, such as are relevant for magnetic relaxation, of $0.5-10 \times 10^{-11}$ sec. Thus the mobility of this water is only slightly lower than in the normal liquid, where the rotational correlation time is 0.26×10^{-11} sec at 25°C. Probably the non-specific hydration in fact consists of several species with a range of correlation times. The rate of exchange between various species of nonspecific hydration, as well as between these water molecules and the bulk liquid, is very fast. The water molecules of the nonspecific hydration are not irrotationally bound to the macromolecule[285] and retain their high mobility down to very low temperatures.

Apart from the specific hydration, which seems to fulfill a structural role, there is no evidence that any appreciable amount of water is strongly immobilized, or occurs as ice or any other crystalline form in macromolecular hydrates. This observation does *not* resolve the question whether water plays a major role in biological systems concerning ion specificity, specific binding properties of cytoplasma, membrane permeability, proton transfer processes, etc. As is evident from the molecular basis of hydrophobic hydration, marked thermodynamic changes may be accompanied by only minor changes in dynamic behavior. The properties of the nonspecific hydration are sufficiently different from the liquid state and the number of molecules involved is sufficiently large to allow for a possible functional role of water in biological processes. Critical and clear-cut experiments, rather than solid-state theories, are needed to evaluate such a role of water in biological systems.

Note added in proof. For a review on the properties of water in biological systems the reader is referred to R. Cooke and I. D. Kuntz, *Ann. Rev. Biophys. Bioengineering* **3**, 95 (1974).

References

1. R. P. Abendroth, *J. Colloid Interface Sci.* **34**, 591 (1970).
2. L. A. Abetsedarskaya, F. G. Miftakhutdinova, and V. D. Fedotov, *Biophysics* **13**, 750 (1968).
3. A. Abragam, "The Principles of Nuclear Magnetism," Oxford (1961).
4. A. W. Adamson, "Physical Chemistry of Surfaces," Interscience Publishers, Easton, Pennsylvania, p. 506 (1960).
5. N. V. Afanas'ev, M. S. Mestik, and V. N. Popova, in "Research in Surface Forces" (B. V. Derjaguin, ed.), Consultants Bureau, New York, Vol. 2, pp. 177, 181 (1966).
6. I. V. Aleksandrov, "The Theory of Nuclear Magnetic Resonance," Academic Press, New York (1966).
7. K. W. Allen and G. A. Jeffrey, *J. Chem. Phys.* **38**, 2304 (1963).
8. L. C. Allen and P. A. Kollman, *J. Colloid Interface Sci.* **36**, 469 (1971).
9. D. M. Anderson, *J. Colloid Interface Sci.* **25**, 174 (1967).
10. D. M. Anderson and P. Hoekstra, *Science* **149**, 318 (1965).
11. D. M. Anderson and P. Hoekstra, *Soil Sci. Soc. Amer. Proc.* **29**, 498 (1965).
12. D. M. Anderson and P. F. Low, *Soil Sci. Soc. Amer. Proc.* **22**, 99 (1958).
13. J. H. Anderson, Jr., and K. A. Wickersheim, *Surface Sci.* **2**, 252 (1964).
14. J. L. Anderson and J. A. Quinn, *J. Chem. Soc. Faraday I* **68**, 744 (1972).
15. Anonymous, *Nature* **230**, 11 (1971).
16. S. G. Ash and G. H. Findenegg, *Special Disc. Faraday Soc.* **1**, 105 (1970).
17. S. G. Ash and G. H. Findenegg, *Trans. Faraday Soc.* **67**, 2122 (1971).
18. E. D. T. Atkins, I. A. Nieduszynski, W. Mackie, K. D. Parker, and E. E. Smolko, *Biopolymers* **12**, 1879 (1973).
19. I. Aviram and A. Schejter, *Biopolymers* **11**, 2141 (1972).
20. P. Bagchi, *J. Colloid Interface Sci.* **41**, 380 (1972).
21. P. Bagchi and R. D. Vold, *J. Colloid Interface Sci.* **38**, 652 (1972).
22. V. A. Bakaev, V. F. Kiselev, and K. G. Krasil'nikov, *Dokl. Akad. Nauk SSSR* **125**, 831 (1959).
23. D. H. Bangham and N. Fakhoury, *J. Chem. Soc.* **1931**, 1324.
24. D. H. Bangham and S. Mossalam, *Proc. Roy. Soc. A* **166**, 558 (1938).
25. D. H. Bangham and R. I. Razouk, *Trans. Faraday Soc.* **33**, 1459 (1937).
26. D. H. Bangham and Z. Saweris, *Trans. Faraday Soc.* **34**, 554 (1938).
27. V. M. Barboi, *Koll. Zh.* **26**, 3 (1964).
28. V. M. Barboi, *Koll. Zh.* **26**, 409 (1964).
29. V. M. Barboi, *Koll. Zh.* **27**, 643 (1965).
30. L. Barclay, A. Harrington, and R. H. Ottewill, *Kolloid Z. u. Z. Polymere* **250**, 655 (1972).

31. L. M. Barclay and R. H. Ottewill, *Special Disc. Faraday Soc.* **1**, 138 (1970).
32. D. Bargeman and F. van Voorst Vader, *J. Electroanal. Chem.* **37**, 45 (1972).
33. P. Barnes, I. Cherry, J. L. Finney, and S. Petersen, *Nature* **230**, 31 (1971).
34. R. M. Barrer and A. V. J. Edge, *Proc. Roy. Soc. (Lond.) A* **300**, 1 (1967).
35. I. Barshad, *Am. Min.* **35**, 225 (1950).
36. W. D. Bascom, *J. Phys. Chem.* **76**, 456 (1972).
37. W. D. Bascom, E. J. Brooks, and B. N. Worthington, *Nature* **228**, 1290 (1970).
38. S. H. Bastow and F. P. Bowden, *Proc. Roy. Soc. A* **134**, 404 (1931).
39. S. H. Bastow and F. P. Bowden, *Proc. Roy. Soc. A* **151**, 220 (1935).
40. P. Becher, "Emulsions: Theory and Practice," Reinhold, New York (1965).
41. P. Becher, *J. Colloid Interface Sci.* **42**, 645 (1973).
42. R. E. Beck and J. S. Schultz, *Biochem. Biophys. Acta* **255**, 273 (1972).
43. G. M. Bell and S. Levine, *Trans. Faraday Soc.* **53**, 143 (1957).
44. G. M. Bell and P. L. Levine, *J. Colloid Interface Sci.* **41**, 275 (1972).
45. J. D. Bell, R. W. Myatt, and R. E. Richards, *Nature Phys. Sci. (Lond.)* **230**, 91 (1971).
46. P. S. Belton, K. J. Packer, and T. C. Sellwood, *Biochim. Biophys. Acta* **304**, 56 (1973).
47. H. A. Benesi and A. C. Jones, *J. Phys. Chem.* **63**, 179 (1959).
48. H. J. C. Berendsen, *J. Chem. Phys.* **36**, 3297 (1962).
49. H. J. C. Berendsen, *Fed. Proc. Fed. Amer. Soc. Exp. Biol.* **25**, 971 (1966).
50. H. J. C. Berendsen, *in* "Theoretical and Experimental Biophysics" (A. Cole, ed.), Vol. 1, pp. 1–76, Marcel Dekker, New York (1967).
51. H. J. C. Berendsen, *in* "Biology of the Mouth," AAAS Symposium, Publ. No. 89, p. 145, Washington, D.C. (1968).
52. H. J. C. Berendsen and H. T. Edzes, *Ann. N. Y. A ad. Sci.* **204**, 459 (1973).
53. H. J. Berendsen and C. Migchelsen, *Ann. N. Y. Acad. Sci.* **125**, 365 (1965).
54. V. N. Beresnev, N. A. Fermor, and N. I. Smirnov, *Zh. Prikl. Khim.* **39**, 1319 (1966).
55. G. Berger, *Chem. Weekbl.* **38**, 42 (1941).
56. P. E. Berghausen, "Adhesion and Adhesives," Wiley, New York, p. 225 (1954).
57. M. S. Bergqvist and E. Forslind, *Acta Chem. Scand.* **16**, 2069 (1962).
58. E. Berlin, P. G. Kliman, and M. J. Pallansch, *J. Colloid Interface Sci.* **34**, 488 (1970).
59. J. D. Bernal and I. Fankuchen, *J. Gen. Physiol.* **25**, 111 (1941).
60. J. D. Bernal and R. H. Fowler, *J. Chem. Phys.* **1**, 515 (1933).
61. Y. G. Bérubé and P. L. J. de Bruyn, *Colloid Interface Sci.* **28**, 92 (1968).
62. G. Beurskens, G. A. Jeffrey, and R. K. McMullan, *J. Chem. Phys.* **39**, 3311 (1963).
63. P. T. Beurskens and G. A. Jeffrey, *J. Chem. Phys.* **40**, 2800 (1964).
64. B. H. Bijsterbosch and J. Lyklema, *J. Colloid Sci.* **20**, 665 (1965).
65. J. J. Bikerman, *in* "Chemistry and Physics of Interfaces," Amer. Chem. Soc. Publ., p. 58, Washington, D.C. (1965).
66. N. Bjerrum, *Fys. Tidsskr.* **48**, 71 (1950).
67. D. J. Blears and S. S. Danyluk, *Biochim. Biophys. Acta* **154**, 17 (1968).
68. J. O'M. Bockris, M. A. V. Devanathan, and K. Müller, *Proc. Roy. Soc. A* **274**, 55 (1963).
69. H. Boehm, *Disc. Faraday Soc.* No. 52 (1972).
70. R. W. Bolander, J. L. Kassner, and J. T. Zung, *Nature* **221**, 1233 (1969).
71. M. Bonamico, S. A. Jeffrey, and R. McMullan, *J. Chem. Phys.* **37**, 2219 (1962).
72. F. Booth, *J. Chem. Phys.* **19**, 391 (1951).

73. H. Boutin, G. J. Safford, and H. R. Danner, *J. Chem. Phys.* **39**, 488 (1963).
74. F. P. Bowden and D. Tabor, "The Friction and Lubrication of Solids," Oxford, p. 55 (1954).
75. E. M. Bradbury, R. E. Burge, J. T. Randall, and G. R. Wilkinson, *Disc. Faraday Soc.* **25**, 173 (1958).
76. R. S. Bradley, *J. Chem. Soc.* **1936**, 1799.
77. W. F. Bradley, *J. Amer. Chem. Soc.* **67**, 975 (1945).
78. W. F. Bradley and J. M. Serratosa, *Clays and Clay Minerals* **7**, 260 (1960).
79. C. B. Bratton, A. L. Hopkins, and J. W. Weinburg, *Science* **147**, 738 (1965).
80. G. Brausse, A. Mayer, T. Nedetzka, P. Schlecht, and H. Vogel, *J. Phys. Chem.* **72**, 3098 (1968).
81. M. M. Breuer and M. G. Kennerley, *J. Colloid Interface Sci.* **37**, 124 (1971).
82. G. W. Brindley (ed.), "X-Ray Identification and Structure of the Clay Minerals," Mineralogical Society of Great Britain Monograph (1951).
83. D. E. Brooks, *J. Colloid Interface Sci.* **43**, 687 (1973).
84. G. Brown, *Clay Min. Bull.* **2**, 109 (1950).
85. G. Brown, "Nomenclature of the Mica Clay Minerals," Mineralogical Society of Great Britain Monograph (G. W. Brindley, ed.), p. 155 (1951).
86. G. Brown, "The X-Ray Identification and Crystal Structures of Clay Minerals," London (1961).
87. S. Brunauer, "The Adsorption of Gases and Vapors," Princeton University Press, Princeton, New Jersey (1945).
88. S. Brunauer, P. H. Emmett, and E. Teller, *J. Amer. Chem. Soc.* **60**, 309 (1938).
89. H. Buckingham and E. Price, "Principles of Electrical Measurements," English University Press, London (1955).
90. R. Bulkley, *Bur. Stand. J. Res.* **6**, 89 (1931).
91. H. B. Bull, *J. Amer. Chem. Soc.* **66**, 1499 (1944).
92. H. B. Bull and K. Breeze, *Arch. Biochem. Biophys.* **149**, 164 (1972).
93. Bull. Int. Assoc. of Engineering Geology No. 5, Paris (June 1972).
94. U. Buontempa, G. Carei, and P. Fasella, *Biopolymers* **11**, 519 (1972).
95. V. M. Burmudez, *J. Phys. Chem.* **74**, 4160 (1970).
96. C. J. Burton, *J. Acoust. Soc. Am.* **20**, 186 (1948).
97. L. Cahn and H. R. Schultz, "Vacuum Microbalance Techniques," Vol. 3, Plenum Press, New York, p. 29 (1966).
98. R. Calvet and R. Prost, *Clays and Clay Minerals* **19**, 187 (1970).
99. H. Y. Carr and E. M. Purcell, *Phys. Rev.* **94**, 630 (1954).
100. A. Carrington and A. D. McLachlan, "Introduction to Magnetic Resonance," Harper and Row, New York (1967).
101. H. B. G. Casimir and D. Polder, *Phys. Rev.* **73**, 360 (1948).
102. J. Cerbon, *Biochim. Biophys. Acta* **144**, 1 (1967).
103. D. C. Chang, C. F. Hazlewood, B. L. Nichols, and H. E. Rorschach, *Nature* **235**, 170 (1971).
104. D. C. Chang, H. E. Rorschach, B. L. Nichols, and C. F. Hazlewood, *Ann. N. Y. Acad. Sci.* **204**, 434 (1973).
105. D. L. Chapman, *Phil. Mag.* **25**, 475 (1913).
106. G. E. Chapman, S. S. Danyluk, and K. A. McLauchlan, *Proc. Roy. Soc. B* **178**, 465 (1971).
107. G. E. Chapman and K. A. McLauchlan, *Proc. Roy. Soc. B* **173**, 223 (1969).

108. J. J. Chessick, A. C. Zettlemoyer, F. H. Healey, and G. J. Young, *Can. J. Chem.* **33**, 251 (1955).
109. Yu. N. Chirgadze and A. M. Ovsepyan, *Biopolymers* **11**, 2179 (1972).
110. H. N. Christenson, *in* "Biological Transport," W. A. Benjamin, New York (1962).
111. K. S. Chua, *Nature* **227**, 834 (1970).
112. N. V. Churaev, V. D. Sobolev, and Z. M. Zorin, *Special Disc. Faraday Soc.* **1**, 213 (1970).
113. R. Cini, G. Loglio, and A. Ficalbi, *Nature* **223**, 1148 (1969).
114. R. Cini, G. Loglio, and A. Ficalbi, *J. Colloid Interface Sci.* **41**, 287 (1972).
115. M. M. Civan and M. Shporer, *Biophys. J.* **12**, 404 (1972).
116. W. F. Claussen, *J. Chem. Phys.* **19**, 259, 662 (1951).
117. E. J. Clayfield and E. C. Lumb, *J. Colloid Interface Sci.* **22**, 269 (1966).
118. E. J. Clayfield and E. C. Lumb, *J. Colloid Interface Sci.* **22**, 285 (1966).
119. E. J. Clayfield and E. C. Lumb, *Macromolecules* **1**, 133 (1968).
120. E. J. Clayfield, E. C. Lumb, and P. H. Mackey, *J. Colloid Interface Sci.* **37**, 382 (1971).
121. J. Clifford, *Chem. Commun.* **1967**, 880.
122. J. Clifford and S. M. A. Lecchini, "Wetting," SCI Monograph No. 25, p. 174 (1967).
123. J. Clifford, J. Oakes, and G. J. Tiddy, *Special Disc. Faraday Soc.* **1**, 175 (1970).
124. J. Clifford and B. A. Pethica, *Trans. Faraday Soc.* **60**, 1483 (1964).
125. J. Clifford and B. A. Pethica, *in* "Hydrogen-Bonded Solvent Systems," (A. K. Covington and P. Jones, eds.) Taylor and Francis, London, p. 169 (1968).
126. J. Clifford, B. A. Pethica, and E. G. Smith, *in* "Membrane Models and the Formation of Biological Membranes," (L. Bolis and B. A. Pethica, eds.) North-Holland Publishing Co., Amsterdam, pp. 19–41 (1968).
127. J. Clifford and B. Sheard, *Biopolymers* **4**, 1057 (1966).
128. J. S. Clunie, J. M. Corkill, and J. F. Goodman, *Disc. Faraday Soc.* **42**, 34 (1966).
129. J. S. Clunie, J. M. Corkhill, J. F. Goodman, and C. P. Ogden, *Trans. Faraday Soc.* **63**, 505 (1967).
130. J. S. Clunie, J. F. Goodman, and B. T. Ingram, *in* "Surface and Colloid Science," (E. Matijevic, ed.), Vol. 3, p. 167 Wiley (1971).
131. J. S. Clunie, J. F. Goodman, and P. C. Symons, *Nature* **216**, 1203 (1967).
132. M. Cohn, *Quart. Rev. Biophys.* **3**, 61 (1970).
133. F. W. Cope, *Biophys. J.* **9**, 303 (1969).
134. J. M. Corkhill, J. F. Goodman, C. P. Ogden, and J. R. Tate, *Proc. Roy. Soc. A* **273**, 84 (1963).
135. J. M. Corkhill, J. F. Goodman, and J. R. Tate, *Trans. Faraday Soc.* **63**, 773 (1967).
136. T. J. Cyr, W. Derbyshire, J. L. Parsons, J. M. V. Blanshard, and R. A. Lawrie, *Trans. Faraday Soc.* **67**, 1887 (1971).
137. T. Dalton and R. S. Snart, *Biochim. Biophys. Acta* **135**, 1059 (1967).
138. B. B. Damaskin, *J. Electroanal. Chem.* **7**, 155 (1964).
139. M. D. Danford and H. A. Levy, *J. Amer. Chem. Soc.* **84**, 3965 (1962).
140. D. W. Davidson, *Can. J. Chem.* **49**, 1224 (1971).
141. J. Davidtz and P. F. Low, *Clays and Clay Minerals* **18**, 325 (1970).
142. D. R. Davies, *Ann. Rev. Biochem.* **36**, 1, 321 (1967).
143. J. T. Davies, *in* "Proc. 2nd Intern. Congr. Surface Activity," Vol. 1, p. 426, Butterworths, London (1957).

144. J. T. Davies, *in* "Recent Progress in Surface Science" (J. F. Danielli, K. G. A. Pankhurst, and A. C. Riddiford, eds.), Vol. 2, p. 129, Academic Press, London (1964).

145. R. F. Davis, *Chem. Eng. News* **1970**, 73.

146. V. Ya. Davydov, A. V. Kiselev, and L. T. Zhuravlev, *Trans. Faraday Soc.* **60**, 2254 (1964).

147. P. Debye, *Trans. Faraday Soc.* **23**, 334 (1927).

148. P. Debye and E. Hückel, *Physik. Z.* **24**, 305 (1923).

149. C. T. Deeds and H. van Olphen, *Adv. Chem. Series* **33**, 332 (1961).

150. R. E. Dehl, *J. Chem. Phys.* **48**, 831 (1968).

151. R. E. Dehl, *Science* **170**, 738 (1970).

152. R. E. Dehl and C. A. J. Hoeve, *J. Chem. Phys.* **50**, 3245 (1969).

153. M. de Paz, A. Pozzo, and M. E. Vallauri, *Chem. Phys. Lett.* **7**, 23 (1970).

154. W. Derbyshire and J. L. Parsons, *J. Mag. Res.* **6**, 344 (1972).

155. B. V. Derjaguin, *J. Phys. Chem.* **3**, 29 (1932).

156. B. V. Derjaguin, *Z. Phys.* **84**, 657 (1933).

157. B. V. Derjaguin, *Trans. Faraday Soc.* **36**, 203 (1940).

158. B. V. Derjaguin, *in* "Research in Surface Forces" (B. V. Derjaguin, ed.), Vol. 2, pp. 3–7 (1966).

159. B. V. Derjaguin, *Disc. Faraday Soc.* **42**, 109 (1966).

160. B. V. Derjaguin, *Pure and Appl. Chem.* **24**, 95 (1970).

161. B. V. Derjaguin and N. V. Churaev, *J. Colloid Interface Sci.* **36**, 415 (1971).

162. B. V. Derjaguin and N. V. Churaev, *Nature* **244**, 430 (1973).

163. B. V. Derjaguin, N. V. Churaev, N. N. Fedyakin, M. V. Talaev, and I. G. Ershova, *Izv. Akad. Nauk SSSR* (*Ser. Khim.*) **1967**(10), 2178.

164. B. V. Derjaguin, I. G. Ershova, B. V. Zheleznyi, and N. V. Churaev, *Dokl. Akad. Nauk SSSR* **172**, 1121 (1967).

165. B. V. Derjaguin, N. N. Fedyakin, and M. V. Talaev, *Dokl. Akad. Nauk SSSR* **167**, 376 (1966).

166. B. V. Derjaguin and R. Greene-Kelly, *Trans. Faraday Soc.* **60**, 449 (1964).

167. B. V. Derjaguin and V. V. Karasev, *Kolloid Z.* **15**, 365 (1953).

168. B. V. Derjaguin and M. Kusakov, *Acta Physicochim. USSR* **10**, 25 (1939).

169. B. V. Derjaguin and M. Kusakov, *Acta Physicochim. USSR* **10**, 153 (1939).

170. B. V. Derjaguin and L. Landau, *Acta Physicochim. USSR* **14**, 633 (1941).

171. B. V. Derjaguin and L. Landau, *Zh. Eksperim. i Teor. Fiz.* **11**, 802 (1941).

172. B. V. Derjaguin, D. S. Lychnikov, K. M. Merzhanov, Y. I. Rabinovich, and N. V. Churaev, *Dokl. Akad. Nauk SSSR* **181**, 823 (1969).

173. B. V. Derjaguin and E. Obuchov, *Colloid J.* **1**, 385 (1935).

174. B. V. Derjaguin, and Ya. I. Rabinovitch, *Koll. Zh.* **31**, 47 (1969).

175. B. V. Derjaguin and L. M. Shcherbakov, *Colloid J. USSR* **23**, 33 (1961).

176. B. V. Derjaguin and A. S. Titijevskaja, *Koll. Zh.* **15**, 416 (1953).

177. B. V. Derjaguin and A. S. Titijevskaja, *Disc. Faraday Soc.* **18**, 24 (1954).

178. B. V. Derjaguin and A. S. Titijevskaja, *in* "Proc. 2nd Internat. Congr. Surface Activity," Vol. 1, p. 211, Butterworths, London (1957).

179. B. V. Derjaguin, B. V. Zheleznyi, Y. I. Rabinovich, C. K. Simonova, M. V. Talaev, and N. V. Churaev, *Dokl. Akad. Nauk SSSR* **190**, 372 (1970).

180. B. V. Derjaguin, B. V. Zheleznyi, N. N. Zakhavaeva, O. A. Kiseleva, A. I. Konovalov, D. S. Lychnikov, Y. I. Rabinovich, M. V. Talaev, and N. V. Churaev, *Dokl. Akad. Nauk SSSR* **189**, 1282 (1969).

181. B. V. Derjaguin and Z. M. Zorin, *Zh. Fiz. Khim.* **29**, 1010 (1955).

182. B. V. Derjaguin and Z. M. Zorin, *in* "Proc. 2nd Int. Congr. Surface Activity London," Vol. 2, 145 (1957).

183. B. V. Derjaguin, Z. M. Zorin, and N. V. Churaev, *Dokl. Akad. Nauk SSSR* **182**, 811 (1969).

184. B. V. Derjaguin, Z. M. Zorin, V. V. Karazev, V. D. Dobolov, E. N. Kromova, and N. V. Churaev, *Dokl. Akad. Nauk SSSR* **187**, 605 (1969).

185. M. A. V. Devanathan, *Proc. Roy. Soc. (Lond.) A* **267**, 256 (1962).

186. O. F. Deveraux and P. L. de Bruyn, *J. Colloid Sci.* **19**, 302 (1964).

187. C. Deverell, *Progress Nuclear Magnetic Resonance (Lond.)* **4**, 235 (1969).

188. C. Devillez, A. Sanfeld, and A. Steinchen, *J. Colloid Interface Sci.* **25**, 295 (1967).

189. A. J. de Vries, *Rec. Trav. Chim.* **77**, 383 (1958).

190. J. J. de Vries and H. J. C. Berendsen, *Nature* **221**, 1139 (1969).

191. R. M. Diamond, *J. Phys. Chem.* **67**, 2513 (1963).

192. M. Dixon and E. C. Webb, "Enzymes," Academic Press, London (1960).

193. A. Doroszkowski and R. Lambourne, *J. Polymer Sci. C* **34**, 253 (1971).

194. A. Doroszkowski and R. Lambourne, *J. Colloid Interface Sci.* **43**, 97 (1973).

195. P. Doty and J. T. Edsall, *Advan. Protein Chem.* **6**, 35 (1951).

196. K. Dransfeld, H. L. Frisch, and E. A. Wood, *J. Chem. Phys.* **36**, 1574 (1962).

197. J. Drenth, J. N. Jansonius, R. Koekoek, and B. G. Wolthers, *Advan. Protein Chem.* **25**, 79 (1971).

198. W. Drost-Hansen, *Ind. Eng. Chem.* **57**, 18 (1965).

199. W. Drost-Hansen, *in* Advances in Chemistry Series, Vol. 67, p. 70 (1967).

200. W. Drost-Hansen, *in* "Proc. First Int. Symp. on Water Desal." Vol. 1, p. 382 (1967).

201. W. Drost-Hansen, *Chem. Phys. Lett.* **2**, 647 (1968).

202. W. Drost-Hansen, *Ind. and Eng. Chem.* **61**, (11), 10 (1969).

203. W. Drost-Hansen, *in* "Chemistry of the Cell Interface," Part B, (H. D. Brown, ed.), p. 1, Academic Press, New York (1971).

204. W. Drost-Hansen, *J. Geophysical Res.* **77**, 5132 (1972).

205. S. S. Dukhin and B. V. Derjaguin, *Dokl. Akad. Nauk SSSR* **159**, 177 (1964).

206. S. S. Dukhin and N. M. Semenikhin, *Koll. Zh.* **32**, 360 (1970).

207. E. M. Duyvis, "The Equilibrium Thickness of Free Liquid Films," Thesis, Utrecht University (1962).

208. E. M. Duyvis and J. Th. G. Overbeek, *Proc. Koninkl. Ned. Akad. Wetenschappen* **B65**, 26 (1962).

209. I. E. Dzyaloshinskii, E. M. Lifshitz, and L. P. Pitaevskii, *Zh. Eksperim. i Teor. Fiz.* **37**, 229 (1959).

210. I. E. Dzyaloshinskii, E. M. Lifshitz, and L. P. Pitaevskii, *Advan. Phys.* **10**, 165 (1959).

211. D. Eagland and G. Pilling, *J. Phys. Chem.* **76**, 1902 (1972).

212. C. H. Edelman, *Verre et silic. industr.* **12**, 3 (1947).

213. C. H. Edelman and J. C. L. Favejee, *Z. Krist.* **102**, 417 (1940).

214. B. E. Eichinger and P. J. Flory, *Trans. Faraday Soc.* **64**, 2035 (1968).

215. R. P. Eischens and J. Jacknow, *in* "Proc. Third Internat. Congress on Catalysis, Amsterdam, 1964," North-Holland, Amsterdam (1965).

216. G. A. H. Elton, *Proc. Roy. Soc. A* **194**, 275 (1948).

217. L. Endom, H. G. Hertz, B. Thuel, and M. D. Zeidler, *Ber. Bunsenges. Phys. Chem.* **71**, 1008 (1967).

218. G. Engel and H. G. Hertz, *Ber. Bunsenges.* **72**, 808 (1968).
219. S. R. Erlander, *Phys. Rev. Lett.* **22**, 177 (1969).
220. S. R. Erlander, *J. Colloid Interface Sci.* **34**, 53 (1970).
221. N. G. Esipova, N. S. Andreeva, and T. V. Gatovskaia, *Bioophys.* **3**, 505 (1958).
222. L. F. Evans, *Ind. Eng. Chem.* **46**, 2420 (1954).
223. R. Evans and D. H. Napper, *Kolloid Z.u.Z. Polymere* **251**, 329 (1973).
224. R. Evans and D. H. Napper, *Kolloid Z.u.Z. Polymere* **251**, 409 (1973).
225. D. H. Everett, J. M. Haynes, and P. J. McElroy, *Nature* **226**, 1033 (1970).
226. D. H. Everett, J. M. Haynes, and P. J. McElroy, *J. Colloid Interface Sci.* **36**, 483 (1971).
227. D. H. Everett, J. M. Haynes, and P. J. McElroy, *Sci. Progr. Oxford* **59**, 279 (1971).
228. D. H. Everett and P. J. McElroy, *J. Colloid Interface Sci.* **36**, 529 (1971).
229. P. A. Faith, "Adsorption and Vacuum Techniques," Institute of Science and Technology, University of Michigan, Ann Arbor (1962).
230. M. Falk, *Canad. J. Chem.* **44**, 1107 (1966).
231. M. Falk, K. A. Hartman, and R. C. Lord, *J. Amer. Chem. Soc.* **84**, 3843 (1962).
232. M. Falk, A. G. Poole, and C. G. Goymour, *Canad. J. Chem.* **48**, 1536 (1970).
233. H. Falkenhagen and G. Kelbg, *Ann. Physik.* **11**, 60 (1952).
234. M. Faraday, *Proc. Roy. Inst.* (*Great Britain*) (1850).
235. N. N. Fedyakin, *Colloid J. USSR* **24**, 425 (1962).
236. N. N. Fedyakin, B. V. Derjaguin, A. V. Novikova, and M. V. Talaev, *Dokl. Akad. Nauk SSSR* **167**, 376 (1966).
237. D. Feil and G. A. Jeffrey, *J. Chem. Phys.* **35**, 1863 (1961).
238. H. Fellner-Feldegg, *J. Phys. Chem.* **73**, 616 (1969).
239. E. D. Finch, J. F. Harmon, and B. H. Muller, *Arch. Biochem. Biophys.* **147**, 299 (1971).
240. E. W. Fischer, *Kolloid Z.* **160**, 120 (1958).
241. I. Fischer and L. Ehrenberg, *Acta Chem. Scand.* **2**, 669 (1948).
242. F. Fister and H. G. Hertz, *Ber. Bunsenges. Phys. Chem.* **71**, 1032 (1967).
243. N. H. Fletcher, *Phil. Mag.* **7**, 255 (1962).
244. N. H. Fletcher, *Phil. Mag.* **8**, 1426 (1963).
245. P. J. Flory, *J. Chem. Phys.* **10**, 51 (1942).
246. P. J. Flory, "Principles of Polymer Chemistry," Cornell University Press, Ithaca (1953).
247. P. J. Flory, "Statistical Mechanics of Chain Molecules," p. 32, Interscience, New York (1969).
248. P. J. Flory and T. G. Fox, Jr., *J. Amer. Chem. Soc.* **73**, 1904 (1951).
249. P. J. Flory and W. R. Krigbaum, *J. Chem. Phys.* **18**, 1086 (1950).
250. C. Folzer, R. W. Hendricks, and A. H. Narten, *J. Chem. Phys.* **54**, 799 (1971).
251. E. Forslind, Swedish Cement and Concrete Research Institute at the Royal Institute of Technology, Stockholm, Proc. No. 16 (1952).
252. E. Forslind, *in* "Proc. 2nd Int. Cong. Rheology," p. 50 (1953).
253. E. Forslind, *Reologi, Särtryck ur Svensk Naturvetenskap* **1966**, 9.
254. E. Forslind, *Quart. Rev. Biophys.* **4**, 325 (1971).
255. H. S. Frank, *Fed. Proc.* **24**(No. 2, Part III), P.S-1 (1965).
256. H. S. Frank and W. Y. Wen, *Disc. Faraday Soc.* **24**, 133 (1957).
257. F. Franks, *Ann. N. Y. Acad. Sci.* **125**, 277 (1965).
258. F. Franks and D. J. G. Ives, *J. Chem. Soc.* (*Lond.*) **1960**, 741.

259. F. Franks, J. R. Ravenhill, and D. S. Reid, *J. Soln. Chem.* **1**, 3 (1972).
260. F. Franks, D. S. Reid, and A. Suggett, *J. Soln. Chem.* **2**, 99 (1973).
261. R. D. B. Frazer and T. P. Macrae, *Nature* **183**, 179 (1959).
262. J. Frenkel, "Kinetic Theory of Liquids," Oxford University Press (1947).
263. J. Frenkel, "Kinetic Theory of Liquids," Dover, New York (1955).
264. G. Frens, Thesis (Utrecht University, 1968).
265. G. Frens, D. J. C. Engel, and J. Th. G. Overbeek, *Trans. Faraday Soc.* **63**, 418 (1967).
266. H. Freundlich, *Z. Phys. Chemie* **44**, 151 (1903).
267. H. Freundlich, "Kapillarchemie," Leipzig (1909).
268. H. Freundlich, "Colloid and Capillary Chemistry," Methuen, London (1926).
269. W. Friese, *Z. Elektrochem.* **56**, 822 (1952).
270. J. J. Fripiat, A. Jelli, G. Poncelet, and J. André, *J. Phys. Chem.* **69**, 2185 (1965).
271. H. Fröhlich, "Theory of Dielectrics," Clarendon Press, Oxford (1958).
272. G. J. C. Frohnsdorff and G. L. Kington, *Proc. Roy. Soc. A* **247**, 469 (1958).
273. A. N. Frumkin, *Acta Physicochim.* **9**, 313 (1938).
274. A. N. Frumkin, *Zh. Fiz. Khim.* **12**, 33 (1938).
275. G. I. Fuks, *Colloid J.* **20**, 705 (1958).
276. N. Fukuta and B. J. Mason, *J. Phys. Chem. Solids* **24**, 715 (1963).
277. B. M. Fung and P. Trautmann, *Biopolymers* **10**, 391 (1971).
278. J. A. Gann, *Kolloid Chem. Beihefte* **8**, 125, 63 (1916).
279. L. Gargallo, L. Sepulveda, and J. Goldfarb, *J. Kolloid Z. u. Z. Polymere* **229**, 51 (1969).
280. M. P. Gingold, *Nature (Phys. Sci.)* **235**, 75 (1972).
281. A. S. Ginzberg, quoted by H. van Olphen, *in* "Clays and Clay Minerals," Proc. 2nd National Conference on Clays and Clay Minerals (A. Swineford and N. Plummer, eds.), p. 429, National Research Council, Washington, D.C. (1953).
282. M. Giskin and J. Hagin, *Israel J. Chem.* **6**, 387 (1968).
283. G. J. Gittens, *J. Colloid Interface Sci.* **30**, 406 (1969).
284. S. H. Glarum, *J. Chem. Phys.* **33**, 1371 (1960).
285. J. A. Glasel, *Nature* **220**, 1124 (1968).
286. Y. M. Glazman, *Disc. Faraday Soc.* **42**, 255 (1966).
287. Y. M. Glazman and I. M. Dykman, *Dokl. Akad. Nauk SSSR* **100**, 299 (1953).
288. Y. M. Glazman and I. M. Dykman, *Koll. Zh.* **18**, 13 (1956).
289. Y. M. Glazman, I. M. Dykman, and E. A. Strel'tsova, *Dokl. Akad. Nauk. SSSR* **117**, 829 (1957).
290. Y. M. Glazman, I. M. Dykman, and E. A. Strel'tsova, *Koll. Zh.* **20**, 149 (1958).
291. J. W. Glen, *Proc. Roy. Soc. A* **228**, 519 (1955).
292. D. N. Glew, *J. Phys. Chem.* **66**, 605 (1962).
293. R. J. Good, *J. Phys. Chem.* **61**, 810 (1957).
294. W. Good, *Nature* **214**, 1250 (1967).
295. G. Gouy, *J. de Phys.* **9**, 457 (1910).
296. J. Graham, *Rev. Pure Appl. Chem.* **14**, 81 (1964).
297. J. Graham, G. F. Walker, and G. W. J. West, *J. Chem. Phys.* **40**, 540 (1964).
298. D. C. Grahame, *Chem. Revs.* **41**, 441 (1947).
299. D. C. Grahame, *J. Amer. Chem. Soc.* **80**, 4201 (1958).
300. D. C. Grahame, and R. Parsons, *J. Amer. Chem. Soc.* **83**, 1291 (1961).
301. E. H. Grant, S. E. Keefe, and S. Takashima, *J. Phys. Chem.* **72**, 4373 (1968).

302. S. J. Gregg, *J. Chem. Soc.* **1942**, 696.
303. J. Gregory, *Advan. Colloid Interface Sci.* **2**, 396 (1969).
304. A. Griffiths, *Phil. Trans. A* **221**, 163 (1920).
305. J. R. Grigera and H. J. C. Berendsen, to be published.
306. J. R. Grigera, K. Hallenga, and H. J. C. Berendsen, to be published.
307. R. E. Grim, "Clay Mineralogy," McGraw-Hill, London (1953).
308. R. E. Grim, "Applied Clay Mineralogy," McGraw-Hill, London (1962).
309. T. B. Grimley and N. F. Mott, *Disc. Faraday Soc.* **1**, 3 (1947).
310. Å. Grudemo, Swedish Cement and Concrete Research Institute at the Royal Institute of Technology, Stockholm, Proc. 22 (1954).
311. G. A. Guderjahn, D. A. Paynter, P. E. Berghausen, and R. J. Good, *J. Chem. Phys.* **28**, 520 (1958).
312. E. A. Guggenheim, "Applications of Statistical Mechanics," Clarendon Press, Oxford (1966).
313. C. Haas, *Phys. Lett.* **3**, 126 (1962).
314. N. Hackerman and W. H. Wade, *J. Phys. Chem.* **68**, 1592 (1964).
315. G. Hägg, Allmän och oorganisk kemi, Almqvist & Wiksell Förlag AB, Stockholm (1963).
316. E. L. Hahn, *Phys. Khim.* **80**, 580 (1950).
317. M. L. Hair, Infrared Spectroscopy in Surface Chemistry, Marcel Dekker, New York (1967).
318. G. L. Haller and R. W. Rice, *J. Phys. Chem.* **74**, 4386 (1970).
319. G. D. Halsey, *J. Chem. Phys.* **16**, 931 (1948).
320. H. C. Hamaker, *Physica* **4**, 1058 (1937).
321. Sir W. B. Hardy, *Proc. Roy. Soc. A* **86**, 610 (1912).
322. W. D. Harkins, *Science* **162**, 292 (1945).
323. W. D. Harkins and G. Jura, *J. Amer. Chem. Soc.* **66**, 1366 (1944).
324. S. C. Harvey, and P. Hoekstra, *J. Phys. Chem.* **76**, 2987 (1972).
325. J. B. Hasted, D. M. Ritson, and C. H. Collie, *J. Chem. Phys.* **16**, 1 (1948).
326. E. A. Hauser, "Colloidal Phenomena. An Introduction to the Science of Colloids," p. 104, McGraw-Hill, New York (1939).
327. H. Hauser, M. C. Phillips, and R. M. Marchbanks, *Biochem. J.* **120**, 329 (1970).
328. A. T. J. Hayward and J. D. Isdale, *Brit. J. Appl. Phys. (J. Phys. D.)* Ser. 2, **2**, 251 (1969).
329. C. F. Hazlewood, B. L. Nichols, and N. F. Chamberlain, *Nature* **222**, 747 (1969).
330. C. F. Hazlewood, B. L. Nichols, D. C. Chang, and B. Brown, *Johns Hopkins Medical J.* **128**, 117 (1971).
331. M. Heberhold, *Chem. Unsere Z.* **1972**, 154.
332. O. Hechter, *Ann. N. Y. Acad. Sci.* **125**, 625 (1965).
333. L. Heller, V. C. Farmer, R. C. Mackenzie, B. D. Mitchell, and H. F. W. Taylor, *Clay Min. Bull.* **5**, 56 (1962).
334. H. Helmholtz, *Wied. Ann.* **7**, 337 (1879).
335. M. A. Hemminga and H. J. C. Berendsen, *J. Mag. Res.* **8**, 133 (1972).
336. S. B. Hendricks, *J. Geol.* **50**, 276 (1942).
337. J. C. Henniker, *Rev. Mod. Phys.* **21**, 322 (1949).
338. D. C. Henry, *Proc. Roy. Soc. (Lond.)* **133**, 106 (1931).
339. L. G. Hepler, *J. Phys. Chem.* **61**, 1426 (1957).
340. H. G. Hertz, *Z. Elektrochemie* **65**, 20 (1961).

341. H. G. Hertz, *Ber. Bunsenges.* **67**, 311 (1963).

342. H. G. Hertz, "Progress in NMR Spectroscopy" (J. W. Emsley, J. Feeney, and L. H. Sutcliffe, eds.), Vol. 3, Pergamon Press, London (1967).

343. H. G. Hertz, *in* "8th Colloquium on NMR Spectroscopy, Aachen 1971."

344. H. G. Hertz, B. Lindman, and V. Siepe, *Ber. Bunsenges.* **73**, 542 (1969).

345. H. G. Hertz and W. Spalthoff, *Z. Elektrochemie* **63**, 1096 (1959).

346. H. G. Hertz and M. D. Zeidler, *Ber. Bunsenges.* **68**, 821 (1964).

347. F. Th. Hesselink, A. Vrij, and J. Th. G. Overbeek, *J. Phys. Chem.* **75**, 2094 (1971).

348. N. E. Hill, S. W. Vaughan, A. H. Price, and M. Davies, "Dielectric Properties and Molecular Behavior," Van Nostrand Reinhold Co., London (1969).

349. T. L. Hill, *J. Chem. Phys.* **17**, 580, 668 (1949).

350. T. L. Hill and G. Jura, *J. Amer. Chem. Soc.* **74**, 1598 (1952).

351. J. C. Hindman, *J. Chem. Phys.* **36**, 1000 (1962).

352. P. V. Hobbs and B. J. Mason, *Phil. Mag.* **9**, 181 (1964).

353. P. Hoekstra and R. D. Miller, *J. Colloid Interface Sci.* **25**, 166 (1967).

354. U. Hofmann, K. Endell, and D. Wilm, *Z. Krist.* **102**, 417 (1933).

355. W. Hol, Thesis Groningen 1971.

356. C. M. Hollabaugh and J. J. Chessick, *J. Phys. Chem.* **65**, 109 (1961).

357. R. A. Horne and R. P. Young, *Electrochim. Acta* **17**, 763 (1972).

358. C. L. Hosler and R. E. Hallgren, *Disc. Faraday Soc.* **30**, 200 (1960).

359. C. L. Hosler, D. C. Jensen, and L. Goldshlak, *J. Meterol.* **14**, 415 (1957).

360. J. Hougardy, J. M. Serratosa, W. Stone, and H. van Olphen, *Special Disc. Faraday Soc.* **1**, 187 (1970).

361. B. F. Howell, *J. Chem. Ed.* **48**, 663 (1971).

362. B. F. Howell and L. Lancaster, *Chem. Commun.* **1971**, 693.

363. H. W. Hoyer and K. S. Birdi, *Biopolymers* **6**, 1507 (1968).

364. E. Hückel, *Physik. Z.* **25**, 204 (1924).

365. M. L. Huggins, *J. Phys. Chem.* **46**, 151 (1942).

366. R. J. Hunter, *J. Colloid Interface Sci.* **22**, 231 (1966).

367. R. J. Hunter, G. C. Stirling, and J. W. White, *Nature, Phys. Sci.* **230**, 192 (1971).

368. N. Ishizaka, *Z. Physik. Chemie* **83**, 97 (1913).

369. A. A. Isirikyan and A. V. Kiselev, *Dokl. Akad. Nauk SSSR* **110**, 1009 (1956).

370. I. B. Ivanov, B. Radoev, E. Manev, and A. Scheludko, *Trans. Faraday Soc.* **66**, 1262 (1970).

371. N. N. Ivanova, D. F. Kurpnova, R. M. Panich, and S. S. Voyutski, *Koll. Zh.* **31**, 63 (1969).

372. I. Iwakimi, *J. Chem. Soc. Nippon Kagaku Zasshi* **80**, 1094 (1959).

373. R. D. Iyengar and A. C. Zettlemoyer, "Solid–Gas Interface," Vol. II, (E. A. Flood, ed.), Marcel Dekker, New York (1967).

374. K. Jäckel, *Kolloid Z.* **197**, 143 (1964).

375. B. Jacobson, *Nature* **172**, 666 (1953).

376. A. Jacobsson and E. Forslind, to be published.

377. K. Jasmund, "Die silikatischen Tonminerale," Angewandte Chemie, Monograph 60 (1955).

378. H. H. G. Jellinek, U. S. Army Snow Ice and Permafrost, Res. Estab. Res. Rept. 23, p. 38 (1957).

379. H. H. G. Jellinek, *J. Colloid Sci.* **14**, 268 (1959).

380. H. H. G. Jellinek, *J. Colloid Interface Sci.* **25**, 192 (1967).

381. H. H. G. Jellinek and S. H. Ibrahim, *J. Colloid Interface Sci.* **25**, 245 (1967).
382. G. A. Johnson, S. M. A. Lecchini, E. G. Smith, J. Clifford, and B. A. Pethica, *Disc. Faraday Soc.* **42**, 120 (1966).
383. S. M. Johnson and A. D. Bangham, *Biochim. Biophys. Acta* **193**, 92 (1969).
384. E. C. Jonas, *in* "Clays and Clay Minerals" Third National Conference (Milligan, ed.), p. 66 (1955).
385. E. C. Jonas and R. E. Grim, *Mineralogical Society of Great Britain* (R. C. Mackenzie, ed.), Chapter 15, p. 389 (1956).
386. G. Jones and M. Dole, *J. Amer. Chem. Soc.* **51**, 2950 (1929).
387. M. N. Jones, K. J. Mysels, and P. C. Scholten, *Trans. Farad. Soc.* **62**, 1336 (1966).
388. G. Jura and T. L. Hill, *J. Amer. Chem. Soc.* **74**, 1598 (1952).
389. J. J. Jurinak, *Proc. Soil Sci. Soc. Amer.* **27**, 269 (1963).
390. J. J. Jurinak, *J. Colloid Sci.* **19**, 477 (1964).
391. H. Kallman and M. Willstätter, *Naturwiss.* **20**, 952 (1932).
392. B. Kamb, *Science* **172**, 231 (1971).
393. V. V. Karasev and Y. M. Luzhnov, *Russ. J. Phys. Chem.* **42**, 1255 (1968).
394. S. Karosaki, *J. Phys. Chem.* **58**, 320 (1954).
395. L. P. Kayashin (ed.), "Water in Biological Systems," Consultants Bureau, New York (1969).
396. W. D. Kingery, *J. Appl. Phys.* **31**, 833 (1960).
397. P. J. Killion, L. H. Reyerson, and B. F. Cameron, *J. Colloid Interface Sci.* **34**, 495 (1970).
398. J. J. Kirkland, *Anal. Chem.* **27**, 1537 (1955).
399. A. V. Kiselev, *Disc. Faraday Soc.* **52**, 14 (1971).
400. A. V. Kiselev, Ya. Koutetski, and I. Chizhek, *Izv. Acad. Sci. USSR Phys. Chem.* **137**, 283 (1961).
401. J. A. Kitchener, *Endeavour* **22**, 118 (1963).
402. J. A. Kitchener, *in* "Recent Progress in Surface Science" (J. F. Danielli, K. G. A. Pankhurst, and A. C. Riddiford, eds.), Vol. 1, p. 51, Academic Press, London (1964).
403. J. A. Kitchener, *Disc. Faraday Soc.* **52**, 379 (1971).
404. J. A. Kitchener and P. R. Mussellwhite, *in* "Emulsion Science," (P. Sherman, ed.), p. 77, Academic Press, London (1968).
405. K. Klier, *J. Opt. Soc. Am.* **62**, 882 (1972).
406. K. Klier, *J. Chem. Phys.* **58**, 737 (1973).
407. K. Klier, J. H. Shen, and A. C. Zettlemoyer, *J. Phys. Chem.* **77**, 1458 (1973).
408. I. M. Klotz, *Fed. Proc.* **24**, S-24 (1965).
409. H. Koelmans and J. Th. G. Overbeek, *Disc. Faraday Soc.* **18**, 52 (1954).
410. S. H. Koenig and W. E. Schillinger, *J. Biol. Chem.* **244**, 3283 (1969).
411. R. A. Kohl, J. W. Cary, and S. A. Taylor, *J. Colloid Sci.* **19**, 699 (1964).
412. L. Korson, W. Drost-Hansen, and F. J. Millero, *J. Phys. Chem.* **73**, 34 (1969).
413. G. Kortum, "Reflectance Spectroscopy," pp. 103–127, 170–216, Springer-Verlag, New York (1969).
414. D. L. Kostin, R. M. Panich, E. G. Lazaryants, L. V. Kosmodem'yanskii, and S. S. Voyutskii, *Koll. Zh.* **31**, 233 (1969).
415. H. R. Kruyt, "Colloid Science," Vol. 1, Elsevier, New York (1952).
416. H. R. Kruyt and M. A. M. Klompe, *Kolloid Beitr.* **54**, 484 (1943).
417. A. Krynicki, *Physica* **32**, 167 (1966).

418. K. Kume, *J. Phys. Soc. Japan* **15**, 1493 (1960).
419. I. D. Kuntz, *J. Amer. Chem. Soc.* **93**, 514 (1971).
420. I. D. Kuntz, T. S. Brassfield, G. D. Law, and G. C. Purcell, *Science* **163**, 1329 (1969).
421. I. Langmuir, *J. Amer. Chem. Soc.* **40**, 1361 (1918).
422. I. Langmuir, *J. Chem. Phys.* **6**, 873 (1938).
423. D. H. Larsen, Calif. Div. *Mines Bull.* **169**, 269 (1955).
424. J. Lawrence, R. Parsons, and R. Payne, *J. Electroanal. Chem. Interfacial Chem.* **16**, 193 (1968).
425. A. G. Leiga, D. W. Vance, and A. T. Ward, *Science* **168**, 114 (1970).
426. R. Lemlich (ed.) "Adsorptive Bubble Separation Techniques," Academic Press, New York (1972).
427. S. Levine, *J. Colloid Interface Sci.* **37**, 619 (1971).
428. S. Levine and G. M. Bell, *J. Colloid Sci.* **17**, 838 (1962).
429. S. Levine and G. M. Bell, *Disc. Faraday Soc.* **42**, 69 (1966).
430. S. Levine, G. M. Bell, and D. Calvert, *Can. J. Chem.* **40**, 518 (1962).
431. S. Levine and J. E. Jones, *Kolloid Z. u. Z. Polymere* **230**, 306 (1969).
432. S. Levine and E. Matijevic, *J. Colloid Interface Sci.* **23**, 188 (1967).
433. H. A. Levy, M. D. Danford, and A. H. Narten, ORNL-3960 (1966).
434. K. E. Lewis and G. D. Parfitt, *Trans. Faraday Soc.* **62**, 204 (1966).
435. E. M. Lifshitz, *Zh. Eksperim. i Teor. Fiz.* **29**, 94 (1955).
436. A. Liljas, K. K. Kannan, P.-C. Bergsten, I. Waara, K. Fridborg, B. Strandberg, U. Carlbom, L. Järup, S. Lövgren, and M. Petef, *Nature, New Biol.* **235**, 131 (1972).
437. B. Lindman, *Diss. LTH* (1971).
438. B. Lindman, S. Forsen, and E. Forslind, *J. Phys. Chem.* **72**, 2805 (1968).
439. G. N. Ling, *Ann. N. Y. Acad. Sci.* **125**, 401 (1965).
440. G. N. Ling, C. Miller, and M. H. Ochsenfeld, *Ann. N. Y. Acad. Sci.* **204**, 6 (1973).
441. J. W. Linnett, *Science* **167**, 1719 (1970).
442. E. R. Lippincot, R. R. Stromberg, W. H. Grant, and G. L. Cessac, *Science* **164**, 1482 (1969).
443. O. C. Lippold, J. G. Nicholls, and J. W. T. Redfearn, *J. Physiol.* **153**, 218 (1960).
444. A. G. Loeb, J. Th. G. Overbeek, and P. H. Wiersema, "The Electrical Double Layer around a Spherical Colloid Particle," Mass. Inst. Tech. Press (1960).
445. H. W. Loeb, G. M. Young, P. A. Quickenden, and A. Suggett, *Ber. Bunsenges. Phys. Chem.* **75**, 115 (1971).
446. F. London, *Z. Physik.* **63**, 245 (1930).
447. F. London, *Trans. Faraday Soc.* **33**, 8 (1937).
448. J. Longuet-Escard, J. Méring, and G. W. Brindley, *in* "Report of the Twenty-First Session Norden. Part XXIV. Proc. of the Internat. Committee for the Study of Clays" (I. Th. Rosenquist and P. Graff-Petersen), p. 17 (1960).
449. D. C. Look and J. J. Lowe, *J. Chem. Phys.* **44**, 2995 (1966).
450. P. F. Low, *Adv. Agron.* **13**, 269 (1961).
451. J. Lucassen and R. S. Hansen, *J. Colloid Interface Sci.* **23**, 319 (1967).
452. J. Lucassen, M. van den Tempel, A. Vrij, and F. Th. Hesselink, *Proc. Koninkl. Ned. Akad. Wetenschappen* **B73**, 109 (1970).
453. W. A. P. Luck, *Ber. Bunsenges. Phys. Chem.* **69**, 626 (1965).
454. B. Luyet, *Biodynamica* **10**, 277 (1969).
455. J. Lyklema, *Trans. Faraday Soc.* **59**, 418 (1963).

456. J. Lyklema, *Disc. Faraday Soc.* **42**, 81 (1966).

457. J. Lyklema and K. J. Mysels, *J. Amer. Chem. Soc.* **87**, 2539 (1965).

458. J. Lyklema and J. Th. G. Overbeek, *J. Coll. Sci.* **16**, 595 (1961).

459. J. Lyklema, P. C. Scholten, and K. J. Mysels, *J. Phys. Chem.* **69**, 116 (1965).

460. L. J. Lynch and A. R. Haly, *Kolloid Z.* **239**, 581 (1970).

461. L. J. Lynch, K. H. Marsden, and E. P. George, *J. Chem. Phys.* **51**, 5673 (1969).

462. S. C. Lyons, Technical Association of the Pulp and Paper Industry, Monograph 20 (1958).

463. D. M. C. MacEwan, *Nature* **154**, 577 (1944).

464. D. M. C. MacEwan, *in* "X-Ray Identification and Crystal Structures of Clay Minerals" (G. W. Brindley, ed.), p. 86 (1951).

465. R. C. Mackenzie, *Geol. Fören. Förh.* **78**, 508 (1956).

466. E. L. Mackor, *J. Colloid Sci.* **6**, 492 (1951).

467. E. L. Mackor and J. H. van der Waals, *J. Colloid Sci.* **7**, 535 (1952).

468. B. Maijgren and E. Forslind, to be published.

469. A. C. Makrides and N. Hackermann, *J. Phys. Chem.* **63**, 594 (1959).

470. J. Malsch, *J. Phys. Z.* **29**, 770 (1928).

471. M. Mandel, Thesis, Université Libre de Bruxelles (1955).

472. W. W. Mansfield, *Trans. Faraday Soc.* **66**, 341 (1970).

473. C. E. Marshall, *Trans. Faraday Soc.* **26**, 173 (1930).

474. C. E. Marshall, *Z. Krist.* (*A*) **91**, 433 (1935).

475. R. T. Martin, *in* "Clays and Clay Minerals, Proc. 6th Conf.," p. 259, Pergamon Press, New York (1959).

476. R. T. Martin, *in* "Clays and Clay minerals, Proc. 9th Conf.," p. 28, Pergamon Press, New York (1962).

477. A. McL. Mathieson and G. F. Walker, *Amer. Min.* **39**, 231 (1954).

478. J. M. Mays and G. W. Brady, *J. Chem. Phys.* **25**, 583 (1956).

479. J. W. McBain and A. M. Baker, *J. Amer. Chem. Soc.* **48**, 690 (1926).

480. E. McCafferty, "The Dielectric Behavior of Adsorbed Water Films on Iron Oxides," Ph.D. Thesis, Lehigh University (1968).

481. E. McCafferty, V. Pravdic, and A. C. Zettlemoyer, *Trans. Faraday Soc.* **66**, 1720 (1970).

482. E. McCafferty and A. C. Zettlemoyer, *Disc. Faraday Soc.* **52**, 239 (1971).

483. J. D. McCowan and R. L. McIntosh, *Can. J. Chem.* **39**, 425 (1961).

484. R. S. McDonald, *J. Phys. Chem.* **62**, 1168 (1958).

485. R. L. McIntosh, "Dielectric Behavior of Physically Adsorbed Gases," Marcel Dekker, New York (1966).

486. V. McKoy and O. Sinanoglu, *J. Chem. Phys.* **38**, 2946 (1953).

487. W. J. McLean and G. A. Jeffrey, *J. Chem. Phys.* **47**, 414 (1967).

488. W. J. McLean and G. A. Jeffrey, *J. Chem. Phys.* **49**, 4556 (1968).

489. R. K. McMullan, M. Bonamico, and G. A. Jeffrey, *J. Chem. Phys.* **39**, 3295 (1963).

490. R. McMullan and G. A. Jeffrey, *J. Chem. Phys.* **31**, 1231 (1959).

491. R. M. McMullan and G. A. Jeffrey, *J. Chem. Phys.* **42**, 2725 (1965).

492. R. K. McMullan, G. A. Jeffrey, and D. Panke, *J. Chem. Phys.* **53**, 3568 (1970).

493. R. K. McMullan, T. H. Jordan, and G. A. Jeffrey, *J. Chem. Phys.* **47**, 1218 (1967).

494. R. K. McMullan, T. C. W. Mak, and G. A. Jeffrey, *J. Chem. Phys.* **44**, 2338 (1966).

495. H. D. Megaw, *Nature* **134**, 900 (1934).

496. D. J. Meier, *J. Phys. Chem.* **71**, 1861 (1967).

497. W. M. Meier, *Z. Krist.* **113**, 430 (1960).

498. J. Méring, *Trans. Faraday Soc.* **42B**, 205 (1946).

499. R. P. Messmer, *Science* **168**, 479 (1970).

500. M. S. Metsik and O. S. Aidanova, in "Research in the Field of Surface Forces" (B. V. Derjaguin, ed.), Vol. 2, p. 169 Consultants Bureau, New York (1966).

501. M. B. M'Ewen and D. L. Mould, *Trans. Faraday Soc.* **53**, 548 (1957).

502. M. B. M'Ewen and M. I. Pratt, *Trans. Faraday Soc.* **53**, 535 (1957).

503. C. Migchelsen and H. J. C. Berendsen, *J. Chem. Phys.* **59**, 296 (1973).

504. C. Migchelsen, H. J. C. Berendsen, and A. Rupprecht, *J. Mol. Biol.* **37**, 235 (1968).

505. C. Migchelsen, K. J. Bienkiewiczs, and H. J. C. Berendsen, to be published.

506. G. Millot, "Geology of Clays," Springer Verlag, Berlin (1970).

507. J. Mingins and B. A. Pethica, *J. Chem. Soc. Faraday Trans. I*, **69**, 500 (1973).

508. V. V. Morariu, R. Mills, and L. A. Woolf, *Nature* **227**, 373 (1970).

509. T. Morimoto, K. Shiomi, and H. Tanaka, *Bull. Chem. Soc. Japan* **37**, 392 (1964).

510. J. N. Mukherjee and R. F. Mitra, *J. Colloid Sci.* **1**, 141 (1946).

511. K. J. Mysels, *Nature* **218**, 265 (1968).

512. K. J. Mysels, K. Shinoda, and S. P. Frankel, "Soap Films, Studies of their Thinning and a Bibliography," Pergamon Press, Oxford (1959).

513. M. Nagao, *J. Phys. Chem.* **75**, 3822 (1971).

514. G. Nagelschmidt, *Z. Krist.* **A93**, 481 (1936).

515. U. Nakaya and A. Matsumoto, U. S. Army Snow Ice and Permafrost Res. Estab. Res. Paper 4 (1953).

516. J. Namy and J. Chaussidon, *Bull. Grp. fr. Argiles* **14**, 101 (1967).

517. P. K. Nandi and D. R. Robinson, *J. Amer. Chem. Soc.* **94**, 1308 (1972).

518. D. H. Napper, *Trans. Faraday Soc.* **64**, 1701 (1968).

519. D. H. Napper, *Ind. Eng. Chem. Prod. Res. Develop.* **9**, 467 (1970).

520. D. H. Napper, *J. Colloid Interface Sci.* **32**, 106 (1970).

521. D. H. Napper, *J. Colloid Interface Sci.* **33**, 384 (1970).

522. D. H. Napper and R. J. Hunter, *in* "Surface Chemistry and Colloids" MTP International Review of Science (M. Kerker, ed.), Vol. 7, p. 241, Butterworths, London (1972).

523. A. H. Narten and S. Lindenbaum, *J. Chem. Phys.* **51**, 1108 (1969).

524. V. E. Nash, *in* "7th Nat. Conf. on Clays and Clay Minerals," p. 328 (1960).

525. G. Navon, R. I. Shulman, B. J. Wylvda, and T. Yamane, *J. Mol. Biol.* **51**, 15 (1970).

526. G. Némethy and H. A. Scheraga, *J. Chem. Phys.* **36**, 3382 (1962).

527. G. Némethy and H. A. Scheraga, *J. Chem. Phys.* **36**, 3401 (1962).

528. P. C. Nicolson, G. U. Yuen, and B. Zaslow, *Biopolymers* **4**, 677 (1966).

529. I. A. Nieduszynski and R. H. Marchessault, *Biopolymers* **11**, 1335 (1972).

530. B. W. Ninham and V. A. Parsegian, *Biophys. J.* **10**, 646 (1970).

531. B. W. Ninham and V. A. Parsegian, *J. Chem. Phys.* **52**, 4578 (1970).

532. B. W. Ninham and V. A. Parsegian, *J. Chem. Phys.* **53**, 3398 (1970).

533. S. Nir, R. Rein, and L. Weiss, *J. Theor. Biol.* **41**, 561 (1973).

534. K. Norrish, *Nature* **173**, 256 (1954).

535. W. J. O'Brien, *Surface Science* **25**, 298 (1971).

536. C. T. O'Konski, *Science* **168**, 1089 (1970).

537. C. T. O'Konski and S. Levine, *J. Colloid Interface Sci.* **36**, 547 (1971).

538. S. Olejnik, G. C. Stirling, and J. W. White, *Special Disc. Faraday Soc.* **1**, 194 (1970).

539. S. Olejnik and J. W. White, *Nature, Phys. Sci.* **236**, 15 (1972).

540. V. Y. Orlov, *Khol. Zh.* **33**, 465 (1971).

541. C. Orr and J. M. Dalla Valle, "Fine Particle Measurement," Macmillan, New York (1959).

542. D. W. J. Osmond, B. Vincent, and F. A. Waite, *J. Colloid Interface Sci.* **42**, 262 (1973).

543. R. H. Ottewill, *in* "Nonionic Surfactants" (M. J. Schick, ed.), p. 627 E. Arnold, London (1967).

544. R. H. Ottewill, *Ann. Rept. Prog. Chem.* **A66**, 183 (1969).

545. R. H. Ottewill and B. J. Vincent, *J. Chem. Soc. Faraday I* **68**, 1533 (1972).

546. R. H. Ottewill and T. Walker, *Kolloid Z. u. Z. Polymere* **227**, 108 (1968).

547. R. K. Outred and E. P. George, *Biophys. J.* **13**, 83 (1973).

548. R. K. Outred and E. P. George, *Biophys. J.* **13**, 97 (1973).

549. J. Th. G. Overbeek, *Adv. Colloid Sci.* **3**, 97 (1950).

550. J. Th. G. Overbeek, *in* "Colloid Science" (H. R. Kruyt, ed.), Vol. 1, Elsevier (1952).

551. J. Th. G. Overbeek, *Pure Appl. Chem.* **10**, 359 (1965).

552. J. Th. G. Overbeek, *Disc. Faraday Soc.* **42**, 7 (1966).

553. J. Th. G. Overbeek, *Disc. Faraday Soc.* **42**, 136 (1966).

554. P. G. Owston and K. Lonsdale, *J. Glaciol.* **1**, 118 (1948).

555. K. J. Packer, *in* "Progress in Nuclear Magnetic Resonance Spectroscopy" (J. W. Emsley, J. Feeney, and L. H. Sutcliffe, eds.) Pergamon Press, Vol. 3, pp. 87–129 (1967).

556. K. J. Packer, C. Rees, and D. J. Tomlinson, *Advances on Molecular Relaxation Processes* **3**, 119 (1972).

557. J. F. Padday, *Special Disc. Faraday Soc.* **1**, 64 (1970).

558. T. F. Page, R. J. Jakobsen, and E. R. Lippincott, *Science* **167**, 51 (1970).

559. L. S. Palmer, A. Cunliffe, and J. M. Hough, *Nature* **170**, 796 (1952).

560. J. G. Paren and J. C. F. Walker, *Nature Phys. Sci.* **230**, 77 (1971).

561. V. A. Parsegian and B. W. Ninham, *Biophys. J.* **10**, 664 (1970).

562. V. A. Parsegian and B. W. Ninham, *J. Colloid Interface Sci.* **37**, 332 (1971).

563. R. Parsons, *J. Electroanal. Chem.* **7**, 136 (1964).

564. R. Parsons, *J. Electroanal. Chem.* **8**, 93 (1964).

565. D. Patterson, *Macromolecules* **2**, 672 (1969).

566. L. Pauling, *J. Amer. Chem. Soc.* **57**, 2680 (1935).

567. L. Pauling, "The Nature of the Chemical Bond," Cornell Univ. Press (1948).

568. L. Pauling and R. E. Marsh, *Proc. Natl. Acad. Sci.* (*U. S.*) **38**, 112 (1952).

569. L. Pauling and R. E. Marsh, *Proc. Natl. Acad. Sci.* (*U. S.*) **38**, 2304 (1953).

570. R. T. Pearson, Ph.D. Thesis Nottingham University (1972).

571. J. B. Peri, *in* "Actes du Deuxième Congrès International de Catalyse, Paris, 1960," Vol. 1, p. 1333, Edition Technip, Paris (1961).

572. G. Peschel and K. H. Aldfinger, *Naturwiss.* **54**, 614 (1967).

573. G. Peschel and K. H. Aldfinger, *Naturwiss.* **56**, 558 (1969).

574. B. A. Pethica, *Special Disc. Faraday Soc.* **1**, 7 (1970).

575. B. A. Pethica, W. K. Thompson, and W. T. Pile, *Nature* **229**, 22 (1971).

576. G. A. Petsko, *Science* **167**, 171 (1970).

577. H. Pezerat and J. Méring, *Clay Min. Bull.* **2**, 156 (1954).

578. H. Pezerat and J. Méring, *C.R. Acad. Sci.* (*Paris*) **265**, 529 (1967).

579. D. Platikanov, N. Rangelova, and A. Sheludko, *God. Sofiisk. Univ. Khim. Fak.* **60**, 293 (1965–66) (*C. A.* **69**, 91093w).

580. J. C. Platteeuw and J. H. van der Waals, *Mol. Phys.* **1**, 91 (1958).

581. M. N. Plooster and S. N. Gitlin, *J. Phys. Chem.* **75**, 3322 (1971).

582. R. E. Powell and W. M. Latimer, *J. Chem. Phys.* **19**, 1139 (1951).

583. M. Prigonine and J. J. Fripiat, *Chem. Phys. Letts.* **12**, 107 (1971).

584. A. Prins and M. van den Tempel, *Special Disc. Faraday Soc.* **1**, 20 (1970).

585. D. Puett, *Biopolymers* **5**, 327 (1967).

586. B. R. Puri, L. R. Sharma, and M. L. Lakhanpal, *J. Phys. Chem.* **58**, 289 (1954).

587. B. R. Puri, D. D. Singh, and Y. P. Myer, *Trans. Faraday Soc.* **53**, 530 (1957).

588. F. A. Quiocho and W. N. Lipscomb, *Advan. Protein Chem.* **25**, 1 (1971).

589. S. W. Rabideau and A. E. Florin, *Science* **169**, 48 (1970).

590. A. Rahman and F. H. Stillinger, *J. Chem. Phys.* **55**, 3336 (1971).

591. G. N. Ramachandran and R. Chandrasekharan, *Biopolymers* **6**, 1649 (1968).

592. M. V. Ramiah and D. A. I. Goring, *J. Polymer Sci.* C **11**, 27 (1965).

593. J. E. B. Randles *Proc. 15th Solvay Conf.* (to be published).

594. L. E. Raraty and D. Tabor, *Proc. Roy. Soc.* A **245**, 184 (1958).

595. I. Ravina and P. F. Low, *Clays and Clay Minerals* **20**, 109 (1971).

596. A. D. Read and J. A. Kitchener, "Wetting," S.C.I. Monograph No. 25, p. 300 (1967).

597. A. D. Read and J. A. Kitchener, *J. Colloid Interface Sci.* **30**, 391 (1969).

598. A. G. Redfield, W. Fite, and H. E. Bleich, *Rev. Sci. Instr.* **39**, 710 (1968).

599. H. A. Resing, *J. Chem. Phys.* **43**, 669 (1965).

600. H. A. Resing, *Advan. Molec. Processes* **3**, 199 (1972).

601. H. A. Resing, J. K. Thompson, and J. J. Krebs, *J. Phys. Chem.* **68**, 1621 (1964).

602. A. D. Roberts and D. Tabor, *Special Disc. Faraday Soc.* **1**, 243 (1970).

603. A. D. Roberts and D. Tabor, *Proc. Roy. Soc.* A **235**, 323 (1971).

604. R. A. Robinson and R. H. Stokes, "Electrolyte Solutions," Butterworths, London (1959).

605. S. Ross, *in* "Chemistry and Physics of Interfaces" *Amer. Chem. Soc. Publ.*, p. 15, Washington, D.C. (1971).

606. S. Ross, *J. Colloid Interface Sci.* **42**, 52 (1973).

607. S. Ross and S. B. Hendricks, *Prof. Pap. H. S. Geol. Surv.* **205B**, 23 (1945).

608. S. Ross and J. P. Olivier, "On Physical Adsorption," Interscience, New York (1964).

609. D. L. Rousseau, *Science* **171**, 170 (1971).

610. D. L. Rousseau and S. P. S. Porto, *Science* **167**, 1715 (1970).

611. A. Rupprecht, *Acta Chem. Scand.* **20**, 494 (1966).

612. E. W. Rushe and W. B. Good, *J. Chem. Phys.* **45**, 4667 (1966).

613. C. Salama and D. A. I. Goring, *J. Phys. Chem.* **70**, 3838 (1966).

614. O. Y. Samoilov, *Zh. Fiz. Khim.* **20**, 1411 (1946).

615. E. T. Samulski and A. V. Tobolski, *Mol. Cryst. Liquid Cryst.* **7**, 433 (1969).

616. A. Sanfeld, and R. Defay, *J. Chim. Phys.* **63**, 577 (1966).

617. A. Sanfeld, C. Devillez, and P. Terlinck, *J. Colloid Interface Sci.* **32**, 33 (1970).

618. D. J. Schiffrin, *Trans. Faraday Soc.* **66**, 2464 (1970).

619. H. Schulze, *J. Prakt. Chem.* **25**, 431 (1882).

620. D. A. Seanor and C. H. Amberg, *Rev. Sci. Instr.* **34**, 917 (1963).

621. W. Senghaphan, G. O. Zimmerman, and C. E. Chase, *J. Chem. Phys.* **51**, 2543 (1969).

622. G. Seytre, J.-F. May, and G. Vallet, *J. Chim. Phys. Physicochim. Biol.* **69**, 959 (1972).

623. A. Sheludko, *Koll. Z.* **155**, 39 (1957).
624. A. Sheludko, *Z. Electrochem.* **61**, 220 (1957).
625. A. Sheludko, *Dokl. Akad. Nauk SSSR* **123**, 1074 (1958).
626. A. Sheludko, *Proc. Koninkl. Ned. Akad. Wetenschap. B* **65**, 86 (1962).
627. A. Sheludko, *Proc. Koninkl. Ned. Akad. Wetenschap. B* **65**, 97 (1962).
628. A. Sheludko, *Advan. Colloid Interface Sci.* **1**, 391 (1967).
629. A. Sheludko, G. Desmirov, and K. Nĩkolov, *God. Sofiisk. Univ. Fiz. Mat. Fak. Vol. 2—Khim.* **49**, 127 (1956).
630. A. Sheludko and D. Exerowa, *Kolloid Z.* **165**, 148 (1959).
631. A. Sheludko and D. Exerowa, *Kolloid Z.* **168**, 24 (1960).
632. A. Sheludko and D. Platikanov, *Kolloid Z.* **175**, 150 (1961).
633. A. Sheludko, D. Platikanov, and E. Manev, *Disc. Faraday Soc.* **40**, 253 (1965).
634. P. Sherman (ed.) "Emulsion Science," Academic Press, London (1968).
635. J. C. Slater and J. G. Kirkwood, *Phys. Rev.* **37**, 682 (1931).
636. A. W. Smith and J. M. Quets, *J. Catalysis* **4**, 163 (1965).
637. I. M. Solomentseva, O. O. Baran, and O. D. Kurylenko, *Dopovidi Akad. Nauk Ukr.SSR* **33**, 1108 (1971).
638. H. Sonntag, *Z. Phys. Chem.* **227**, 248 (1964).
639. H. Sonntag and H. Klare, Jr., *Kolloid Z.* **195**, 35 (1964).
640. H. Sonntag, J. Netzel, and H. Klare, Jr. *Kolloid Z.* **211**, 121 (1966).
641. H. Sonntag, J. Netzel, and B. Unterberger, *in* "Thin Liquid Films and Boundary Layers," *Special Disc. Faraday Soc.* **1**, 57 (1970).
642. H. Sonntag and K. Strenge, "Coagulation and Stability of Disperse Systems," Halstead Press (1970).
643. M. J. Sparnaay, *Rec. Trav. Chim.* **81**, 395 (1962).
644. R. B. Spooner and P. W. Selwood, *J. Amer. Chem. Soc.* **71**, 2184 (1949).
645. G. Sposito and K. L. Babcock, *in* "Proc. 14th Nat. Conf. on Clays and Clay Minerals," p. 133 (1966).
646. G. Srinivasan, J. J. Chessick, and A. C. Zettlemoyer, *J. Phys. Chem.* **66**(10), 1819 (1962).
647. E. O. Stejskal, *Advan. Molecular Relaxation Processes* **3**, 27 (1972).
648. O. Stern, *Z. Elektrochem.* **30**, 508 (1924).
649. F. H. Stillinger, Jr. and A. Ben-Naim, *J. Chem. Phys.* **47**, 4431 (1967).
650. F. H. Stillinger and A. Rahman, *J. Chem. Phys.* **57**, 1281 (1972).
651. W. H. Stockmayer, *J. Polymer Sci.* **15**, 595 (1955).
652. A. Suggett, *in* "Dielectric and Related Molecular Processes," Vol. 1, p. 100, Chemical Society, London (1972).
653. A. Suggett, P. A. Mackness, M. J. Tait, H. W. Loeb, and G. M. Young, *Nature* **228**, 456 (1970).
654. M. V. Sussman and L. Chin, *Science* **151**, 324 (1966).
655. T. J. Swift and R. E. Connick, *J. Chem. Phys.* **37**, 307 (1962); Erratum, *J. Chem. Phys.* **41**, 2553 (1964).
656. A. Szent Györgyi, "Bioenergetics," Academic Press, New York (1957).
657. M. J. Tait and F. Franks, *Nature* **230**, 91 (1971).
658. M. J. Tait, A. Suggett, F. Franks, S. Ablett, and P. A. Quickenden, *J. Soln. Chem.* **1**, 131 (1972).
659. C. Tanford, "Physical Chemistry of Macromolecules," p. 197, Wiley, New York (1963).

660. J. E. Tanner and S. O. Stejskal, *J. Chem. Phys.* **49**, 1768 (1968).

661. N. Tcheurekdjian, A. C. Zettlemoyer, and J. J. Chessick, *J. Phys. Chem.* **68**, 773 (1964).

662. J. W. Telford and J. S. Turner, *Phil. Mag.* **8**, 527 (1963).

663. J. Timmerman and H. Bodson, *Compt. Rend.* **204**, 1804 (1937).

664. J. Thompson, *Proc. Roy. Soc.* **11**, 473 (1861).

665. W. Thompson, *Phil. Mag.* **42**, 448 (1871).

666. W. K. Thompson, *J. Chem. Soc.* **1964**, 3658.

667. W. K. Thompson, *Trans. Faraday Soc.* **61**, 2635 (1965).

668. J. M. Thorne and H. Slaughter, *Thermochim. Acta* **3**, 181 (1972).

669. B. H. Torrie, I. D. Brown, and H. E. Petch, *Can. J. Phys.* **42**, 229 (1964).

670. A. J. Tursi and E. R. Nixon, *J. Chem. Phys.* **52**, 1521 (1970).

671. A. J. Tyler, J. A. G. Taylor, B. A. Pethica, and J. A. Hockey, *Trans. Faraday Soc.* **67**, 483 (1971).

672. M. van den Tempel, *in* "Proc. 2nd Intern. Congr. Surface Activity," Vol. 1, p. 439, Butterworths, London (1957).

673. M. van den Tempel, *J. Colloid Sci.* **13**, 125 (1958).

674. J. W. Vanderhoff, *in* "Clean Surfaces, Their Preparation and Characterization for Interfacial Studies" (G. Goldfinger, ed.), Dekker, New York (1970).

675. G. E. van Gils, *J. Colloid Interface Sci.* **30**, 272 (1969).

676. H. van Olphen, *J. Inst. Petrol.* **36**, 223 (1950).

677. H. van Olphen, *Disc. Faraday Soc.* **11**, 82 (1951).

678. H. van Olphen, Clay Colloid Chemistry, p. 53, Interscience, New York (1963).

679. C. Veder, "Grouts and Drilling Muds in Engineering Practice," Butterworths, London (1963).

680. E. J. W. Verwey and J. Th. G. Overbeek, "Theory of the Stability of Lyophobic Colloid," Elsevier, Amsterdam (1948).

681. B. Vincent, *J. Colloid Interface Sci.* **42**, 270 (1973).

682. B. Vincent, B. H. Bijsterbosch, and J. Lyklema, *J. Colloid Interface Sci.* **37**, 171 (1971).

683. J. Visser, *Advan. Colloid Interface Sci.* **3**, 331 (1972).

684. M. J. Vold, *J. Colloid Sci.* **16**, 1 (1961).

685. M. von Smoluchowski, *Bull. Acad. Sci. Cracovie* **1903**, 182.

686. M. von Stackelberg, *Naturwiss.* **36**, 327, 359 (1949).

687. A. Vrij, *Disc. Faraday Soc.* **42**, 23 (1966).

688. A. Vrij, F. Th. Hesselink, J. Lucassen, and M. van den Tempel, *Proc. Koninkl. Nederl. Akad. Wetenschappen* **B 73**, 124 (1970).

689. W. H. Wade and N. Hackerman, *J. Phys. Chem.* **65**, 1681 (1961).

690. K. W. Wagner, *Arch. Electrotechn.* **2**, 371 (1914).

691. G. F. Walker, *Clay Min. Bull.* **3**, 302 (1958).

692. G. F. Walker, *Min. Soc. London*, 297 (1961).

693. G. E. Walrafen, *J. Chem. Phys.* **36**, 1035 (1962).

694. G. E. Walrafen, *J. Chem. Phys.* **40**, 3249 (1964).

695. G. E. Walrafen, *J. Chem. Phys.* **44**, 1546 (1966).

696. G. E. Walrafen, *J. Chem. Phys.* **52**, 4176 (1970).

697. G. E. Walrafen, *J. Chem. Phys.* **55**, 768 (1971).

698. J. A. Walter and A. B. Hope, *Aust. J. Biol. Sci.* **24**, 497 (1971).

699. J. A. Walter and A. B. Hope, *Prog. Biophys. Mol. Biol.* **23**, 1 (1971).

700. J. R. Watts-Tobin, *Phil. Mag.* **8**, 333 (1963).
701. R. A. Weiler, Thesis, Institut Agronomique, Université Catholique de Louvain (1966).
702. A. Weiss and J. Russow, *in* "14th Internat. Clay Conf. Proc., Stockholm," p. 203 (1963).
703. A. Weiss and J. Russow, *in* "21st Intern. Clay Conf. Proc., Stockholm," p. 69 (1963).
704. P. V. Wells, *Ann. Phys.* **16**, 69 (1921).
705. J. W. Whalen, *J. Phys. Chem.* **66**, 511 (1962).
706. O. Wiener, *Abhandl. math. phys. Klasse sächs. Akad. Wiss. (Leipzig)* **32**, 256 (1912).
707. P. H. Wiersema, A. L. Loeb, and J. Th. G. Overbeek, *J. Colloid Interface Sci.* **22**, 78 (1966).
708. G. R. Wiese, R. O. James, and T. W. Healy, *Disc. Faraday Soc.* **52**, 302 (1971).
709. E. Willis, G. K. Rennie, C. Smart, and B. A. Pethica, *Nature* **222**, 159 (1969).
710. B. H. Wilsdon, *J. Soc. Chem. Ind.* **53**, 397T (1934).
711. H. Winkler, *in* "Proc. 10th Colloq. Ampère Leipzig," p. 219 (1961).
712. H. Winkler, *Wiss. Zeit. Karl-Marx Univ. Leipzig Math. Naturwiss.* **4**, 913 (1965).
713. D. E. Woessner, *J. Chem. Phys.* **35**, 41 (1961).
714. D. E. Woessner and B. S. Snowden, Jr., *J. Chem. Phys.* **50**, 1516 (1969).
715. D. E. Woessner and B. S. Snowden, Jr., *Ann. N. Y. Acad. Sci.* **204**, 113 (1973).
716. D. E. Woessner, B. S. Snowden, Jr., and G. H. Meyer, *J. Chem. Phys.* **51**, 2968 (1969).
717. D. E. Woessner, B. S. Snowden, Jr., and G. H. Meyer, *J. Colloid Interface Sci.* **34**, 43 (1970).
718. H. Wroblowa, Z. Kovac, and J. O'M. Bockris, *Trans. Faraday Soc.* **61**, 1523 (1965).
719. H. W. Wyckoff, D. Tsernoglou, A. W. Hanson, J. R. Knox, B. Lee, and F. M. Richards, *J. Biol. Chem.* **245**, 305 (1970).
720. D. J. C. Yates, *J. Phys. Chem.* **65**, 746 (1961).
721. A. Yonath and W. Traub, *J. Mol. Biol.* **43**, 461 (1969).
722. G. J. Young, *J. Colloid Sci.* **13**, 67 (1958).
723. G. C. Yukhnevich, A. V. Karyakin, N. I. Khitarov, and E. E. Senderov, *Geokhimiya* **1961**, 849.
724. A. C. Zettlemoyer, *Ind. Eng. Chem.* **57**, 27 (1965).
725. A. C. Zettlemoyer, *J. Colloid Interface Sci.* **28**, 343 (1968).
726. A. C. Zettlemoyer and J. J. Chessick, *in* "Advances in Chemistry Series," No. 43, p. 88 (1964).
727. A. C. Zettlemoyer, F. J. Micale, and Y. K. Lui, *Ber. Bunsenges. Phys. Chem.* **71**, 286 (1967).
728. A. C. Zettlemoyer, G. J. Young, and J.J. Chessick, *J. Phys. Chem.* **59**, 962 (1955).
729. V. Zhurblis, *Khimiya i Zhizn.* **1969**(12), 37.
730. B. H. Zimm, *J. Chem. Phys.* **16**, 1093 (1948).
731. H. Zocher, *Z. Anorg. Chem.* **147**, 91 (1925).
732. Z. M. Zorin and N. V. Churaev, *Colloid J. USSR* **30**, 279 (1968).

Author Index

351

Subject Index

Compound Index